T0305591

Risk Management Applications Used to Sustain Quality in Projects

This practical guide covers the steps necessary to sustain quality in a project from start to finish. The book shows how to identify risks at different processes, phases, and stages and offers directions on how to mitigate and reduce risks using analysis, evaluation, and monitoring.

Risk Management Applications Used to Sustain Quality in Projects: A Practical Guide focuses on applying risk management principles to manage quality in all project management processes, stages, and phases. The book discusses the potential risks that may occur at the different phases of the project life cycle, their effects on projects, and how to prevent them. It explores all the process elements and activities of risk management and provides steps on how to make the project more qualitative, competitive, and economical. Risk management processes are discussed at each project management process and project life cycle phase/stage to help the reader understand how various risks can occur and how to mitigate and reduce them.

The main audience for this book is project management professionals, quality managers, systems engineers, construction managers, and risk management professionals, as well as industrial engineers, academics, and students.

Quality Management and Risk Series

Series Editor: Abdul Razzak Rumane, Senior Engineering Consultant, Kuwait

This new series will include the latest and innovative books related to quality management and risk related topics. The definition of quality relating to manufacturing, processes, and the service industries, is to meet the customer's need, satisfaction, fitness for use, conforms to requirements, and a degree of excellence at an acceptable price. With globalization and competitive markets, the emphasis on quality management has increased. Quality has become the most important single factor for the survival and success of any company or organization. The demand for better products and services at the lowest costs have put tremendous pressure on organizations to improve the quality of products, services, and processes, in order to compete in the marketplace. Because of these changes, the ISO 9001 now lists that risk-based thinking must be incorporated into the management system by considering the context of the organization. Quality management and risk management now play an important role in the over all quality management system. This means that books which cover quality, need to also cover risk to update practices/processes, tools, and techniques, per ISO 9001. The goal of this new series is to include the books that will meet this need and demand.

Quality Management in Oil and Gas Projects
Abdul Razzak Rumane

Risk Management Applications Used to Sustain Quality in Projects
A Practical Guide
Abdul Razzak Rumane

For more information on this series, please visit: www.routledge.com/Quality-Management-and-Risk-Series/book-series/CRCQMR

Risk Management
Applications Used to
Sustain Quality in Projects
A Practical Guide

Abdul Razzak Rumane

CRC Press
Taylor & Francis Group
Boca Raton London New York

CRC Press is an imprint of the
Taylor & Francis Group, an **informa** business

First edition published 2023
by CRC Press
6000 Broken Sound Parkway NW, Suite 300, Boca Raton, FL 33487–2742

and by CRC Press
4 Park Square, Milton Park, Abingdon, Oxon, OX14 4RN

CRC Press is an imprint of Taylor & Francis Group, LLC

© 2023 Taylor & Francis Group, LLC

ISBN: 978-1-032-15771-9 (hbk)
ISBN: 978-1-032-15772-6 (pbk)
ISBN: 978-1-003-24561-2 (ebk)

DOI: 10.1201/9781003245612

Typeset in Times
by Apex CoVantage, LLC

To
my parents
for their prayers and love.
My prayers are always for my father and my mother, who
encouraged and inspired me all the time.

I wish they would have been here to see this book and give me
blessings.
To
my wife.
I miss my wife, who stood with me all the time during the writing of
my earlier books.

Contents

Foreword

I am highly honored to be invited to write the foreword for Dr. Rumane's latest book, which is a highly commendable addition to his prolific outputs over the past several years. I was present when his first 2010 book entitled "Quality Management in Constructions Projects" was launched officially in Kuwait City in 2011. Since then, Dr. Rumane has published half dozen books, mainly in the field of quality in construction projects, which represent a prolific rate of about one book every two years. The intellectuality embodied in Dr. Rumane's writing is something I have never witnessed in my over four decades of affiliation with the book publishing industry. His stream of publishing successes is not a fluke. I first met Dr. Rumane at an international conference in Detroit, Michigan, in 2009, when he intimated that he was writing a construction management book. This piqued my interest, and we struck several collaborative discussions. Although he was already discussing with several publishers at that time, I encouraged and invited him to write the book for my Taylor & Francis Group book series on Industrial Innovation (later changed to Systems Innovation Book Series), which was still in its infancy at that time. Within one year, Dr. Rumane had completed the manuscript, well ahead of the contractual deadline. This is a direct demonstration of Dr. Rumane's commitment to the practice of compliance with the schedule, quality, and cost expectations that are typical in the construction industry. Dr. Rumane doesn't just write about the tools and techniques for the construction industry, he also practices what he preaches. This is evidenced by the quick succession of his successful books. He is now a worldwide leading authority on the topics of quality management and delivery in the construction industry. In fact, his literary products in the construction industry are now highly coveted in other industries as well. This is a huge societal contribution that many authors never achieve. Dr. Rumane now has his own Taylor & Francis Group Book Series. Some of his books have already been translated into multiple languages. Beyond his first book in 2010, Dr. Rumane has written several other relevant books, including Quality Management in Oil and Gas Projects, Quality Auditing in Construction Projects, Handbook of Construction Management, Quality Tools for Managing Construction Projects, and others.

This brings us to Dr. Rumane latest intellectual output entitled "**Risk Management Applications Used to Sustain Quality in Projects: A Practical Guide**". As with his previous books, Dr. Rumane covered all the bases involved in any comprehensive project, not just in the construction industry, but also in other industries. Chapters in the book are Overview of Risk Management, Overview of Quality of Projects, Project Management Process, Risk Management in Quality of Project Processes, and Risk Management in Quality of Project Life Cycle Phases. Without resorting to a verbatim duplication of the book's preface or the table of contents, I can assure all readers that this book is what is needed to understand and tackle all the nuances of risk management in all projects, small or large, private or corporate, domestic or global. Multinational companies, in particular, will derive great benefits from this book, written by an experienced, dedicated, and world-centric professional. I strongly recommend the book for all facets of project pursuits.

Adedeji B. Badiru, Ph.D., PE, PMP
Dean, College of Engineering and Management
Airforce Institute of Technology, Ohio, the United States

Preface

ISO 9001:2015 emphasizes risk-based thinking. One of the key changes in the revision is to establish systematic approach to consider risk, rather than treating "prevention" as a separate component of quality system.

Projects are of a non-repetitive nature. A project is a plan or program carried out by the people with the assigned resources to achieve an objective within a finite duration. They have an abundance of risks affecting most participants in certain portion of project processes during the life cycle of the project.

Risk management is the process of identifying, assessing, prioritizing different kinds of risks; planning risk mitigation; implementing mitigation plan; and controlling the risks. It is a process of thinking systematically about the possible risks, problems, or disasters before they happen and setting up the procedure that will avoid the risk or minimize the impact or cope with the impact. The objectives of project risk management are to increase the probability and impacts of positive events and decrease the probability and impacts of events adverse to the project objectives. Risk is the probability that the occurrence of an event may turn into undesirable outcome (loss, disaster). It is virtually anything that threatens or limits the ability of an organization to achieve its objectives. It can be unexpected and unpredictable events which have the potential to damage the functioning of organization in terms of money, or, in worst scenario, it may cause the business to close.

Risk management process is designed to reduce or eliminate the risk of certain kind of events happening (occurring) or having an impact on the project. The risk process consists of following steps:

1. Identify the potential sources of risk on the project.
2. Analyze their impact on the project.

   a) Qualitative
   b) Quantitative

3. Select those with a significant impact on the project.

   a) Prioritization

4. Determine how the impact of risk can be reduced.

   a. Avoidance
   b. Transfer
   c. Reduction
   d. Retention (Acceptance)

5. Select best alternative
6. Develop and implement mitigation plan
7. Monitor and control the risks by implementing risk response plan, tracking identified risks, identifying new risks and evaluating the risk impact.

Projects have many varying risks due to the involvement of people and resources. Risk management throughout the life cycle of the project is important and essential to prevent unwanted consequences and effects on the project. Projects have the involvement of many stakeholders such as project owners, financial institutions funding the project, planners, executors, and monitoring and controlling agencies who are affected by the risk. Each of these parties has involvement with certain portion of overall project risk; however, the owner has a greater share of risks as the owner is involved from the inception until completion of project and beyond. The owner must also take initiatives to develop risk consciousness and awareness among all the parties emphasizing upon the importance of explicit consideration of risk at each stage of the project as the owner is ultimately responsible for overall project cost.

Risk management is an ongoing process. In order to reduce the overall risk in quality of the projects, the risk assessment (identification, analysis, and evaluation) process must start as early as possible to maximize project benefits. There are number of risks that can be identified at each stage of the project. Early risk identification can lead to better estimation of the cost in the project budget, whether through contingencies, contractual or insurance. Risk identification is most important function in construction projects.

It is essential that risks are identified in each process, activity of the project, phase/stage of the life cycle of the project, and mitigate the risk and effect of risk on the project.

The book is developed to provide all the related information to project professional practitioners; risk management professionals involved to sustain the quality in projects to know about various risks that span occurrence from the initiation of the project through the closing of the project, identifying major risks at different process groups, phases/stages of project life cycle, their probable effects on the project, ownership of the risk, and risk treatment/control measures to mitigate/reduce the risk by analysis, evaluation, and monitoring to make the project most qualitative, competitive, and economical.

For the benefit of professionals, the book contains many valuable figures that can be applied for risk treatment and take control measures to mitigate and avoid the occurrence of risk in the project.

For the sake of proper understanding, the book is divided into five chapters, and each chapter is divided into a number of sections covering risk management processes at each of the phases of construction projects.

Chapter 1 is about overview of risk management. It discusses risk management process that includes the development of risk management plan and risk management process for identifying, assessing, prioritizing different kinds of risks, planning risk mitigation, implementing mitigation plan, and controlling the risks.

Chapter 2 is an overview of quality of projects. It covers brief definition of project, quality of construction projects, and principles of quality in construction projects developed on the ISO principles CLIPSCFM.

Chapter 3 is about project management processes. It discusses project management processes to manage and control various processes and activities that are performed in a project from project initiation to the closing of the project.

Management Processes discussed in this chapter covers briefs about Integration Management; Stakeholder Management; Scope Management; Schedule Management; Cost Management; Quality Management; Resource Management; Communication Management; Risk Management; Contract Management; Health, Safety, and Environment Management (HSE); Financial Management; and Claim Management. The contents of this chapter are for the benefits of project management and risk management professionals who are not familiar or need more information about the management processes, for activities at different stages/phases of projects in which the risk is likely to occur and has to be considered for risk-based decision-making to achieve and sustain quality in project processes and for activities for risk identification and control.

Chapter 4 details risk management in quality of project process groups that include initiation, planning, execution, monitoring, and control and closing. The chapter discusses the identification of potential risks in the major activities of process groups, probable effects on the project, ownership of the risks, and risk treatment and control measures to avoid and mitigate occurrence of risk(s) in the project.

Chapter 5 details risk management in quality of project life cycle phases/stages from the initiation of the project till the closing of the project. The chapter discusses the identification of potential risks in the major activities/elements of project life cycle phases, probable effects on the project, ownership of the risks, and risk treatment and control measures to avoid, mitigate the occurrence of risk(s) in the project.

The books, I am certain, will meet the requirements of construction professionals, quality professionals, project owners, students, and academics and satisfy their needs.

Acknowledgment

SHARING THE KNOWLEDGE WITH OTHERS
IS THE MOTTO OF THIS BOOK

Many of colleagues and friends extended help while preparing the book by arranging reference material, and hence many thanks to all of them for their support.

I thank publishers and authors, whose writings are included in this book, for extending their support by allowing me to reprint their material.

I thank reviewers, from various professional organizations, for their valuable input to improve my writing. I thank members of ASQ Audit Division, ASQ Design & Construction Division, The Institution of Engineers (India), IEI, Kuwait Chapter, Kuwait Society of Engineers, and ASQ Kuwait Section for their support to bring out this book.

I thank Dr. Adedeji B. Badiru for his nicely worded thought-provocative Foreword and support and best wishes all the time.

I thank Cindy Renee Carelli, executive editor of CRC Press, the senior editorial assistant, and other CRC staff for their support and contribution to make this construction related book a reality.

I thank Mr. Cliff Moser, former chairman, ASQ Design and Construction Division; Mr. John F. Mascaro, chairman, ASQ Design and Construction Division and ASQ Audit Division; and Mr. Raymond R. Crawford of former chairman, ASQ Design and Construction Division for their best wishes all the time.

I thank Engr. Adel Kharafi, former chairman, Kuwait Society of Engineers, and former president of World Federation of Engineering Organizations (WFEO) for his good wishes all the time. I thank Engr. Ahmad Alkandari, director, Kuwait Municipality, for his support and good wishes. I thank Engr. Ahmad Almershed, former Undersecretary, MSNA, Kuwait, for his good wishes all the time. I thank Dr. Fadel Safer, former minister of Public Works, Kuwait, for his support and good wishes. I thank Engr. Faisal D. Alatel, chairman, Kuwait Society of Engineers and president of Federation of Arab Engineers, for his support and good wishes. I thank Prof. Mohammed Aichouni, University of Hail, KSA, for his support and good wishes all the time. I thank Dr. Mohammad Ben Salamah, former chair, ASQ Kuwait Section, for his support and best wishes. I thank Dr. N. N. Murthy of Jagruti Kiran Consultants for his support and good wishes. I thank Dr. Othman Alshamrani, Imam Abdulrehman Bin Faisal University, KSA, for his support and good wishes all the time. I thank Ms. Rima Al Awadhi chair, ASQ Kuwait Section and Team Leader at Kuwait Oil Company for her support and good wishes. I thank Engr. Wael Aljasem of Kuwait Project Management Society for his support and good wishes. I thank Engr. Yaseen Farraj, former director, Ministry of Public Works, Kuwait, for his support and good wishes.

I also thank Mr. Bashir Ibrahim Parkar of Dar SSH International and Engr. Ganeshan Swaminathan, quality consultant, for their valuable input and support.

I extend my thanks to Dr. Ted Coleman, professor and department chairman, California State University, San Bernardino and former chancellor, KW University, for his everlasting support.

My special thanks to H.E. Sheikh Rakan Nayef Jaber Al Sabah for his support and good wishes.

I thank all my well-wishers whose inspiration helped me to complete this book.

Most of the data discussed in this book is from author's practical and professional experience and are accurate to the best of author's knowledge and ability. However, if any discrepancies are observed in the presentation, I would appreciate communicating them to me.

The contributions of my son Ataullah, my daughter Farzeen, and daughter-in-law Masum are worth mentioning here. They encouraged me and helped me in my preparatory work to achieve the final product. I thank my brothers, sisters, and all the family members for their support, encouragement, and good wishes all the time.

Abdul Razzak Rumane

About the Author

Abdul Razzak Rumane (Ph.D.) is Chartered Quality Professional-Fellow of The Chartered Quality Institute (UK) and certified consultant engineer in the field of Electrical Engineering and Project Management. He obtained a Bachelor of Engineering (Electrical) degree from Marathwada University (now Dr. Babasaheb Ambedkar Marathwada University), India, in 1972 and received his Doctor of Philosophy (Ph.D.) from Kennedy Western University, USA (now Warren National University), in 2005. His dissertation topic was "Quality Engineering Applications in Construction Projects".

Dr. Rumane has been honored by The International University of Ministry and Education, Missouri, USA, which awarded honorary degree of Doctor of Philosophy (D.Phil.) for expertise in Quality Management in the year 2021, and The Yorker International University, the United States, awarded an Honorary Doctorate of Engineering in the year 2007. The Albert Schweitzer International Foundation, Spain, honored him with gold medal for "Outstanding contribution in the field of Quality in Construction Projects"; and World Quality Congress awarded him "Global Award for Excellence in Quality Management and Leadership".

Dr. Rumane is an accomplished engineer. He is associated with a number of professional organizations. He is a fellow of The Chartered Quality Institute (UK), Senior Member of American Society for Quality, fellow of The Institution of Engineers (India) and has an Honorary Fellowship of Chartered Management Association (Hong Kong). He is senior member of Institute of Electrical and Electronics Engineers (USA) and member of Kuwait Society of Engineers, among others.

Dr. Rumane has attended many international conferences and has made technical presentations at various conferences. Dr. Rumane is an author of six useful books entitled "Quality Management in Construction Projects" (First Edition, 2010), "Quality Tools for Managing Construction Projects" (2013), "Quality Management in Construction Projects" (Second Edition, 2017), "Quality Management in Oil and Gas Projects" (2021), and editor of book entitled "Handbook of Construction Management: Scope, Schedule, and Cost Control" (2016). All of these books are published by CRC Press (Taylor & Francis Group), USA. A book entitled "Quality Auditing in Construction Projects: A Handbook" (2019) is published by Routledge, UK (Taylor & Francis Group). His book "Quality Management in Construction Projects" is translated into Korean language.

Presently, he is treasurer of ASQ Kuwait Section and International Liaison committee chair ASQ (Design and Construction Division) for the year 2022. He served as secretary ASQ GC for the year 2019 and Secretary ASQ LMC, Kuwait, for the years 2017 and 2018. He was honorary chairman of The Institution of Engineers (India), Kuwait chapter for the years 2016–17, 2013–14, and 2005 to 2007.

Dr. Rumane's professional career exceeds 50 years including 10 years in manufacturing industries and over 40 years in construction projects. Presently he is associated with SIJJEEL Co., Kuwait, as a Advisor and director, Construction Management.

Abbreviations

AACE	American Association of Cost Engineers
ASCE	American Society of Civil Engineers
ASHRAE	American Society of Heating, Refrigeration, and Air-conditioning Engineers
ASQ	American Society for Quality
BMS	Building Management System
CII	Construction Industry Institute
CDM	Construction (Design and Management)
CEN	European Committee for Standardization
CMAA	Construction Management Association of America
CSC	Construction Specifications, Canada
CSI	Construction Specification Institute
FEED	Front End Engineering Design
FIDIC	Federation International des Ingeneurs-Counceils
HAZID	Hazard Identification
HAZOP	Hazard and Operability
HSE	Health, Safety and Environment
ICE	Institute of Civil Engineers (UK)
IEC	International Electrotechnical Commission
IEEE	Institute of Electrical and Electronics Engineers
IoT	Internet of Things
IP	Ingress Protection
ISO	International Organization for Standardization
OH&S	Occupational Health and Safety
PHSER	Procedure for Project HSE Review
PCM	Planning and Control Manager
PMC	Project Management Consultant
PMI	Project Management Institute
PMBOK	Project Management Book of Knowledge
QMS	Quality Management System
QS	Quantity Surveyor
RE	Resident Engineer
TIC	Total Investment Cost

SYNONYMS

Owner	Client, Employer
Consultant	Architect/Engineer (A/E), Designer, Design Professionals, Designer, Consulting Engineers, Supervision Professional Specialist Consultant
Engineer	Resident Project Representative

Engineer's Representative	Resident Engineer
Project Charter	Terms of Reference (TOR), Client Brief, Definitive Project Brief
Project Manager	Construction Manager
Contractor	Constructor, Builder, EPC Contractor
Quantity Surveyor	Cost Estimator, Contract Attorney, Cost Engineer, Cost and Works Superintendent
Main Contractor	General Contractor

1 Overview of Risk Management

1.1 RISK MANAGEMENT

Risk is the probability that the occurrence of an event may turn into undesirable outcome (loss, disaster). It is virtually anything that threatens or limits the ability of an organization to achieve its objectives. It can be unexpected and unpredictable events which have the potential to damage the functioning of organization in terms of money, or, in worst scenario, it may cause the business to close.

Risk is any unexpected event that may occur suddenly thus affecting the organization's strategic and operational objectives. Risk is the possibility of events or activities impeding the achievement of an organization's strategic and operational objectives.

According to ISO 31000, risk is the effect of uncertainty on objectives resulting a positive or negative deviation from what is expected.

ISO 9001:2015 emphasizes risk-based thinking to achieve conformity and customer satisfaction while developing quality management system. Risk-based thinking enables an organization to determine the factors, under different clauses of ISO 9000: 2015 Clauses, which could cause its processes and its quality management system to deviate from the planned results/objectives to put in place preventive control to minimize negative effects and to make maximum use of opportunities as they arise.

Risk management is a set of coordinated activities to direct and control the risk. Risk management is the process of identifying, analyzing, assessing, and prioritizing different kinds of risks; responding to any risk by planning risk mitigation; implementing mitigation plan; and controlling the risks. It is a process of thinking systematically about the possible risks, problems, or disasters before they happen and setting up the procedure that will avoid the risk or minimize the impact or cope with the impact. The objectives of project risk management are to increase the probability and impacts of positive events and decrease the probability and impacts of events adverse to the project objectives.

There are many categories of risks. Table 1.1 lists major categories of risks that may occur in a project.

1.2 ISO 31000 RISK MANAGEMENT STANDARD

ISO is the world's largest developer and publisher of international standards. ISO (the International Organization for Standardization) is a worldwide federation of national standards bodies (ISO member bodies). The work of preparing International Standards is normally carried out through ISO technical committees. Each member body interested in a subject for which a technical committee has been established has

DOI: 10.1201/9781003245612-1

TABLE 1.1
Typical Categories of Risks in a Project

Serial Number	Category
1	Management
2	Statutory/Regulatory
3	Contracting (project contracting system)
4	Project study
5	Owner (stakeholder) requirements/project scope
6	Technical (design)
7	Technology (new technology)
8	Planning of project (schedule)
9	Financial (project budget)
10	Resources/Skilled workforce
11	Project activities (construction)
12	Health, safety, and environment (HSE)
13	Operational (performance)
14	Economical
15	Commercial
16	Legal
17	Political/Social
18	Natural/Environmental
19	Geological
20	Supply and demand (market)
21	Force majeure

the right to be represented on that committee. It is a nongovernmental organization that forms a bridge between the public and private sectors. ISO has more than 21,000 international standards. Of all the standards produced by ISO, the ones that are most widely known are the ISO 9000 and ISO 14000 series. ISO 9000 has become an international reference for quality requirements in business-to-business dealings, and ISO 14000 looks to achieve at least as much, if not more, in helping organizations to meet their environmental changes. ISO 9000 and ISO 14000 families are known as "generic management system standards".

ISO 9001:2015 emphasizes risk-based thinking to achieve conformity and customer satisfaction while developing quality management system. In order to effectively meet the quality management system's goal, ISO 90001:2015 requires the organizations to consider the risks as part of their quality management system's plan. It required to systematically consider the risk in all the clauses of ISO 9001:2015.

ISO 31000 is an international standard published in 2009 that provides principles and guidelines for an effective risk management. It outlines a generic approach to risk management, which can be applied to different types of risks (financial, safety, project risks) and used by any type of organization. The standard provides a uniform vocabulary and concepts for discussing risk management. It provides guidelines and principles that can help to undertake a critical review of your organization's risk management process. ISO 31000 is intended to serve as a guide for the design, implementation, and maintenance of risk management. ISO 31000:2009 describes a systematic and logical process, during which organizations manage risk by identifying it, analyzing, and then evaluating whether the risk should be modified by risk treatment in order to satisfy their risk criteria.

ISO standards are updated periodically. ISO 31000:2009 was revised in 2018. The book discusses the guidelines intended to streamline risk management for organizations and the clauses described in ISO 31000:2018. Table 1.2 lists the clauses of ISO 31000:2018.

Risk is involved in any activity of an organization. Risk management can be applied to an entire organization, at its many areas and levels, at any time, as well as to specific functions, projects, and activities.

Risk management is application specific. In some circumstances, it can therefore be necessary to supplement the vocabulary in this Guide. Where terms related to the management of risk are used in a standard, it is imperative that their intended meanings within the context of the standard are not misinterpreted, misrepresented, or misused. ISO 31000 focuses on best practice principles for implementing, maintaining, and improving a framework for risk management.

TABLE 1.2
ISO 31000:2018 Sections (Clauses)

Section (Clause) No.	Relevant Clause in 31000:2018	Description
1	Scope	
2	Normative references	
3	Terms and references	
	3.1	Risk
	3.2	Risk management
	3.3	Stakeholder
	3.4	Risk source
	3.5	Event
	3.6	Sequence
	3.7	Likelihood
	3.8	Control
4	Principles	
	4.1	Integrated
	4.2	Structured and comprehensive
	4.3	Customized
	4.4	Inclusive
	4.5	Dynamic
	4.6	Best available information
	4.7	Human and cultural factors
	4.8	Continual improvement

(Continued)

TABLE 1.2
(Continued)

Section (Clause) No.	Relevant Clause in 31000:2018	Description
5	**Framework**	
	5.1	General
	5.2	Leadership and commitment
	5.3	Integration
	5.4	Design
	5.4.1	Understanding the organization and its context
	5.4.2	Articulating risk management commitment
	5.4.3	Assisting organizational roles, authorities, responsibilities, and accountabilities
	5.4.4	Allocating resources
	5.4.5	Establishing communication and consultation
	5.5	Implementation
	5.6	Evaluation
	5.7	Improvement
	5.7.1	Adapting
	5.7.2	Continually improving
6	**Process**	
	6.1	General
	6.2	Communication and consultation
	6.3	Scope, context, and criteria
	6.3.1	General
	6.3.2	Defining the scope
	6.3.3	External and internal context
	6.3.4	Defining risk criteria
	6.4	Risk assessment
	6.4.1	General
	6.4.2	Risk identification
	6.4.3	Risk analysis
	6.4.4	Risk evaluation
	6.5	Risk treatment
	6.5.1	General
	6.5.2	Selection of risk treatment options
	6.5.3	Preparing and implementing risk treatment plans
	6.6	Monitoring and review
	6.7	Recording and reporting

ISO 31000 is a set of guidelines, not requirements, whereas ISO 9001 and ISO 14001 are requirements, which means they compose a strict set of specifications that can be certified to. ISO 31000 is not for certification purposes. It's simply a set of best practice guidelines.

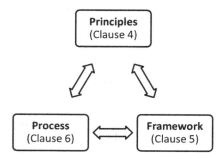

FIGURE 1.1 Risk management clauses' relationship.

ISO 31000 is an effort to acknowledge that business operations always contain a degree of uncertainty and, therefore, risk. No matter what our business goals, there's always a chance that things might go wrong.

ISO 31000 family consists of:

1. ISO 31000:2018 (Principles and Guidelines on Implementation) ISO/IEC
2. 31010:2009 (Risk Assessment Techniques)
3. ISO Guide 73:2009 (Risk Management Vocabulary)

ISO 31000 aims to simplify risk management into a set of clearly understandable and actionable guidelines that should be straightforward to implement, regardless of the size, nature, or location of a business.

Risk management is an ongoing process that continues throughout the life cycle of a project. Managing risk is based on Principles, Framework, and Process outlined under Clauses 4, 5, and 6 of ISO 31000 Risk Management System. These activities have coordinated relationship. Figure 1.1 shows the relationship between these activities.

1.2.1 RISK MANAGEMENT PRINCIPLES

As per ISO 31000, there are eight principles of Risk Management. The purpose of Risk Management principles is for Value Creation and Protection. Figure 1.2 illustrates eight principles of risk management.

These principles provide guidance on the characteristics of effective and efficient risk management, communicating its value and explaining its intention and purpose. The principles are foundation for risk managements and should be considered for establishing the risk management framework and processes.

These principles are further explained as follows:

1. Integrated

- The organization should integrate its risk management efforts into all parts and activities of the organization.
- Risk management should be considered as companywide activity.

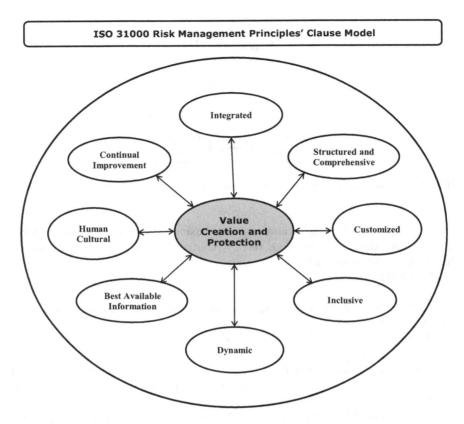

FIGURE 1.2   Risk management principles clause model.

2. Structured and comprehensive

- A structured and comprehensive approach that leads to most consistent, effective, and desirable risk management outcomes

3. Customized

- The risk management framework and process should be customized and proportionate to meet organization's external and internal context related to achieve the goals and objectives of the organization.

4. Inclusive

- The risk management process should encourage timely involvement and participation of all stakeholders to make the process most effective. The risk management process should encourage the use of reliable information, views, and perception of all stakeholders and implement into risk management efforts.

5. Dynamic

- The risk management should be dynamic and capable of managing organizational changes, including its external and internal context and adapting the changes. A risk management process should help the organization anticipate, identify, acknowledge, and respond to the changes and expectations in an appropriate and timely manner.

6. Best available information

- Consider the information from the historical, present, as well future expectations. Risk management should explicitly take into account any limitations and uncertainties associated with such information and expectations. Information should be timely, clear, and available to relevant stakeholders in a timely and clear manner.

7. Human and cultural factors

- Human and cultural factors significantly influence all aspects of risk management at each level and stage. Ensure that risk management process should consider human and cultural factors.

8. Continual improvement

- Continually improve the risk management process through lesson learned and experience.

1.2.2   RISK MANAGEMENT FRAMEWORK

The purpose of risk management framework is to assist the organization in integrating risk management into significant activities and functions. The effectiveness of risk management will depend on its integration with the governance of the organization, including decision-making. The involvement and support of stakeholders, particularly top management, is required for establishment of risk management framework. Figure 1.3 illustrates PDCA cycle for ISO 3100 risk management framework clause activities.

PDCA cycle for establishment of framework considers all the activities such as integrating, designing, implementing, evaluating, and improving risk management across the organization as per following actions:

1. Plan

a) Establishment of risk management framework

- Ask stakeholders to support the establishment of a framework
- Identify gaps in the existing risk management practices and processes
- Evaluate existing risk management practices and processes
- Design requirements to develop risk management framework

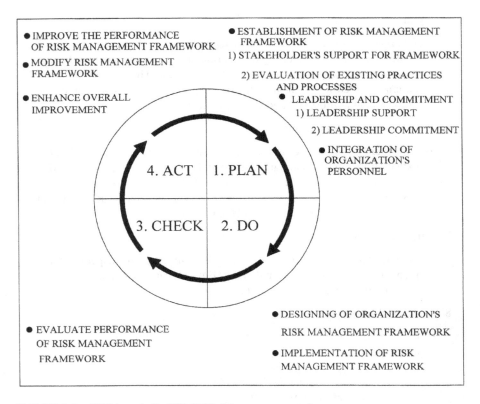

FIGURE 1.3   PDCA cycle for ISO 3100 risk management clause.

    b) Show leadership by making a commitment to risk management, which needs

- Support from organization leadership for risk management framework
- Commitment from top management for risk management framework
- Commitment from oversight bodies for risk management framework
- To determine policy for the establishment of risk management approach, plan, or course of action
- Communication plan to inform the risk to all the stakeholders

    c) Make organization's personnel responsible for risk management

- Make risk management an integral part of the organization's culture.
- Make everyone in the organization to be responsible for managing risk.
- Make organization's leader to commit for customizing and implementing all of the framework.
- Ensure that the necessary resources are allocated to manage risk.

- Involve internal and external stakeholders at every stage of risk management process.
- Assign authority, responsibility, and accountability at appropriate levels within the organization.

2. Do

a) Design organization's customized risk management framework to satisfy organization's need.
b) Design risk management framework to fill the gaps that are existing in current practices and processes.
c) Design risk management framework considering organization's external and internal context.
d) Assign and allocate resources for risk management activities.
e) Determine and define the confidence level.
f) Establish communication and consultation system.
g) Implement risk management framework by developing an appropriate plan including time and resources.

3. Check

a) Evaluate performance of risk management framework.
b) Periodically measure risk management framework performance against the organization's implementation plans, indicators, and its behavior.

4. Act

a) Improve the performance of risk management framework.
b) Modify risk management framework.
c) Enhance overall improvement,

1.2.3   RISK MANAGEMENT PROCESS

The risk management process involves the systematic application of policies, procedures, and practices to the activities of communicating and consulting, establishing the context and assessing, treating, monitoring, reviewing, recording, and reporting risk. Figure 1.4 illustrates organizational structure of ISO 31000 process clause.

The risk management process is an integral part of management and decision-making and integrated into the structure, operation, and processes of the organization. It can be applied at strategic, operational, program, or project level.

ISO 31000 risk management process clause consists of following elements:

1. General
2. Communication and consultation
3. Scope, context, and criteria
4. Risk assessment
5. Risk treatment
6. Monitoring and review
7. Recording and reporting

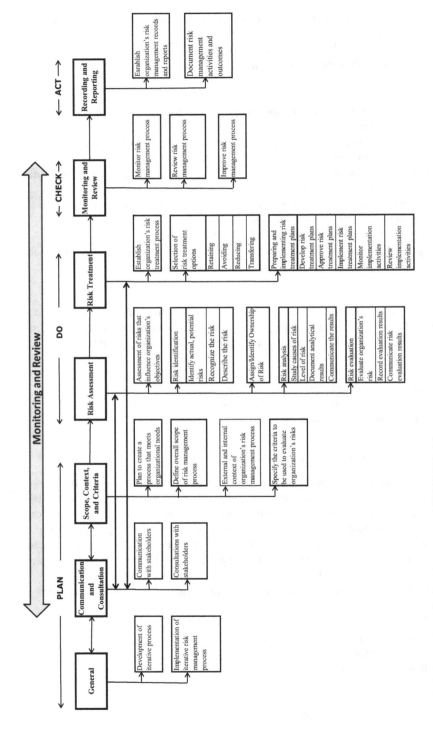

FIGURE 1.4 Organizational structure of ISO 31000 risk management process clause.

1.2.3.1 General

Risk management process is an iterative process. Therefore, risk management process also has to be developed as an iterative process. It is an ongoing process that continues throughout the life cycle of the process. The activities are sequential but iterative and monitored at each stage.

1.2.3.2 Communication and Consultations

The purpose of communication and consultation is to keep informed the relevant stakeholders, both internal and external, to understand the risk, the basis on which decisions are made, and the reasons why particular action is required. The communication method is to be clearly defined based on the standard policies and procedures of the organization. Communication seeks to promote awareness and understanding of risk, whereas consultation involves obtaining feedback and information from the relevant stakeholders to take decisions. Communication and consultations are to be coordinated closely in order to facilitate factual, timely, relevant, accurate and understandable exchange of information among the relevant stakeholders taking into consideration the confidentiality and integrity of the information as well as the privacy right of individuals. Communication and consultations with relevant stakeholders, both internal and external, should take place within and throughout all the stages of risk management process.

The purpose of communication and consultation is to

- Bring together experts from different areas to participate at each step of risk management process
- Involve internal and external stakeholders at each step of risk management process
- Ensure that different views from experts are considered while defining risk criteria and when evaluating the risk
- Provide sufficient information to facilitate risk oversight and decision-making
- Build a sense of inclusiveness and ownership among those affected by the risk

1.2.3.3 Scope, Context, and Criteria

The purpose of establishing scope, context, and criteria is to plan and create customize risk management process to meet organizational needs. Following activities are to be considered while creating a process that meets the organization's customized needs.

- The organization should define the scope of the risk management activities.
- The organization to clearly specify the scope of the organization's risk management activities.
- The organization to clarify the external and internal context of organization's risk management process. The external and internal context is the environment in which the organization seeks to define and achieve its objectivities.

- The organization to specify the criteria that the organization plans to use to evaluate the organization's risks.
- The organization should specify the amount and type of risks that the organization may or may not take to achieve its objectives.

1.2.4 Risk Assessment

Risk assessment is the act of determining the probability that risk will occur and the impact that event would have, should it occur. Risk assessment is designed to reduce or eliminate the risk of certain kind of events happening (occurring) or having an impact on the project. Risk assessment consists of following major elements:

1. Risk identification
2. Risk analysis
3. Risk evaluation

Risk assessment should be conducted systematically, iteratively and collaboratively, drawing on the knowledge and views of stakeholders. It should use the best available information collected through communication and consultation among the stakeholders, supplemented by further enquiry as necessary.

1.2.4.1 Risk Identification

The purpose of risk identification is to find, recognize, and registration of the risks that might help or prevent an organization in achieving its objectives. It is a process of research, recognition, and registration of risk. Risk identification involves determining the source and type of risk which may affect the project. It is also to determine how often (frequency) it occurs.

The organization can use a range of techniques for identifying uncertainties that may affect one or more objectives. Following tools and techniques are used to identify risks.

1. Benchmarking
2. Brainstorming
3. Delphi technique
4. Interviews
5. Past database, historical data from similar projects
6. Questionnaires
7. Risk breakdown structure
8. Workshops
9. Root Cause Analysis
10. SWOT Analysis

A brief description of some of these tools is given below.

1. Brainstorming

Benchmarking is the process of measuring the actual performance of the organization's products, processes, services and compare to the best-known industry standards to assist the organization in improving the performance of their products, processes, and services. Benchmarking involves analyzing an existing situation, identifying and measuring factors critical to the success of the product or services, comparing them with other businesses, analyzing the results, and implementing an action plan to achieve better performance. Following is the process for benchmarking.

1. Collect internal and external data on work, process, method, product characteristics, and system selected for benchmarking.
2. Analyze data to identify performance gaps and determine cause and differences.
3. Prepare action plan to improve the process in order to meet or exceed the best practices in the industry.
4. Search for the best practices among market leaders, competitors, and non-competitors, which lead to their superior performance.
5. Improve the performance by implementing these practices.

Figure 1.5 illustrates benchmarking process.

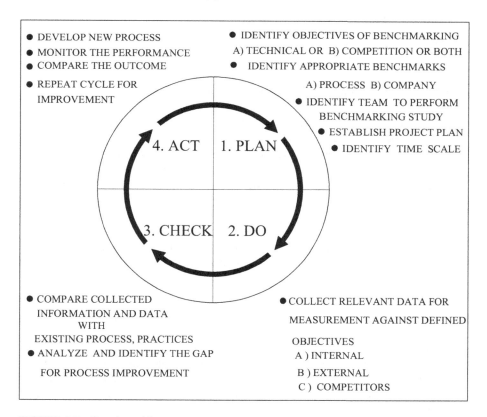

FIGURE 1.5   Benchmarking process.

Source: Abdul Razzak Rumane. (2013). *Quality Tools for Managing Construction Projects*. Reprinted with permission of Taylor & Francis Group.

2. Brainstorming

Brainstorming is listing all the ideas put forth by a group in response to a given question or problem. It is a process of creating ideas by storming some objectives. In 1939, a team led by advertising executive Alex Osborn coined the term "brainstorm". According to Osborn, brainstorm means using the brain to storm a creative problem. Classical brainstorming is the most well-known and often-used technique for idea generation in a short period of time. It is based on the fundamental principles of deferment of judgment and that quantity breeds quality. It involves questions such as

- Does the item have any design features that are not necessary?
- Can two or more parts be combined together?
- How can we cut down the weight?
- Are these nonstandard parts that can be eliminated?

There are four rules for successful brainstorming:

1. Criticism is ruled out.
2. Freewheeling is welcomed.
3. Quantity is wanted.
4. Contribution and improvement are sought.

A classical brainstorming session has the following basic steps:

- Preparation: The participants are selected, and a preliminary statement of the problem is circulated.
- Brainstorming: A warm-up session with simple unrelated problems is conducted; the relevant problem and the four rules brainstorming are presented; and ideas are generated and recorded using checklists and other techniques if necessary.
- Evaluation: The ideas are evaluated relative to the problem.

Generally, a brainstorming group should consist of four to seven people, although some suggest larger group. Figure 1.6 illustrates a Brainstorming process.

3. Delphi Technique

Delphi technique is intended to determine a consensus among experts on a subject matter. The goal of Delphi technique is pick brains of experts in the subject area, treating them as contributors to create ideas. It is a measure and method for consensus building by using questionnaire and obtaining responses from the panel of experts in the selected subjects. Delphi technique employs multiple iterations designed to develop consensus opinion about the specific subject. The selected expert group answers questions by facilitator. The responses are summarized and further circulated for group comments to reach the consensus. The

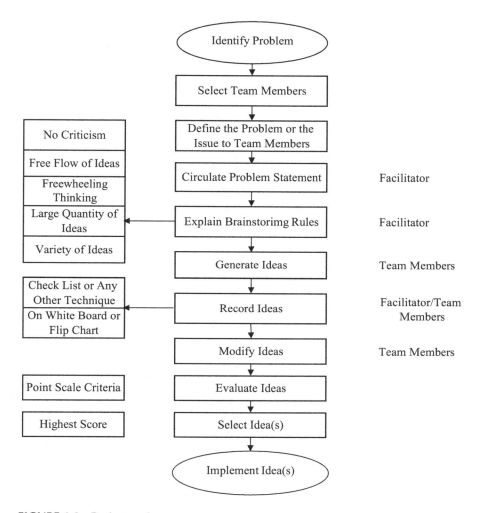

FIGURE 1.6 Brainstorming process.

Source: Abdul Razzak Rumane. (2013). *Quality Tools for Managing Construction Projects*. Reprinted with permission of Taylor & Francis Group.

iteration/feedback process allows the team members to reassess their initial judgment and change or modify the earlier suggestions. Figure 1.7 illustrates a Delphi Technique Process.

4. Questionnaires

Quality tools such as 5W2H and 5Why can be used for risk assessment.

   a) 5W2H is about asking the questions to understand about a process or problem.

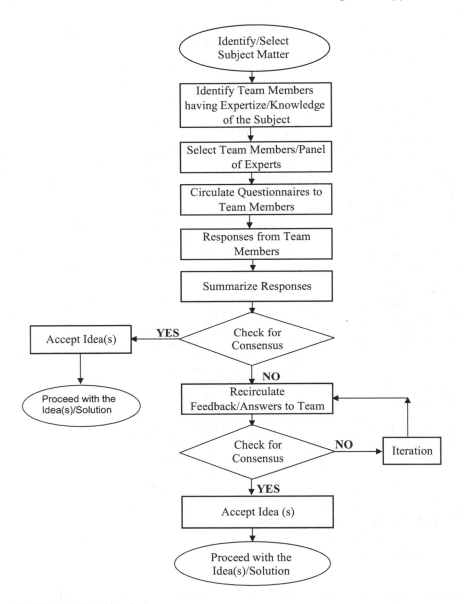

FIGURE 1.7   Delphi technique process.

Source: Abdul Razzak Rumane. (2013). *Quality Tools for Managing Construction Projects*. Reprinted with permission of Taylor & Francis Group.

The five Ws are:

1. Why
2. What

3. When
4. Where
5. Who

And two Hs are:

1. How
2. How much

b) 5Why is used to analyze and solve any problem where the root cause is unknown.

5. Risk Breakdown Structure

Systems engineering starts from the complexity of the large-scale problem as a whole and moves toward structural analysis and the partitioning process until the questions of interest are answered. This process of decomposition is called a Work Breakdown Structure (WBS). The WBS is a hierarchical representation of system levels. Being a family tree, the WBS consists of a number of levels, starting with the complete system at level 1 at the top and progressing downward through as many levels as necessary to obtain elements that can be conveniently managed.

Benefits of systems engineering applications are:

- Reduction in cost of system design and development, production/construction, system operation and support, system retirement, and material disposal
- Reduction in system acquisition time
- More visibility and reduction in the risks associated with the design decision-making process. Risk breakdown system is similar to work breakdown system (WBS). The WBS is a hierarchical representation of system levels. Being a family tree, the WBS consists of a number of levels, starting with the complete system at level 1 at the top and progressing downward through as many levels as necessary to obtain elements that can be conveniently managed.

Risk Breakdown Structure (RBS) represents a systematic and logical breakdown of the risk into its components (activities). It is constructed by dividing the risk into major elements with each of these being divided into sub-elements. This is done till a breakdown is done in terms of manageable units of risk for which responsibility can be defined and risk owner can be assigned. Figure 1.8 illustrates a sample Risk Breakdown Structure. (For more details about systems engineering and RBS, please read Section 3.1 of Chapter 3.)

6. Root Cause Analysis

It is used to organize a graphical display of multiple causes with a particular effect. Figure 1.9 illustrate root cause and effect analysis.

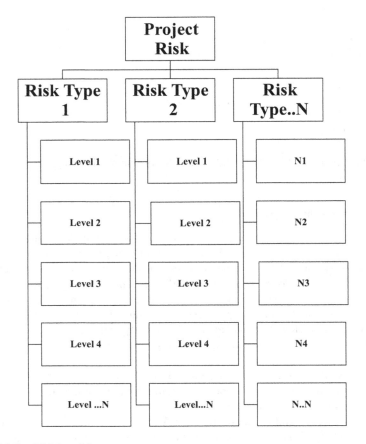

FIGURE 1.8 Risk breakdown structure.

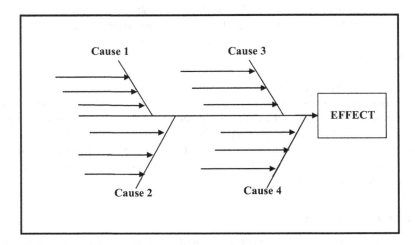

FIGURE 1.9 Cause and effect diagram.

7. SWOT Analysis

SWOT is an analysis to measure the strength, weakness, opportunities, and threat the risk has to the project.

While identifying the risk, the following factors and the relationship between these factors should be considered.

- Tangible and intangible source of risk
- Causes and events
- Vulnerabilities and capabilities
- Changes in the internal and external context
- Indicators of emerging risks
- The nature and value of assets and resources
- Consequences and their impact on objectives
- Limitations of knowledge and reliability of information
- Time-related factors
- Biases, assumptions, and beliefs of those involved

The identified risks are classified into following.

1. Internal
2. External

These are further divided into following main categories.

- Management
- Project (Contract)
- Technical
- Construction
- Physical
- Logistic
- Health, Safety, and Environmental
- Statutory/Regulatory
- Financial
- Commercial
- Economical
- Political
- Legal
- Natural

Each identified risk is documented in a risk register.

1.2.4.1.1   Risk Register

Risk register is a document recording details of all the identified risks at the beginning of the project and during the life cycle of the project in a format which consists of comprehensive lists of significant risks along with the actions and cost estimated

TABLE 1.3
Risk Register

Project Name
Risk Register

Serial Number	Risk Identification Number (Risk ID)	Description of Risk	Owner of Risk	Estimated Likelyhood of Risk	Impact	Estimated Severity	Prioritization	List of Activities Influenced	Leading Indicators for Risk	Risk Mitigation Plan	Risk Mitigation Plan on Leading Indicator	Timeline for Mitigation Action	Tracking of Leading Indicators (Monitoring)	Date of Review/ Update	Forecasting Risk Happenings	Action to be Taken in Future

SAMPLE FORM

with the identified risks. Risk register is updated every time a new risk is identified or relevant actions are taken. Table 1.3 illustrates an example risk register.

1.2.4.2 Risk Analysis

Risk analysis is the process to analyze the listed risks. The purpose of risk analysis is to comprehend the nature of risk and its characteristics including, where appropriate, the level of risk. Table 1.4 illustrates risk levels normally assumed for Probability Impact Matrix. The % Probability of Occurrence shown in the table is indicative. The organization can determine the probability level as per the nature of business.

Risk analysis involves a detailed consideration of uncertainties, risk resources, consequences, likelihood, events, scenarios, controls, and their evidences. There are two methods of analyzing risks. These are

1. Qualitative analysis
2. Quantitative analysis

1.2.4.2.1 Qualitative Analysis

Qualitative analysis is a process to assess the probability of occurrence (likelihood) of the risk and its impact (consequence). This process prioritizes risks according to their potential effect on project objectives. Qualitative risk analysis requires that the probability and consequences of the risks be evaluated using established qualitative-analysis methods and tools.

TABLE 1.4
Risk Probability Levels

Serial Number	Value	Definition	Meaning	% Probability of Occurrence
1	Level 5	Very high (frequent)	• Almost certain that the risk will occur. • Frequency of occurrence is very high.	40–80
2	Level 4	High (likely)	• It's likely to happen • Frequency of occurrence is less	20–40
3	Level 3	Moderate (occasional)	• Its occurrence is occasional.	10–20
4	Level 2	Low (unlikely)	• It's unlikely to happen	5–10
5	Level 1	Very Low (rare)	• The probability to occur is rare.	0–5

Source: Abdul Razzak Rumane. (2016). *Handbook of Construction Management: Scope, Schedule, and Cost Control.* Reprinted with permission of Taylor & Francis Group.

Following tools and techniques are used for qualitative analysis.

- Failure Mode and Effects Analysis (FMEA)
- Group discussion (workshop)
- Pareto diagram
- Probability and Impact assessment
- Probability levels
- Risk categorization

1.2.4.2.2   Quantitative Analysis

Quantitative analysis is a process to quantify the probability of risk and its impact based on numerical estimation. The quantitative risk analysis process aims to analyze numerically the probability of each risk and its consequences on project objectives, as well as the extent of overall project risk.

Following tools and techniques are used for quantitative analysis.

- Event tree analysis
- Probability analysis
- Sensitivity analysis
- Simulation techniques (Monte Carlo Simulation)

These techniques

- Determine the probability of achieving a specific project objective
- Quantify the risk exposure for the project and determine the size of schedule and cost contingency reserves that may be needed
- Identify the risk requiring the most attention by quantifying their relative contribution to project risk
- Identify realistic and achievable scope, schedule, and cost targets

Risk analysis should consider factors such as:

- The likelihood of events and consequences
- The nature and magnitude of consequences
- Complexity and connectivity
- Time-related factors and volatility
- The effectiveness of existing controls
- Sensitivity and confidence level

Similarly, the risk impact is analyzed for each of the identified risk that may be classified into different levels such as

- Very high
- High
- Substantial

- Possible
- Slight

Based on the Impact level, Probability and Impact Matrix is prepared. Table 1.5 is a sample Risk Probability and Impact Matrix.

Risk analysis provides an input to risk evaluation, to decisions on whether risks need to be treated and how, and on the most appropriate risk treatment strategy and methods. The results provide insights for decisions, where choices are being made, and options involve different types of levels of risk.

1.2.4.3 Risk Evaluation

The purpose of risk evaluation is to support decisions. Risk evaluation involves comparing the results of the risk analysis with the established risk criteria to determine where additional action is required. This can lead to a decision to:

- do nothing further
- consider risk treatment options
- undertake further analysis to better understand the risk
- maintain existing controls
- reconsider objectives

Decisions should take account of the wider context and the actual and perceived consequences to external and internal stakeholders. The outcome of risk evaluation should be recorded, communicated, and then validated at appropriate levels of the organization.

TABLE 1.5
Risk Probability and Impact Matrix

Probability Severity			Probability				
			Level 5	Level 4	Level 3	Level 2	Level 1
			Very High (Frequent)	High (Likely)	Moderate (Occasional)	Low (Unlikely)	Very Low (Rare)
Impact/Severity	Level 5	Very High					
	Level 4	High					
	Level 3	Substantial			SAMPLE FORM		
	Level 2	Possible					
	Level 1	Slight					
			Risk Assessment Matrix for; Scope/Schedule/Cost/Quality				

The results of risk assessment/evaluation are used to prioritize risks to establish very high to very low ranking. Prioritization of risks depends on following factors:

1. Probability (occurrence)
2. Impact (consequences)
3. Urgency
4. Proximity
5. Manageability
6. Controllability
7. Responsiveness
8. Variability
9. Ownership ambiguity

The prioritization list helps the project manager to plan actions and assign the resources to mitigate the realization of high-value probability.

1.2.5 RISK TREATMENT

The purpose of risk treatment is to select and implement options for addressing risk.
    Risk treatment involves an iterative process of:

- formulating and selecting risk treatment options
- planning and implementing risk treatment
- assessing the effectiveness of that treatment
- deciding whether the remaining risk is acceptable

if not acceptable, then taking further treatment.
    Selecting the most appropriate risk treatment option(s) involves balancing the potential benefits derived in relation to the achievement of the objectives against costs, effort, or disadvantages of implementation.

1.2.5.1 Plan Risk Response

Plan Risk Response is a process that determines what action (if any) will be taken to address the identified and assessed risks which are listed under Risk Register on prioritization basis. Risk response process is used for developing options and actions to enhance opportunities and reduce the threats to the identified risk activities in the project.

For each identified risk, a response must be identified. The risk owner and project team have to select the risk response for each of the identified risk.

The probability of the risk event occurring and the impacts (threats) is the basis for evaluating the degree to which the response action is to be evolved. Based on the Risk Probability Impact Matrix, Risk Response Strategy for scope, schedule, cost, quality, and HSE can be listed as per Table 1.6.

TABLE 1.6
Risk Response Strategy

Potential Risk		Probability Level	Impact/Severity Level	Risk Response Strategy
Scope	1			
	2			
Schedule	1			
	2			
		SAMPLE TABLE		
Cost	1			
	2			
Quality	1			
	2			
HSE	1			
	2			

Generally, risk response strategies for impact (consequences) on the project fall into one of the following categories:

1. Avoidance
2. Transfer
3. Mitigation (reduction)
4. Acceptance (retention)

1. Avoidance

Avoidance is changing the project scope, objectives, or plan to eliminate the risk or to protect the project objectives from the impact (threat).

2. Transfer

Transfer is transferring the risk to someone else who will be responsible to manage the risk. Transferring the threat does eliminate the threat, it still exists; however, it is owned and managed by other party.

3. Mitigation

Mitigation is reduction in the probability and/or impact to an acceptable threshold. It is done by taking series of control actions.

4. Acceptance

It is acceptance of consequences after response actions, understanding the risk impact should it occur.

Risk treatment options are not necessarily mutually exclusive or appropriate in all circumstances. Options for treating risk may involve one or more of the following:

- avoiding the risk by deciding not to start or continue with the activity that gives rise to the risk
- taking or increasing the risk in order to pursue an opportunity
- removing the risk source
- changing the likelihood
- changing the consequences
- sharing the risk (e.g., through contracts, buying insurance)
- retaining the risk by informed decision

Justification for risk treatment is broader than solely economic considerations and should take into account all of the organization's obligations, voluntary commitments, and stakeholder views. The selection of risk treatment options should be made in accordance with the organization's objectives, risk criteria, and available resources.

1.2.5.2   Approve Risk Response

When selecting risk treatment options, the organization should consider the values, perceptions, and potential involvement of stakeholders and the most appropriate ways to communicate and consult with them. Though equally effective, some risk treatments can be more acceptable to some stakeholders than to others.

Risk treatments, even if carefully designed and implemented, might not produce the expected outcomes and could produce unintended consequences. Monitoring and review need to be an integral part of the risk treatment implementation to give assurance that the different forms of treatment become and remain effective.

Risk treatment can also introduce new risks that need to be managed.

If there are no treatment options available or if treatment options do not sufficiently modify the risk, the risk should be recorded and kept under ongoing review.

Decision-makers and other stakeholders should be aware of the nature and extent of the remaining risk after risk treatment. The remaining risk should be documented and subjected to monitoring, review, and, where appropriate, further treatment.

1.2.5.3   Implementing of Risk Treatment Plan

The purpose of risk treatment plans is to specify how the chosen treatment options will be implemented, so that arrangements are understood by those involved, and progress against the plan can be monitored. The treatment plan should clearly identify the order in which risk treatment should be implemented.

Treatment plans should be integrated into the management plans and processes of the organization in consultation with appropriate stakeholders.

The information provided in the treatment plan should include:

- the rationale for selection of the treatment options, including the expected benefits to be gained
- those who are accountable and responsible for approving and implementing the plan
- the proposed actions
- the resources required, including contingencies
- the performance measures
- the constraints
- the required reporting and monitoring
- when actions are expected to be undertaken and completed

1.2.6 MONITORING AND REVIEWING

The purpose of monitoring and review is to assure and improve the quality and effectiveness of process design, implementation, and outcomes. Ongoing monitoring and periodic review of the risk management process and its outcomes should be a planned part of the risk management process, with responsibilities clearly defined.

It is a systematic process of tracking identified risks, monitoring residual risks, identifying new risks, execution of risk response plan, and evaluating the effectiveness of implementation of actions against established levels of risk in the area of scope, time, cost, and quality throughout the life cycle of the project. It involves timely implementation of risk response to identified risk to ensure the best outcome for a risk to a project. It is an ongoing process for the life cycle of the project.

Monitoring and review should take place in all stages of the process. Monitoring and review includes planning, gathering, and analyzing information; recording results (update risk register); and providing feedback.

Risk monitoring process provides information that assists with making effective decisions in advance of the risk occurring. Communication to all project stakeholders is needed to assess periodically the acceptability of the level of risk on the project. It helps monitor status and trends continually (scope, schedule, cost estimates, quality of product, etc.).

The results of monitoring and review should be incorporated throughout the organization's performance management, measurement, and reporting activities.

Risk monitoring process helps to determine if:

- Risk treatment and risk responses have been implemented as planned
- Risk treatment/response actions are effective as expected, or if new treatment plan should be developed
- Project assumptions are still valid
- Risk exposure has changed from its prior state, with analysis of trends
- A risk trigger has occurred
- Proper policies and procedures are followed
- Risks have occurred or arisen that were not previously identified

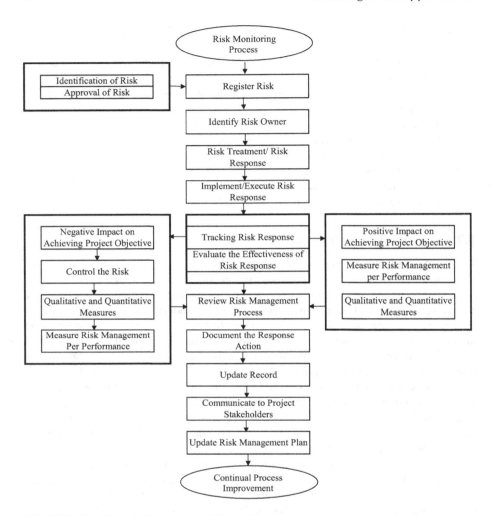

FIGURE 1.10 Typical flowchart for risk monitoring process.

Figure 1.10 illustrates a typical flowchart for Risk Monitoring Process.

1.2.7 RECORDING AND REPORTING

The risk management process and its outcomes should be documented and reported through appropriate mechanisms. Recording and reporting aims to:

- communicate risk management activities and outcomes across the organization
- provide information for decision-making
- improve risk management activities
- assist interaction with stakeholders, including those with responsibility and accountability for risk management activities

TABLE 1.7
Risk Monitoring and Control Form

Serial Number	Description	Action			
1	Risk ID				
2	Description of risk				
3	Response	Sample form			
4	Strategy of response	Avoidance	Transfer	Mitigation	Acceptance
5	Monitoring and control	Risk owner	Review date	Critical issue	
6	Estimated impact on the project				
7	Actual impact on the project				
8	Revised response				
9	Record update date				
10	Communication to stakeholders				

Decisions concerning the creation, retention, and handling of documented information should take into account, but not be limited to, their use, information sensitivity, and the external and internal context.

Reporting is an integral part of the organization's governance and should enhance the quality of dialogue with stakeholders and support top management and oversight bodies in meeting their responsibilities. Factors to consider for reporting include, but are not limited to,

- differing stakeholders and their specific information needs and requirements
- cost, frequency, and timeliness of reporting
- method of reporting
- relevance of information to organizational objectives and decision-making.

Table 1.7 illustrates risk monitoring and control form.

Figure 1.11 is typical flowchart for risk management procedure developed based on ISO 31000 Risk Management Clauses.

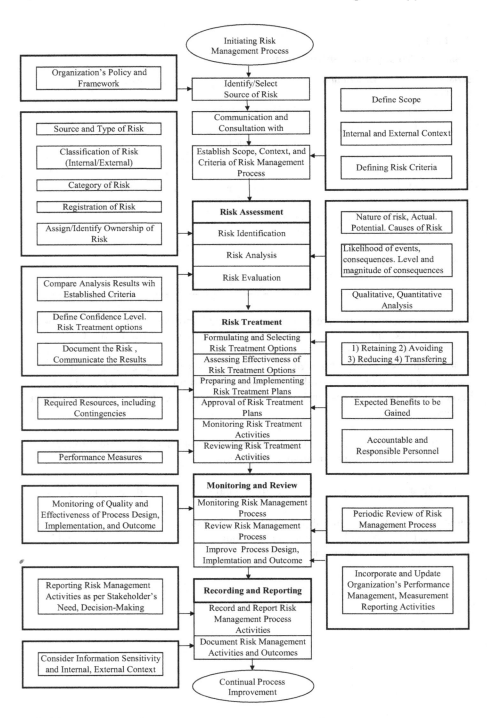

FIGURE 1.11 Typical flowchart for risk management procedure.

1.3 BENEFITS OF RISK MANAGEMENT SYSTEM

Some of the benefits of risk management system are as follows:

- likelihood of increase in achieving the objectives of the organization
- improvement of management system
- encourage proactive management
- improve governance
- improve stakeholder confidence and trust
- improve the ability to identify threats and opportunities
- identifying both positive and negative opportunities to improve decision-making and allocation of resources
- improve controls
- improve operational efficiency and effectiveness
- awareness among employees and stakeholders about the importance of risk management
- adherence to compliance with regulatory requirements
- basis for planning and decision-making
- enhance documentation and reporting system
- compliance with codes and standards
- proper management of stakeholders' expectations
- minimize the possibility and/or impact of project scope creep
- creating realistic project schedule estimate
- creating realistic project cost estimate
- proper management of project resources

2 Overview of Quality of Projects

2.1 OVERVIEW OF QUALITY

Quality issues have been of great concern throughout the recorded history of humans. During the New Stone Age, several civilizations emerged, and some 4000–5000 years ago, considerable skills in construction were acquired. The pyramids in Egypt were built in approximately 2589–2566 BCE. Hammurabi, the king of Babylonia (1792–1750 bce), codified the law, according to which, during the Mesopotamian era, builders were responsible for maintaining the quality of buildings and were given the death penalty if any of their construction collapsed and their occupants were killed. The extension of Greek settlements around the Mediterranean after 200 BCE left records showing that temples and theaters were built using marble. India had strict standards for working in gold in the 4th century BCE.

China's recorded quality history can be traced back to earlier than 200 BCE. China had instituted quality control in its handicrafts during the Zhou dynasty between 1100 and 250 bce. During this period, the handicraft industry was mainly engaged in producing ceremonial artifacts. This industry survived the long succession of dynasties that followed up to 1911 CE.

During the Middle Ages, guilds took the responsibility for quality control upon themselves. Guilds and governments carried out quality control; consumers, of course, carried out informal quality inspection throughout the history.

The guilds' involvement in quality was extensive. All craftsmen living in a particular area were required to join the corresponding guild and were responsible for controlling the quality of their own products. If any of the items was found defective, then the craftsman discarded the faulty items. The guilds also initiated punishments for members who turned out shoddy products. They maintained inspections and audits to ensure that artisans followed quality specifications. The guild hierarchy consisted of three categories of workers: apprentice, journeyman, and master. The guilds had established specifications for input materials, manufacturing processes, and finished products, as well as methods of inspection and testing. They were active in managing quality during Middle Ages until the Industrial Revolution marginalized their influence.

2.1.1 Definition of Quality

Quality has different meanings for different people. The American Society for Quality (ASQ) glossary defines quality as

A subjective term for which, each person has his or her own definition. In technical usage, quality can have two meanings:

DOI: 10.1201/9781003245612-2

1. The characteristics of a product or service that bear on its ability to satisfy stated or implied needs.
2. A product or service free of deficiencies.

It further states that it is

- Based on customers' perceptions of a product's design and how well the design matches the original specifications.
- The ability of a product and service to satisfy stated or implied needs.
- Achieved by conforming to established requirements within an organization.

The International Organization for Standardization (ISO, 1994a) defines quality as "the totality of characteristics of an entity that bears on its ability to satisfy stated or implied needs".

Based on various definitions, it is possible to evolve a common definition of quality, which is mainly related to the manufacturing, processes, and service industries as follows:

- Meeting the customer's need
- Fitness for use
- Conforming to requirements

2.1.2 EVOLUTION OF QUALITY MANAGEMENT SYSTEM

The Industrial Revolution began in Europe in the mid-19th century. It gave birth to factories, and the goals of the factories were to increase productivity and reduce costs. Prior to the Industrial Revolution, items were produced by individual craftsman for individual customers, and it was possible for workers to control the quality of their products. Working conditions then were more conducive to professional pride. Under the factory system, the tasks needed to produce a product were divided among several or many factory workers. Under this system, large groups of workmen were performing similar types of work, and each group was working under the supervision of a foreman who also took on the responsibility of controlling the quality of the work performed. Quality in the factory system was ensured by means of skilled workers, and the quality audit was done by inspectors.

The broad economic result of the factory system was mass production at low costs. The Industrial Revolution changed the situation dramatically with the introduction of a new approach to manufacturing.

The beginning of the 20th century marked the inclusion of process in quality practices. During World War I, the manufacturing process became more complex. Production quality was the responsibility of quality control departments. The introduction of mass production and piecework created quality problems as workmen were interested in earning more money by the production of extra products, which in turn led to bad workmanship. This situation made factories introduce full-time quality inspectors, which marked the real beginning of inspection quality control and thus the introduction of quality control departments headed by superintendents.

Walter Shewhart introduced statistical quality control in the process. His concept was that quality is not relevant to the finished product but to the process that created the product. His approach to quality was based on the continuous monitoring of process variation. The statistical quality control concept freed manufacturers from the time-consuming 100% quality control system because it accepted that variation is tolerable up to certain control limits. Thus, the quality control focus shifted from the end of line to the process.

The systematic approach to quality in industrial manufacturing started during the 1930s when some attention was given to the cost of scrap and rework. With the impact of mass production, which was required during World War II, it became necessary to introduce a more stringent form of quality control. This was instituted by manufacturing units and was identified as Statistical Quality Control (SQC). SQC made a significant contribution in that it provided a sampling rather than 100% product inspection. However, SQC was instrumental in exposing the under appreciation of the engineering of product quality.

From the foregoing writings and many others on the history of quality, it is evident that the quality system in its different forms has moved through distinct quality eras such as

1. Quality inspection
2. Quality control
3. Quality assurance
4. Quality engineering
5. Quality management

2.1.2.1   Quality Inspection

Prior to the Industrial Revolution, items were produced by an individual craftsman, who was responsible for material procurement, production, inspection, and sales. In case any quality problems arose, the customer would take up issues directly with the producer. The Industrial Revolution provided the climate for continuous quality improvement. In the late 19th century, Fredrick Taylor's system of Scientific Management was born. It provided the backup for the early development of quality management through inspection. At the time when goods were produced individually by craftsmen, they inspected their own work at every stage of production and discarded faulty items. When production increased with the development of technology, scientific management was born out of a need for standardization rather than craftsmanship. This approach required each job to be broken down into its component tasks. Individual workers were trained to carry out these limited tasks, making craftsmen redundant in many areas of production. The craftsmen's tasks were divided among many workers. This also resulted in mass production at lower cost, and the concept of standardization started resulting in interchangeability of similar types of bits and pieces of product assemblies. One result of this was a power shift away from workers and toward management.

With this change in the method of production, inspection of the finished product became the norm rather than inspection at every stage. This resulted in wastage because defective goods were not detected early enough in the production process. Wastage added costs that were reflected either in the price paid by the consumer or in reduced profits. Due to the competitive nature of the market, there was pressure on manufacturers to reduce the price for consumers, which in turn required cheaper input prices and lower production costs. In many industries, emphasis was placed on automation to try to reduce the costly mistakes generated by workers. Automation led to greater standardization, with many designs incorporating interchanges of parts. The production of arms for the 1914–1918 war accelerated this process.

An inspection is a specific examination, testing, and formal evaluation exercise and overall appraisal of a process, product, or service to ascertain if it conforms to established requirements. It involves measurements, tests, and gauges applied to certain characteristics in regard to an object or an activity. The results are usually compared to specified requirements and standards for determining whether the item or activity is in line with the target. Inspections are usually nondestructive. Some of the nondestructive methods of inspection are:

- Visual
- Liquid dyed penetrant
- Magnetic particle
- Radiography
- Ultrasonic
- Eddy current
- Acoustic emission
- Thermography

The degree to which inspection can be successful is limited by the established requirements. Inspection accuracy depends on:

1. Level of human error
2. Accuracy of the instruments
3. Completeness of the inspection planning

Human errors in inspection are mainly due to:

- Technique errors
- Inadvertent errors
- Conscious errors
- Communication errors

Most construction projects specify that all the contracted works are subject to inspection by the owner/consultant/owner's representative.

2.1.2.2 Quality Control

The quality control era started at the beginning of the 20th century. The Industrial Revolution had brought about the mechanism and marked the inclusion of process in quality practices. The ASQ termed the quality control era as process orientation that consists of product inspection and statistical quality control.

The quality control can be defined as a process of analyzing data collected through statistical techniques to compare with actual requirements and goals to ensure its compliance with some standards.

A control chart is a graphical representation of the mathematical model used to detect changes in a parameter of the process. Charting statistical data is a test of the null hypothesis that the process from which the sample came has not changed. A control chart is employed to distinguish between the existence of a stable pattern of variation and the occurrence of an unstable pattern. If an unstable pattern of variation is detected, action may be initiated to discover the cause of the instability. Removal of the assignable cause should permit the process to return to stable state.

There are a variety of methods, tools, and techniques that can be applied for quality control and the improvement process. These are used to create an idea, engender planning, analyze the cause, analyze the process, foster evaluation, and create a wide variety of situations for continuous quality improvement. These tools can also be used during various stages of a construction project.

2.1.2.3   Quality Assurance

Quality assurance is the third era in the quality management system.

The ASQ defines quality assurance as "all the planned and systematic activities implemented within the quality system that can be demonstrated to provide confidence a product or service will fulfill requirements for quality".

ASQ details this era:

After entering World War II in December 1941, the United States enacted legislation to help gear the civilian economy to military production. At that time, military contracts were typically awarded to manufacturers who submitted the lowest competitive bid. Upon delivery, products were inspected to ensure conformance to requirements.

During this period, quality became a means to safety. Unsafe military equipment was clearly unacceptable, and the armed forces inspected virtually every unit of product to ensure that it was safe for operation. This practice required huge inspection forces and caused problems in recruiting and retaining competent inspection personnel. To ease the problems without compromising product safety, the armed forces began to utilize sampling inspection to replace unit-by-unit inspection. With the aid of industry consultants, particularly the Bell Laboratories, they adapted sampling tables and published them in a military standard: Mil-Std-105. The tables were incorporated into the military contracts themselves. In addition to creating military standards, the armed forces helped their suppliers improve their quality by sponsoring training courses in Shewhart's statistical quality control (SQC) techniques. While the training led to quality improvements in some organizations, most companies had little motivation to truly integrate the techniques. As long as government contracts paid the bills, organizations' top priority remained meeting production deadlines. Most SQC programs were terminated once the government's contracts came to an end.

According to ISO 9000 (or BS 5750), quality assurance is "those planned and systematic actions necessary to provide adequate confidence that product or service will satisfy given requirements for quality". ISO 8402–1994 defines quality assurance as "all the planned and systematic activities implemented within the quality system, and demonstrated as needed, to provide adequate confidence that an entity will fulfill requirements for quality".

The third era of quality management saw the development of quality systems and their application principally to the manufacturing sector. This was due to the impact of the following external environment upon the development take-up of quality systems at this time:

- Growing and, more significantly, maturing populations
- Intensifying competition

These converging trends contributed greatly to the demand for more, cheaper, and better-quality products and services. The result was the identification of quality assurance schemes as the only solution to meet this challenge.

Quality assurance is the activity of providing evidence to establish confidence among all concerned that quality-related activities are being performed effectively. All these planned or systematic actions are necessary to provide adequate confidence that a product or service will satisfy given requirements for quality.

Quality assurance covers all activities from design, development, production/ construction, installation, and servicing to documentation and also includes regulations of the quality of raw materials, assemblies, products, and components; services related to production; and management, production, and inspection processes.

Quality assurance in construction projects covers all activities performed by the design team, contractor, and quality controller/auditor (supervision staff) to meet owners' objectives as specified and to ensure that the project/facility is fully functional to the satisfaction of the owners/end users.

2.1.2.4 Quality Engineering

Feigenbaum (1991) defines quality engineering technology as "the body of technical knowledge for formulating policy and for analyzing and planning product quality in order to implement and support that quality system which will yield full customer satisfaction at minimum cost".

2.1.2.5 Quality Management

The ASQ glossary defines quality management as "the application of quality management system in managing a process to achieve maximum customer satisfaction at the lowest overall cost to the organization while continuing to improve the process".

However, quality actually emerged as a dominant thinking only since World War II, becoming an integral part of overall business system focused on customer satisfaction and becoming known in recent times as "Total Quality Management", with its three constitutive elements:

- Total: Organization-wide
- Quality: Customer satisfaction
- Management: Systems of managing

2.1.2.6   Total Quality Management

The Total Quality Management (TQM) concept was born following World War II. It was stimulated by the need to compete in the global market where higher quality, lower cost, and more rapid development are essential to market leadership. Today, TQM is considered a fundamental requirement for any organization to compete, let alone lead, in its market. It is a way of planning, organizing, and understanding each activity of the process and removing all the unnecessary steps routinely followed in the organization. TQM is a philosophy that makes quality values the driving force behind leadership, design, planning, and improvement in activities. Table 2.1 summarizes periodical changes in the quality system.

2.1.2.6.1   Changing Views of Quality

The failure to address the culture of an organization is frequently the reason for management initiatives either having limited success or failing altogether. To understand the culture of the organization and using that knowledge to implement, cultural changes are important elements of TQM. The culture of good teamwork and cooperation at all levels in an organization is essential to the success of TQM. Table 2.2 describes cultural changes needed in an organization to meet Total Quality Management.

TABLE 2.1
Periodical Changes in Quality System

Period	System
• Middle Ages (1200–1799)	• Guilds: Skilled craftsmen were responsible to control their own products.
• Mid-18th century (Industrial Revolution)	• Establishment of factories; increase in productivity; mass production; assembly lines; several workers were responsible to produce a product; production by skilled workers and quality audit by inspectors.
• Early 19th century	• Craftsmanship model of production.
• Late 19th century (1880s)	• Fredrick Taylor and "Scientific Management"; quality management through inspection.
• Beginning of 20th century (1920s)	• Walter Shewhart introduced Statistical Process Control; introduction of full-time quality inspection and quality-control department; quality management.
• 1930s	• Introduction of sampling method.
• 1950s	• Introduction of Statistical Quality Process in Japan.
• Late 1960s	• Introduction of Quality Assurance
• 1970s	• Total Quality Control
	• Quality Management
• 1980s	• Total Quality Management (TQM)
• Beginning of 21st century	• Integrated Quality Management (IQM)

Source: Abdul Razzak Rumane. (2017). *Quality Management in Construction Projects*, second edition. Reprinted with permission of Taylor & Francis Group.

TABLE 2.2
Cultural Changes Required to Meet Total Quality Management

From	To
• Inspection orientation	• Defect prevention
• Meet the specification	• Continuous improvement
• Get the product out	• Customer satisfaction
• Individual input	• Cooperative efforts
• Sequential engineering	• Team approach
• Quality control department	• Organizational involvement
• Departmental responsibility	• Management commitment
• Short-term objective	• Long-term vision
• People as cost burden	• Human resources as assets
• Purchase of products or services on price-alone basis	• Purchase on total cost minimization basis
• Minimum cost suppliers	• Mutual beneficial supplier relationship

Source: Abdul Razzak Rumane. (2017). *Quality Management in Construction Projects*, second edition. Reprinted with permission of Taylor & Francis Group.

The prominence of product quality in the public mind has resulted in quality becoming a cardinal priority for most organizations. The identification of quality as a core concern has evolved through a number of changing business conditions. These include

1. Competition
2. The customer-focused organization
3. Higher levels of customer expectation
4. Performance improvement
5. Changes in organization forms
6. Changing workforce
7. Information revolution
8. Electronic commerce
9. Role of a quality department

2.2 PROJECT DEFINITION

The authors of *A Guide to the Project Management Body of Knowledge* (PMBOK 2000) define the word *project* in term of its distinctive characteristics: "A project is a temporary endeavor undertaken to create a unique product or service". "Temporary" means that every project has a definite beginning and a definite end. "Unique" means that the product or service is different in some distinguishing way from all similar products or services.

It further states that projects are often critical components of the performing organization business strategy. Examples of projects include

- Developing a new product or service
- Effecting a change in structure, staffing, or style of an organization
- Designing a new transportation vehicle/aircraft
- Developing or acquiring a new or modified information system
- Running a campaign for political office
- Implementing a new business procedure or process
- Constructing a building or facility

The duration of a project is finite; projects are not ongoing efforts, and the project ceases when its declared objectives have been attained. Among other shared characteristics, projects are

1. Performed by people
2. Constrained by limited resources
3. Planned, executed, and controlled

Based on aforementioned definitions, the project can be defined as follows: "A project is a plan or program performed by the people with assigned resources to achieve an objective within a finite duration".

2.3 CONSTRUCTION PROJECTS

Construction is translating owner's goals and objectives, by the contractor, to build the facility as stipulated in the contract documents, plans, specifications within budget, and on schedule.

Construction has a history of several thousand years. The first shelters were built from stone or mud and the materials collected from the forests to provide protection against cold, wind, rain, and snow. These buildings were primarily for residential purposes, although some may have had some commercial function.

During the New Stone Age, people introduced dried bricks, wall construction, metal working, and irrigation. Gradually, people developed the skills to construct villages and cities, and considerable skills in building were acquired. This can be seen from the great civilizations in different parts of the world—some 4,000–5,000 years ago. During the early period of Greek settlement, which was about 2000 BCE, the buildings were made of mud using timber frames. Later, temples and theaters were built from marble. Some 1,500–2,000 years ago, Rome became the leading center of world culture, which extended to construction.

Marcus Vitruvius Pollo, the 1st-century military and civil engineer, penned in Rome the world's first major treatise on architecture and construction. It dealt with building materials, the styles and design of building types, the construction process, building physics, astronomy, and building machines.

During the Middle Ages (476–1492 AD), improvements happened in agriculture and artisanal productivity and exploration, and, as a consequence, the broadening of commerce took place and in the late Middle Ages, building construction became a major industry. Craftsmen were given training and education in order to develop skills and to raise their status. At this time, guilds came up to identify true craftsmen and set standards for quality.

The 15th century brought a "renaissance" or renewal in architecture, building, and science. Significant changes occurred during the 17th century and thereafter due to the increasing transformation of construction and urban habitat.

The scientific revolution of the 17th and 18th centuries gave birth to the great Industrial Revolution of the 18th century. After some delay, construction followed these developments in the 19th century.

The first half of the 20th century witnessed the construction industry becoming an important sector throughout the world, employing many workers. During this period, skyscrapers, long-span dams, shells, and bridges were developed to satisfy new requirements and marked the continuing progress of construction techniques. The provision of services such as heating, air conditioning, electrical lighting, water mains, and elevators in buildings became common. The 20th century has seen the construction and building industry transforming into a major economic sector. During the second half of the 20th century, the construction industry began to industrialize, introducing mechanization, prefabrication, and system building. The design of building services systems changed considerably in the last 20 years of the 20th century. It became the responsibility of designers to follow health, safety, and environmental regulations while designing any building.

Building and commercial—traditional A/E type—construction projects account for an estimated 25% of the annual construction volume. Building construction is a labor-intensive endeavor. Every construction project has some elements that are unique. No two construction or R&D projects are alike. Though it is clear that many building projects are more routine than research and development projects, some degree of customization is a characteristic of the projects.

Construction projects involve a cross-section of many different participants. These both influence and depend on each other in addition to the "other players" involved in the construction process. Figure 2.1 illustrates the concept of the traditional construction project organization.

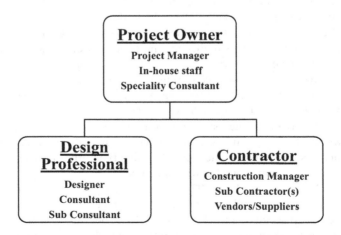

FIGURE 2.1 Traditional construction project.

Traditional construction projects involve three main groups. These are

1. Owners—A person or an organization that initiates and sanctions a project. He/she outlines the needs of the facility and is responsible for arranging the financial resources for creation of the facility.
2. Designers (A/E)—This group consists of one or more architects or engineers and consultants. They are the owner's appointed entities accountable for converting the owner's conception and need into a specific facility with detailed directions through drawings and specifications adhering to the economic objectives. They are responsible for the design of the project and in certain cases its supervision.
3. Contractors—A construction firm engaged by the owner to complete the specific facility by providing the necessary staff, work force, materials, equipment, tools, and other accessories to the satisfaction of the owner/end user in compliance with the contract documents. The contractor is responsible for implementing the project activities and for achieving the owner's objectives.

Construction projects are executed on the basis of a predetermined set of goals and objectives. With traditional construction projects, the owner heads the team, designating a project manager. The project manager is a person/member of the owner's staff or independently hired person/firm with overall or principal responsibility for the management of the project as a whole.

In certain cases, owners engage a professional firm, called a construction manager, trained in the management of construction processes to assist in developing bid documents and overseeing and coordinating the project for the owner. The basic construction management concept is that the owner assigns a contract to a firm that is knowledgeable and capable of coordinating all the aspects of the project to meet the intended use of the project by the owner. In the construction management type of construction projects, the consultants (architect/engineer) prepare complete design drawings and contract documents, then the project is put for competitive bid, and the contract is awarded to the competitive bidder (contractor). Next, the owner hires a third-party (construction manager) to oversee and coordinate the construction.

There are numerous types of construction projects:

• Process Type Projects

  • Liquid chemical plants
  • Liquid/solid plants
  • Solid process plants
  • Petrochemical plants
  • Petroleum refineries

• Non-process Type Projects

  • Power plants
  • Manufacturing plants

- Support facilities
- Miscellaneous (R&D) projects
- Civil construction projects
- Commercial A/E projects

Civil construction projects and commercial A/E projects can further be categorized into four somewhat arbitrary but generally accepted major types of construction. These are:

1. Residential construction
2. Building construction (institutional and commercial)
3. Industrial construction
4. Heavy engineering construction

Residential construction: Residential construction includes single-family homes, multiunit town houses, garden, apartments, high-rise apartments, and villas.

Building construction: Building construction includes structures ranging from small retail stores to urban redevelopment complexes, from grade schools to new universities, hospitals, commercial office towers, theaters, government buildings, recreation centers, warehouses, and neighborhood centers.

Industrial construction: Industrial construction includes petroleum refineries, petroleum plants, power plants, heavy manufacturing plants, and other facilities essential to our utilities and basic industries.

Heavy engineering construction: Heavy engineering construction includes dams and tunnels, bridges, railways, airports, highways and urban rapid transit system, ports and harbors, water treatment and distribution, sewage and storm water collection, treatment and disposal system, power lines, and communication network.

Table 2.3 illustrates types of construction projects.

2.4 QUALITY OF CONSTRUCTION PROJECTS

Construction projects are mainly capital investment projects. They are customized and non-repetitive in nature. Construction projects have become more complex and technical, and the relationships and the contractual grouping of those who are involved are also more complex and contractually varied. The products used in construction projects are expensive, complex, immovable, and long-lived. Generally, a construction project is composed of building materials (civil), electro-mechanical items, finishing items, and equipment. These are normally produced by other construction-related industries/manufacturers. These industries produce products as per their own quality management practices complying with certain quality standards or against specific requirements for a particular project. The owner of the construction project or his representative has no direct control over these companies unless he/his representative/appointed contractor commits to buy their product for use in their facility. These organizations may have their own quality management program. In manufacturing or service industries, the quality management of all in-house manufactured products is performed by manufacturer's own team, under the control of

TABLE 2.3
Types of Construction Projects

1	**Process Type Projects**			
1.1	Liquid chemical plants			
1.2	Liquid/solid plants			
1.3	Solid process plants			
1.4	Petrochemical plants			
1.5	Petroleum refineries			
2	**Non-process**		**Type Projects**	
2.1	Power plants			
2.2	Manufacturing plants			
2.3	Support facilities			
2.4	Miscellaneous (R&D) projects			
2.5	Civil construction projects	**Categories of civil construction projects and**	Residential construction	Family homes, multiunit town houses, garden, apartments, condominiums, high-rise apartments, villas.
2.6	Commercial A/E projects	**Commercial A/E projects**	Building construction (institutional and commercial)	schools, universities, hospitals, commercial office complexes, shopping malls, banks, theatres, stadiums, government buildings, warehouses, recreation centers, amusement parks, holiday resorts, neighborhood centers
			Industrial construction	Petroleum refineries, petroleum plants, power plants, heavy manufacturing plants, steel mills, chemical processing plants
			Heavy engineering	Dams, tunnels, bridges, highways, railways, airports, urban rapid transit system, ports, harbors, power lines and communication network
			Environmental	Water treatment and clean water distribution, sanitary and sewage system, waste management

Source: Abdul Razzak Rumane. (2013). *Quality Tools for Managing Construction Projects.* Reprinted with permission of Taylor & Francis Group.

same organization having jurisdiction over their manufacturing plants at different locations. Quality management of vendor-supplied items/products is carried out as stipulated in the purchasing contract as per the quality control specifications of the buyer.

Quality has different meanings to different people. The definition of quality relating to manufacturing, processes, and service industries is as follows:

- Meeting the customer's need
- Customer satisfaction
- Fitness for use
- Conforming to requirements
- Degree of excellence at an acceptable price

The International Organization for Standardization (ISO) defines quality as "the totality of characteristics of an entity that bears on its ability to satisfy stated or implied needs".

However, the definition of quality for construction projects is different to that of manufacturing or services industries as the product is not repetitive but a unique piece of work with specific requirements.

Quality in construction project is not only the quality of product and equipment used in the construction of facility, it is also the total management approach to complete the facility. Quality of construction depends mainly upon the control of construction, which is the primary responsibility of contractor.

Quality in manufacturing passes through a series of processes. Material and labor are input through a series of process out of which a product is obtained. The output is monitored by inspection and testing at various stages of production. Any nonconforming product identified is either repaired, reworked, or scrapped, and proper steps are taken to eliminate problem causes. Statistical process control methods are used to reduce the variability and to increase the efficiency of process. In construction projects, the scenario is not the same. If anything goes wrong, the nonconforming work is very difficult to rectify, and remedial action are sometimes not possible.

Quality management in construction projects is different to that of manufacturing. Quality in construction projects is not only the quality of products and equipment used in the construction, it is also the total management approach to complete the facility as per the scope of works to customer/owner satisfaction within the budget and to be completed within specified schedule to meet owner's defined purpose. The nature of the contracts between the parties plays a dominant part in the quality system required from the project, and the responsibility for achieving them must therefore be specified in the project documents. The documents include plans, specifications, schedules, bill of quantities, and so on. Quality control in construction typically involves insuring compliance with minimum standards of material and workmanship in order to ensure the performance of the facility according to the design. These minimum standards are contained in the specification documents. For the purpose of insuring compliance, random samples and statistical methods are commonly used as the basis for accepting or rejecting work completed and batches of materials. Rejection of a batch is based on nonconformance or violation of the relevant design specifications.

FIGURE 2.2   Construction project quality trilogy.

Based on aforementioned points, quality of construction projects can be defined as the following: Construction Project quality is fulfillment of owner's needs as per defined scope of works within a budget and specified schedule to satisfy owner's/ user's requirements. The phenomenon of these three components can be called as "Construction Project Trilogy" and is illustrated in Figure 2.2.

Thus, quality of construction projects can be evolved as follows:

1) Properly defined scope of work
2) Owner, project manager, design team leader, consultant, and constructor's manager are responsible to implement the quality.
3) Continuous improvement can be achieved at different levels as follows:

   a) Owner: Specifies the latest needs
   b) Designer: Provides specification to include latest quality materials, products, and equipment
   c) Constructor: Uses latest construction equipment to build the facility

4) Establishment of performance measures

   a) Owner:

      I) To review and ensure that designer has prepared the contract documents which satisfy his needs
      II) To check the progress of work to ensure compliance with the contract documents

   b) Consultant:

      I) As a consultant designer to include the owner's requirements explicitly and clearly defined in the contact documents
      II) As a supervision consultant supervise contractor's work as per contract documents and the specified standards

   c) Contactor: To construct the facility as specified and use the materials, products, and equipment which satisfy the specified requirements

5) Team Approach: Every member of the project team should know principles of Total Quality Management (TQM) knowing that TQM is collaborative efforts, and everybody should participate in all the functional areas to improve the quality of project works. They should know that it is a collective effort by all the participants to achieve project quality.

6) Training and Education: Consultant and contractor should have customized training plans for their management, engineers, supervisors, office staff, technicians, and labors.

7) Establish Leadership: Organizational leadership should be established to achieve the specified quality. The staff and labors should be encouraged and helped to understand the quality to be achieved for the project.

These definitions, when applied to construction projects, relate to the contract specifications or owner/end user requirements to be constructed in such a way that construction of the facility is suitable for owner's use or it meets the owner requirements. Quality in construction is achieved through complex interaction of many participants in the facilities' development process.

The quality plan for construction projects is part of the overall project documentation consisting of:

1. Well-defined specification for all the materials, products, components, and equipment to be used to construct the facility.
2. Detailed construction drawings.
3. Detailed work procedure.
4. Details of the quality standards and codes to be complied.
5. Cost of the project.
6. Manpower and other resources to be used for the project.
7. Project completion schedule.

Participation involvement of all three parties at different levels of construction phases is required to develop quality system and application of quality tools and techniques. With the application of various quality principles, tools, and methods by all the participants at different stages of construction project, rework can be reduced, resulting in savings in the project cost and making the project qualitative and economical. This will ensure the completion of construction and making the project most qualitative, competitive, and economical.

2.5 PRINCIPLES OF QUALITY IN CONSTRUCTION PROJECTS

An ISO document has listed eight quality management principles on which the quality management system standards of the revised ISO 9000:2000 series are based. These are as follows:

Principle 1—Customer focus
Principle 2—Leadership
Principle 3—Involvement of people

Principle 4—Process approach
Principle 5—System approach to management
Principle 6—Continual improvement
Principle 7—Factual approach to design making
Principle 8—Mutual beneficial supplier relationship

Table 2.4 illustrates quality principles of construction projects evolved taking into consideration ISO quality principles (CLIPSCFM)

TABLE 2.4
Principles of Quality in Construction Projects

Principle	Construction Projects' Quality Principle	
Principle 1 (Customer Focus)	1.1	Designer, Consultant are responsible to provide Owner's requirements explicitly and clearly defining the standards of the end products and their compliance in the contract documents.
	1.2	Engineering design should include the process, process equipment, engineering systems' requirements clearly and without any ambiguity for the ease of operation
	1.3	The project and end products should satisfy owner's need, requirements and should be suitable for intended usage
Principle 2 (Leadership)	2.1	Owner, Designer, Consultant, Contractor are fully responsible for the application of quality management system to meet customer requirements and strive to exceed customer expectations by complying with defined scope of work in the contract documents
	2.2	Every member of the project team should exert collaborative efforts in all the functional areas to improve the quality of project
Principle 3 (Involvement/ Engagement of people)	3.1	Each member of project team should participate and fully involve as per their abilities in all the functional areas by adhering to team approach and coordination to continuously improve the quality of the project
Principle 4 (Process Approach)	4.1	Contractor to build the facility as stipulated in the contract documents, plan, specifications as per the approved schedule, and within agreed-upon budget to meet owner's objectives
	4.2	Contractor should study all the documents during tendering/bidding stage and submit his proposal taking into consideration all the requirements specified in the contract documents and identifying, understanding, and managing interrelated processes as a system in achieving the specified product output
	4.3	Contractor is responsible to provide all the resources, manpower, material, equipment, etc., to build the facility as per specifications to produce the specified products
	4.4	Contractor to check executed/installed works to confirm that works have been performed/executed as specified, using specified/approved materials, approved shop drawings, installation methods, and specified references, codes, standards to meet intended use

(Continued)

TABLE 2.4
(Continued)

Principle	Construction Projects' Quality Principle
Principle 5 (System Approach to Management)	5.1 Contractor to prepare contractor's quality control plan (CQCP) and follow the same to ensure meeting the performance standards specified in the contract documents.
	5.2 Method of payments (Work progress, material, equipment, etc.) to be clearly defined in the contract documents. Rate analysis of Bill of Quantities (BOQ) or Bill of Materials (BOM) item to be agreed before signing of contract
	5.3 Contract documents should include a clause to settle the dispute arising during construction stage
Principle 6 (Continual Improvement)	6.1 Contractor shall follow the submittal procedure specified in the contract documents for detailed design, procurements, checklists, inspection, and testing procedures as per the communication matrix. The contents of transmittals and executed works should be reviewed prior to submission for approval
Principle 7 (Factual Approach to Evidence-Based Design Making)	7.1 Contractor shall follow an agreed-upon quality assurance and quality control plan. Consultant and PMC shall be responsible to oversee the compliance with contract documents and specified standards and codes
	7.2 Contractor is responsible to construct the facility to produce the products as specified and use the material, products, systems, equipment, and methods which satisfy the specified requirements (factual approach to design making)
Principle 8 (Mutual Beneficial Relationship) Relationship Management	8.1 Contractor/All team members should participate and put collective efforts to perform the works as per agreed-upon construction program and hand over the project as per contracted schedule to meet the owner's requirements
	8.2 All team members should focus on participative management and strong operational accountability at the individual contributory level to follow principles of Total Quality Management

2.6   RISKS IN CONSTRUCTION PROJECTS

Construction projects have many varying risks. Risk management throughout the life cycle of the project is important and essential to prevent unwanted consequences and effects on the project. Construction projects have the involvement of many stakeholders such as project owners, developers, design firms (consultants), contractors, banks, and financial institutions funding the project, who are affected by the risk. Each of these parties have involvement with certain portion of overall construction project risk; however, the owner has a greater share of risks as the owner is involved from the inception until completion of project and beyond. The owner must take initiatives to develop risk consciousness and awareness among all the parties emphasizing upon the importance of explicit consideration of risk at each stage of the project as the owner is ultimately responsible for overall project const. Traditionally,

1. Owner/Client is responsible for the investment/finance risk
2. Designer (Consultant) is responsible for design risk
3. Contractors, sub-contractors are responsible for construction risk.

Construction projects are characterized as very complex projects, where uncertainty comes from various sources. Construction projects involve a cross-section of many different participants. They have varying project expectations. Those both influence and depend on each other in addition to the "other players" involved in the construction process. The relationships and the contractual groupings of those who are involved are also more complex and contractually varied. Construction projects often require large amounts of materials and physical tools to move or modify these materials. Most items used in construction projects are normally produced by other construction-related industries/manufacturers. Therefore, risk in construction projects is multifaceted. Construction projects inherently contain a high degree of risk in their projection of cost and time as each is unique. No construction project is without any risk. Risk management in construction projects is mainly focused on delivering the project with:

1. What was originally accepted (as per defined Scope)
2. Agreed-upon time (as per the schedule without any delay)
3. Agreed-upon budget (no overruns to accepted cost)

Risk management is an ongoing process. In order to reduce the overall risk in construction projects, the risk assessment (identification, analysis, and evaluation) process must start as early as possible to maximize project benefits. There are number of risks which can be identified at each stage of the project. Early risk identification can lead to better estimation of the cost in the project budget, whether through contingencies, contractual or insurance. Risk identification is most important function in construction projects.

Risk factors in construction projects can be categorized into a number of ways according to level of details or selected viewpoints. These are categorized based on various risks factors and source of risk. Contractor has to identify related risks affecting the construction, analyze these risks, evaluate the effects on the contract, and evolve the strategy to counter these risks, before bidding for a construction contract. Construction project risks mainly relate to following:

- Scope and change management
- Schedule/time management
- Budget/cost management
- Quality management
- Resources and manpower management
- Communication management
- Procurement/contract management
- Health, safety, and environmental management

Table 2.5 illustrates typical categories of risks in construction projects.

TABLE 2.5

Typical Categories of Risks in Construction Projects

Serial Number	Category	Types
1	Management	Selection of project delivery and contracting system
		Selection of Project Management Consultant
		Selection of Designer
		Selection of Contractor/EPC Contractor
2	Contract (Project)	Scope/Design changes
		Schedule
		Cost
		Conflict resolution
		Delay in change order negotiations
3	Statutory	Statutory/regulatory delay
4	Technical	Incomplete design
		Incomplete scope of work
		Design changes
		Design mistakes
		Errors and omissions in contract documents
		Incomplete specifications
		Ambiguity in contract documents
		Inconsistency in contract documents
		Inappropriate schedule/plan
		Inappropriate construction method
		Conflict with different trades
		Improper coordination with regulatory authorities
		Inadequate site investigation data
5	Technology	New technology
6	Construction	Delay in mobilization
		Delay in transfer of site
		Different site conditions to the information provided
		Changes in scope of work
		Resource (labor) low productivity
		Equipment/plant productivity
		Insufficient skilled workforce
		Union and labor unrest
		Failure/Delay of machinery and equipment
		Quality of material
		Failure/Delay of material delivery
		Delay in approval of submittals
		Extensive subcontracting
		Subcontractor's subcontractor
		Failure of project team members to perform as expected
		Information flow breaks
7	Physical	Damage to equipment
		Structure collapse
		Damage to stored material

Serial Number	Category	Types
		Leakage of hazardous material
		Theft at site
		Fire at site
8	Logistic	Resources' availability
		Spare parts' availability
		Consistent fuel supply
		Transportation facility
		Access to worksite
		Unfamiliarity with local conditions
9	Health, safety, and environment	Injuries
		Health and safety rules
		Environmental protection rules
		Pollution rules
		Disposal of waste
10	Financial	Inflation
		Recession
		Fluctuations in exchange rate
		Availability of foreign exchange (certain countries)
		Availability of funds
		Delays in payment
		Local taxes
11	Economical	Variation of construction material price
		Sanctions
12	Commercial	Import restrictions
		Custom duties
13	Legal	Permits and licenses
		Professional liability
		Litigation
14	Political	Change in laws and regulations
		Constraints on employment of expatriate workforce
		Use of local agent and firms
		Civil unrest
		War
15	Natural/ Environmental	Flood
		Earthquake
		Cyclone
		Sandstorm
		Landslide
		Heavy rains
		High humidity
		Fire
16	Geological	Estimation of accessible reserves
17	Supply and demand	Financial crises and macroeconomic factors
18	Operational	Availability of skilled workers

Source: Abdul Razzak Rumane. (2013). *Quality Tools for Managing Construction Projects.* Reprinted with permission of Taylor & Francis Group.

3 Project Management Process

3.1 SYSTEMS ENGINEERING

Systems are pervasive throughout the universe in which we live. This world can be divided into the natural world and the human-made world. Systems appeared first in natural forms and subsequently with the appearance of human beings. Systems were created based on components, attributes, and relationships.

Systems engineering and analysis, when coupled with new emerging technologies, reveal unexpected opportunities for bringing new improved systems and products into being that will be more competitive in the world economy. Product competitiveness is desired by both commercial and public-sector producers worldwide to meet consumer expectations. These technologies and processes can be applied to construction projects. The systems engineering approach to construction projects help us understand the entire process of project management in order to manage its activities at different levels of various phases to achieve economical and competitive results. The cost effectiveness of the resulting technical activities can be enhanced by giving more attention to what they are to do, before addressing what they are composed of. To ensure economic competitiveness regarding the product, engineering must become more closely associated with economics and economic facilities. This is best accomplished through the life cycle approach to engineering.

Experience in recent decades indicates that properly coordinated and functioning human-made systems will result in a minimum of undesirable side effects through the application of this integrated, life-cycle-oriented "systems" approach. The consequences of not applying systems engineering in the design and development and/or reengineering of systems have been disruptive and costly.

The systems approach is a technique, which represents a broad-based systematic approach to problems that may be interdisciplinary. It is particularly useful when problems are affected by many factors, and it entails the creation of a problem model that corresponds as closely as possible to reality. The systems approach stresses the need for the engineer to look for all the relevant factors, influences, and components of the environment that surround the problem. The systems approach corresponds to a comprehensive attack on a problem and to an interest in, and commitment to, formulating a problem in the widest and fullest manner that can be professionally handled.

System Definition

There are many definitions of *system*. One dictionary definition calls it "a group or combination of interrelated, independent or interacting elements forming a collective

DOI: 10.1201/9781003245612-3

entity". A system is an assembly of components or elements having a functional relationship to achieve a common objective for a useful purpose. A system is composed of components, attributes, and relationships. These are described as follows:

1. Components are the operating parts of the system consisting of input, process, and output. Each system component may assume a variety of values to describe a system state, as set by some control action and one or more restrictions.
2. Attributes are the properties or discernible manifestations of the components of a system. These attributes characterize the system.
3. Relationships are the links between components and attributes.

The properties and behavior of each component of the set have an effect on the properties and behavior of the set as a whole and depend on the properties and behavior of at least one other component on the list. The components of the system cannot be divided into independent subsets. A system is more than the sum of its components and parts. Not every set of items, facts, methods, or procedures is a system. To qualify the system, it should have a functional relationship, interaction between many components and useful purpose. The purposeful action performed by a system is its function. A basic behavioral concept of a system is that it is a device, which accepts one or more inputs and generates one or more outputs from them. This simple behavioral approach to systems is generally known as the Black Box and is represented schematically in Figure 3.1. The Black Box system phenomenon establishes the functional relationship between system inputs and outputs.

Every system is made up of components and components that can be broken down into similar components. If two hierarchical levels are involved in a given system, the lower is conveniently called a subsystem. The designation of system, subsystem, and components are relative because the system at one level in the hierarchy is the component at another level. Everything that remains outside the boundaries of the system is considered to be environmental. Material, energy, and/or information that pass through the boundaries are called "inputs" to the system. In reverse, material, energy, and/or information that pass from the system to the environment are called outputs.

Accordingly, a system is an assembly of components or elements having a functional relationship to achieve a common objective for a useful purpose.

3.1.1 Systems Engineering Approach in Projects

The life cycle of a project begins with the identification of need and extends through conceptual and preliminary design, detail design, and development, production and/

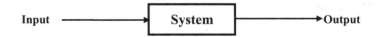

FIGURE 3.1   Black box.

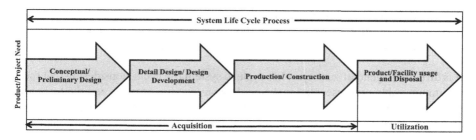

FIGURE 3.2 Product/project system life cycle.

or construction, product use, phase-out, and disposal. The program phases are classified as acquisition and utilization to recognize procedure and customer activities. This classification represents a generic approach. Sometimes, the acquiring process may involve both the customer and the producer (or contractor), whereas acquiring may include a combination of contractor and consumer (or ultimate user) activities.

In general, engineering has focused mainly on product performance as the main objective rather than on development of overall system of which the product is a part. Application of a systems engineering process leads to reduction in the cost of design development, production/construction, and operation and hence results in reduction in life cycle cost of the product; thus, the product becomes more competitive and economical. Systems engineering provides the basis for a structural and logical approach. The need for systems engineering increases with the size of projects.

The product/project system life cycle process is illustrated in Figure 3.2 and is fundamental to the application of system engineering.

The life cycle begins with the identification of need and extends through conceptual and preliminary design, detail design, and development, production and/or construction, product use, phase-out, and disposal. The program phases are classified as acquisition and utilization to recognize procedure and customer activities.

A systems engineering approach to projects helps to understand the entire process of project management and to manage and control its activities at different levels of various phases to ensure timely completion of the project with economical use of resources to make the construction project most qualitative, competitive, and economical.

Systems engineering starts from the complexity of the large-scale problem as a whole and moves toward the structural analysis and partitioning process until the questions of interest are answered. This process of decomposition is called a work breakdown structure (WBS). The WBS is a hierarchical representation of system levels. Being a family tree, the WBS consists of a number of levels, starting with the complete system at level 1 at the top and progressing downward through as many levels as necessary to obtain elements that can be conveniently managed.

Benefits of systems engineering applications are

- Reduction in the cost of system design and development, production/construction, system operation and support, system retirement, and material disposal

- Reduction in system acquisition time
- More visibility and reduction in the risks associated with the design deci-sion-making process

It is difficult to generalize project life cycle to system life cycle, considering that there are innumerable processes that make up the construction process, and the technolo-gies and processes, as applied to systems engineering, can also be applied to con-struction projects. However, the number of phases shall depend on the complexity of the project.

Most construction projects are divided into five phases depending on the size and complexity of the project. In certain case, it may have seven phases. Each phase can further be subdivided into the WBS principle to reach a level of complexity where each element/activity can be treated as a single unit that can be conveniently man-aged. WBS represents a systematic and logical breakdown of the project phase into its components (activities). It is constructed by dividing the project into major elements with each of these being divided into sub-elements. This is done until a breakdown is done in terms of manageable units of work for which responsibility can be defined. WBS involves envisioning the project as a hierarchy of goal, objectives, activities, sub-activities, and work packages. The hierarchical decomposition of activities con-tinues until the entire project is displayed as a network of separately identified and nonoverlapping activities. Each activity will be single purposed, of a specific time duration, and manageable; its time and cost estimates are easily derived, deliverables clearly understood, and responsibility for its completion clearly assigned. The WBS helps in:

- Effective planning by dividing the work into manageable elements, which can be planned, budgeted, and controlled
- Assignment of responsibility for work elements to project personnel and outside agencies
- Development of control and information system

WBS facilitates the planning, budgeting, scheduling, and control activities for the project manager and its team. By the application of WBS phenomenon, the construc-tion phases are further divided into various activities. Division of these phases will improve the control and planning of the construction project at every stage before a new phase starts. From the perspective of risk management, it will be easy to identify risk in each activity of the project life cycle.

3.1.2 Systems Engineering Approach in Risk Management

The risk management process involves the systematic application of policies, proce-dures, and practices to the activities of communicating and consulting, establishing the context and assessing, treating, monitoring, reviewing, recording, and reporting risk (please refer Figure 1.4).

The system life cycle is fundamental to the application of systems engineer-ing. A systems engineering approach helps to understand the entire process of product/project life cycle and to manage and control its activities at different levels of various phases to ensure timely completion of the product/project.

Systems engineering starts from the complexity of the large-scale problem as a whole and moves toward structural analysis and the partitioning process until the questions of interest are answered. This process of decomposition is called a WBS. The WBS is a hierarchical representation of system levels. Being a family tree, the WBS consists of a number of levels, starting with the complete system at level 1 at the top and progressing downward through as many levels as necessary to obtain elements that can be conveniently managed.

The system life cycle is fundamental to the application of systems engineering. A systems engineering approach to risk management process helps to understand the entire process of risk management process activities and to manage and control its activities at different levels of various elements and breaking down each activity on the principle of WBS to achieving the objectives of the project.

Risk Breakdown Structure (RBS) is similar to WBS. Both are organized as hierarchical breakdowns, and each progressively goes downward decomposing the element/activity/item to manageable level. With the RBS, risks are organized according to risk categories. Figure 3.3 illustrated RBS for technical risk category in a project.

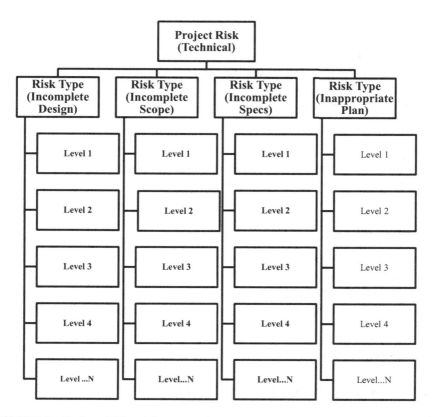

FIGURE 3.3 Project risk breakdown structure.

3.1.3 PROJECT LIFE CYCLE PHASES

The system life cycle is the fundamental to the application of systems engineering. A systems engineering approach to construction projects helps to understand the entire process of project management and to manage and control its activities at different levels of various phases. The project elements/activities derived with systems engineering approach help conveniently manage the project and also manage risk management process to achieve the project objectives.

Though it is difficult to generalize project life cycle to system life cycle. However, considering that there are innumerable processes that make up the construction process, the technologies and processes, as applied to systems engineering, can also be applied to construction projects. The number of phases shall depend on the complexity of the project. Duration of each phase may vary from project to project. There are mainly two categories of construction projects.

1. Process Type of projects
2. Non-process Type of projects

The components/activities of process types of project life cycle phases divided on WBS principle are listed as follows:

1. Feasibility Study
2. Concept Design
3. Front End Engineering Design (FEED)
4. Bidding and Tendering
5. Engineering, Procurement, and Construction (EPC)

   • Detailed engineering
   • Procurement
   • Construction

6. Testing, Commissioning, and Hand Over

The components/activities of non-process types of project life cycle phases divided on WBS principle are listed as follows:

1. Conceptual Design
2. Schematic Design
3. Design Development
4. Construction Documents
5. Bidding and Tendering
6. Construction
7. Testing, Commissioning, and Hand Over

3.2 MANAGEMENT PROCESS GROUPS

Project Management Body of Knowledge (PMBOK® Guide) published by Project Management Institute describes application of Project Management Processes

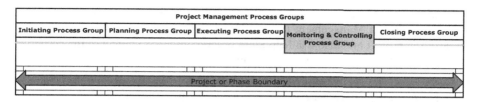

Project Management Process Groups				
Initiating Process Group	Planning Process Group	Executing Process Group	Monitoring & Controlling Process Group	Closing Process Group

Project or Phase Boundary

FIGURE 3.4 Project management process groups.

during the life cycle of projects to enhance the chances of success over a wide range of projects. PMBOK® Guide-Fifth Edition identifies and describes five Project Management Process Groups required for successful completion of any project. These are:

1. Initiating Process Group
2. Planning Process Group
3. Executing Process Group
4. Monitoring and Controlling Process Group
5. Closing Process Group

Figure 3.4 illustrates an overview of Project Management Process Groups.

In order to conveniently manage the project, each of these phases are treated as a project itself with all the five process groups operating as they do for overall project, and each phase is composed of activities, elements having functional relationship to achieve a common objective for useful purpose.

These process groups are independent of application areas or industry focus. These groups consist of 47 Project Management Processes and are further grouped into 13 separate Knowledge Areas. These are:

1. Integration Management
2. Stakeholder Management
3. Scope Management
4. Schedule Management
5. Cost Management
6. Quality Management
7. Resource Management
8. Communication Management
9. Risk Management
10. Contract Management
11. Health, Safety, and Environment Management (HSE)
12. Financial Management
13. Claim Management

The life cycle of a product/project begins with needs that result from business case which suggests production of a product or a new project/facility.

The major activities evolved based on the Management Processes are described under five Project Management Process Groups. These activities help manage and control the project and also can help identify risks involved in each of the activities and treat the risks to achieve project objectives.

3.2.1   INITIATING PROCESS GROUP

Table 3.1 illustrates major construction activities relating to Project Initiating Process Group.

3.2.2   PLANNING PROCESS GROUP

Table 3.2 illustrates major construction activities relating to Project Planning Process Group.

3.2.3   EXECUTING PROCESS GROUP

Table 3.3 illustrates major construction activities relating to Project Executing Process Group.

TABLE 3.1
Major Project Activities Relating to Initiating Process Group

Serial Number	Management Processes	Activities		Elements	
1	Integration Management	1.1	Develop Project Charter	1.1.1	Project Inception
				1.1.2	Problem Statement/ Need Identification
				1.1.3	Need Analysis
				1.1.4	Need Statement
				1.1.5	Need Feasibility
				1.1.6	Project Goals and Objectives
				1.1.7	Project Deliverables
				1.1.8	Design Deliverables
		1.2	Develop Preliminary Scope Statement	1.2.1	Project Terms of Reference (TOR)
				1.2.2	Contract Documents
2	Stakeholder Management	2.1	Identify Stakeholders	2.1	Project Delivery System
				2.2	Project Life Cycle
				2.3	Project Team Members
				2.4	Other Parties

Source: Abdul Razzak Rumane. (2016). *Handbook of Construction Management: Scope, Schedule, and Cost Control*. Reprinted with permission of Taylor & Francis Group.

TABLE 3.2

Major Project Activities Relating to Planning Process Group

Serial Number	Management Processes	Activities		Elements	
1	Integration Management	1.1	Project baseline plan	1.1.1	Preliminary plans
2	Stakeholder Management	2.1	Responsibilities matrix	2.1.1	Owner, Designer, Contractor, Other Stakeholders
		2.2	Stakeholders requirement (work progress)	2.2.1	Design Progress
				2.2.2	Construction progress
				2.2.3	Testing, commissioning, and handover
		2.3	Change reporting	2.3.1	Updated schedule
				2.3.2	Variation report
				2.3.3	Cost variation
		2.4	Project updates		
		2.5	Status reports	2.5.1	Status logs
				2.5.2	Performance reports
				2.5.3	Issue log
		2.6	Meetings	2.6.1	Kick-off meeting
				2.6.2	Progress meetings
				2.6.3	Coordination meetings
				2.6.4	Other meetings
		2.7	Payments	2.7.1	Payment status
3	Scope Management	3.1	Establish scope baseline plan		
		3.2	Collect requirements	3.2.1	Need statement
				3.2.2	Project goals and objectives
				3.2.3	Project Terms of Reference (TOR)
				3.2.4	Owner's preferred requirements
		3.3	Project scope documents	3.3.1	Design development
				•	Concept design
				•	Schematic design
				•	Detail design
				3.3.2	Final design
				3.3.3	Bill of Quantity
				3.3.4	Project specifications
				3.3.5	Construction documents
				3.3.6	Project deliverables

(Continued)

TABLE 3.2
(Continued)

Serial Number	Management Processes	Activities		Elements	
		3.4	Organizational breakdown structure	3.4.1	Project delivery system
				3.4.2	Organizing
				3.4.3	Staffing
				3.4.4	Project design
		3.5	Work Breakdown Structures	3.5.1	Project life cycle
				3.5.2	Work packages
4	Schedule Management	4.1	Bill of Quantity	4.1.1	Quantities take-off
				4.1.2	Sequencing of activities
				4.1.3	Estimate activity resources
				4.1.4	Estimate duration of activity
		4.2	Identify project assumption	4.2.1	Dependencies
				4.2.2	Risks and constraints
				4.2.3	Milestone
		4.3	Develop baseline schedule		
		4.4	Develop schedule	4.4.1	Pre-design stage
				4.4.2	Design development
				•	Concept design
				•	Schematic design
				•	Detail design
				4.4.3	Contract documents
				4.4.4	Bidding/tendering and contract award
				4.4.5	Construction phase
				4.4.6	Testing, commissioning, and handover
		4.5	Construction schedule	4.5.1	Contractor's construction schedule
5	Cost Management	5.1	Estimate cost	5.1.1	Conceptual estimate
				5.1.2	Preliminary estimate
				5.1.3	Detail estimate
				5.1.4	Definitive estimate
		5.2	Estimate budget	5.2.1	Prepare budget
		5.3	Determine project cost baseline	5.3.1	S-Curve
				5.3.2	Cost loading
				5.3.3	Resource loading

Serial Number	Management Processes	Activities		Elements	
		5.4	Estimate cost	5.4.1	Estimate project resources cost
				5.4.2	Estimate project material cost
				5.4.2	Estimate project equipment cost
				5.4.3	Bill of Quantities
				5.4.4	BOQ price analysis
		5.5	Contracted project value	5.5.1	Progress payments
		5.6	Change order Procedure	5.6.1	Change order
				5.6.2	Cost variation
6	Quality Management	6.1	Project quality management plan	6.1.1	Quality codes and standards to be compiled
				6.1.2	Design criteria
				6.1.3	Design procedure
				6.1.4	Quality matrix (design stage)
				6.1.5	Well-defined specification
				6.1.6	Detailed construction drawings
				6.1.7	Quality matrix (construction phase)
				6.1.8	Construction process
				6.1.9	Detailed work procedures
				6.1.10	Quality matrix (inspection, testing during execution)
				6.1.11	Defect prevention/ rework
				6.1.12	Quality matrix (testing and handing over-start up)
				6.1.13	Regulatory requirements
				6.1.14	Quality assurance/ quality control procedures
				6.1.15	Reporting quality assurance/quality Control problems
				6.1.16	Stakeholders quality requirements

(Continued)

TABLE 3.2
(Continued)

Serial Number	Management Processes	Activities		Elements	
7	Resource Management	7.1	Project human resources	7.1.1	Construction/project manager
				7.1.2	Designer's team
				7.1.3	Supervision team
		7.2	Construction resources	7.2.1	Contractor's core team
				7.2.2	Construction material
				7.2.3	Construction equipment
				7.2.3	Construction labor
				7.2.4	Subcontractor (s)
8	Communication Management	8.1	Communication plan	8.1.1	Communication matrix
		8.2	Communication methods	8.2.1	Design progress
				8.2.2	Work progress
				8.2.3	Project issues
				8.2.4	Project variations
				8.2.5	Authorities
		8.3	Submittal procedures	8.3.1	Submittal procedure
				8.3.1	Progress payments
				8.3.2	Progress reports
				8.3.1	Minutes of meetings
				8.3.2	Other meetings
		8.4	Documents	8.4.1	Design documents
				8.4.2	Contract documents
				8.4.3	Construction documents
				8.4.4	As-built documents
				8.4.5	Authority-approved documents/drawings
		8.5	Logs	8.5.1	Issue log
				8.5.2	Correspondence with stakeholders
				8.5.3	Correspondence with team members
				8.5.4	Regulatory authorities
9	Risk Management	9.1	Risk identification	9.1.1	During inception
				9.1.2	During design
				9.1.3	During bidding
				9.1.4	During construction
				9.1.5	During testing and commissioning
				9.1.6	During handing over
		9.2	Managing risk	9.2.1	Risk register
				9.2.2	Risk analysis
				9.2.3	Risk response

Serial Number	Management Processes	Activities	Elements
10	Contract Management	10.1 Project delivery system	10.1.1 Selection of CM
			10.1.2 Selection of designer
		10.2 Bidding and tendering	10.2.1 Pre-qualification of contractors
			10.2.2 Issue tender documents
			10.2.3 Acceptance of tender
11	Health, Safety and Environment	11.1 Environmental compatibility	
		11.2 Safety management plan	11.2.1 Safety consideration in design
			11.2.2 HSE plan for construction site safety
			11.2.3 Emergency evacuation plan
		11.3 Waste management plan	
12	Financial Management	12.1 Financial planning	12.1.1 Payments to Designer (Consultant), Construction/Project Manager, Contractor
			12.1.2 Material procurement
			12.1.3 Equipment procurement
			12.1.4 Project staff salaries
			12.1.5 Bonds, insurance, guarantees
			12.1.6 Cash flow
13	Claim Management	13.1 Claim identification	13.1.1 Design errors
			13.1.2 Additional works
			13.1.3 Delays in payment
		13.2 Claim Quantification	13.2.1 Change order procedures
			• Cost
			• Time

Source: Abdul Razzak Rumane. (2016). *Handbook of Construction Management: Scope, Schedule, and Cost Control.* Reprinted with permission of Taylor & Francis Group.

TABLE 3.3
Major Project Activities Relating to Project Executing Group

Serial Number	Management Processes	Activities		Elements	
1	Integration Management	1.1	Design development	1.1.1	Concept design
				1.1.2	Schematic design
				1.1.3	Detail design
		1.2	Construction	1.2.1	Notice to proceed
				1.2.2	Mobilization
				1.2.3	Submittals
				1.2.4	Execution
				1.2.5	Corrective actions
				1.2.6	Project deliverables
		1.3	Implement changes	1.3.1	Approved changes
				1.3.1	Preventive actions
				1.3.2	Defect repairs
				1.3.3	Rework
				1.3.4	Update scope
				1.3.5	Update plans
				1.3.6	Update contract documents
2	Stakeholder Management	2.1	Project status/ performance report	2.1.1	Updated plans
		2.2	Payments	2.2.1	Progress payments
		2.3	Change requests	2.3.1	Site work instruction
				2.3.2	Change orders
				2.3.3	Schedule
				2.3.2	Materials
		2.4	Conflict resolution		
		2.5	Issue log		
3	Scope Management				
4	Schedule Management				
5	Cost Management				
6	Quality Management	6.1	Quality Assurance	6.1.1	Design compliance to TOR
				6.1.2	Design coordination with all disciplines
				6.1.3	Material approval
				6.1.4	Shop drawing approval
				6.1.5	Method approval
				6.1.6	Method statement
				6.1.7	Mock-up
				6.1.8	Quality audit
				6.1.9	Functional and technical compatibility

Serial Number	Management Processes	Activities		Elements	
7	Resource Management	7.1	Project staff	7.1.1	Project/ construction manager staff
				7.1.2	Supervision staff
		7.2	Project manpower	7.2.1	Core staff
				7.2.2	Site staff
				7.2.3	Workforce
		7.3	Team management	7.3.1	Team behavior
				7.3.2	Conflict resolution
				7.3.3	Demobilization project workforce
		7.4	Construction resources	7.4.1	Material
				7.4.2	Equipment
				7.4.3	Subcontractor (s)
8	Communication Management	8.1	Submittals	8.1.1	Shop drawings
				8.1.2	Material
				8.1.3	Change orders
				8.1.4	Payments
		8.2	Documentation	8.2.1	Status log
				8.2.3	Issue log
				8.2.4	Minutes of meetings
				8.2.5	Contract documents
				8.2.6	Specifications
				8.2.7	Payments
		8.3	Correspondence	8.3.1	Stakeholders
				8.3.2	Regulatory authorities
				8.3.3	Correspondence among team members
9	Risk Management	9.1	Manage risk	9.1.1	Risk register
				9.1.2	Risk response
10	Contract Management	10.1	Contract documents	10.1.1	Notice to proceed
		10.2	Selection of subcontractor(s)		
		10.3	Selection of materials, systems, and equipment		
		10.4	Execution of works		
11	Health, Safety, and Environment	11.1	HSE management plan	11.1.1	Site safety

(Continued)

TABLE 3.3
(Continued)

Serial Number	Management Processes	Activities	Elements
			11.1.2 Preventive and mitigation measures
			11.1.2 Temporary Firefighting
			11.1.3 Environmental protection
			11.1.4 Waste management
			11.1.5 Safety hazards
12	Financial Management		
13	Claim Management		

Source: Abdul Razzak Rumane. (2016). *Handbook of Construction Management: Scope, Schedule, and Cost Control.* Reprinted with permission of Taylor & Francis Group.

3.2.4 MONITORING AND CONTROLLING PROCESS GROUP

Table 3.4 illustrates major construction activities during Project Monitoring and Controlling Process Group.

TABLE 3.4
Major Project Activities Relating to Monitoring and Controlling Process Group

Serial Number	Management Processes	Activities	Elements
1	Integration Management	1.1 Project Performance	1.1.1 Design Performance
			1.1.2 Construction Performance
			1.1.3 Project Start-Up
			1.1.4 Forecasted Schedule
			1.1.4 Forecasted Cost
			1.1.3 Issues
		1.2 Change Management System	1.2.1 Design Changes
			1.2.2 Design Errors
			1.2.3 Change Requests
			1.2.4 Scope Change
			1.2.5 Variation Orders
			1.2.6 Site Work Instruction
			1.2.7 Alternate Material
			1.2.8 Specs/Methods

Serial Number	Management Processes	Activities		Elements	
		1.3	Change Analysis	1.3.1	Review, Evaluate Changes
				1.3.2	Approve, Delay, Reject Changes
				1.3.3	Corrective Actions
				1.3.4	Preventive Actions
		1.4	Compliance to Contract Documents		
2	Stakeholder Management	2.1	Project Performance	2.1.1	Progress Reports
				2.1.2	Updates
				2.1.3	Safety Report
				2.1.4	Risk Report
		2.2	Project Updates	2.2.1	Contract Documents
		2.3	Payments	10.3.1	Payment Certificate
		2.4	Change Requests	2.4.1	Site Work Instruction
				2.4.2	Change Orders
		2.5	Issue Log	2.5.1	Anticipated Problems
		2.6	Minutes of Meetings	2.6.1	Progress Meetings
				2.6.2	Other Meetings
3	Scope Management (Contract Documents)	3.1	Validate Scope	3.1.1	Conformance to TOR
				3.1.2	Review of Design Documents
				3.1.3	Conformance to Contract Documents
				3.1.4	Approval of Changes
				3.1.5	Authorities Approval of Deliverables
				3.1.6	Stakeholders Approval of Deliverables
				3.1.7	Quality Audit
		3.2	Scope Change Control	3.2.1	Variation Orders
				3.2.1	Change Orders
		3.3	Performance Measures		
4	Schedule Management	4.1	Schedule Monitoring	4.1.1	Project Status

(Continued)

TABLE 3.4
(Continued)

Serial Number	Management Processes	Activities		Elements	
		4.2	Schedule Control	4.2.1	Progress Curve
		4.3	Schedule Changes	4.3.1	Approved Changes
		4.4	Progress Monitoring	4.4.1	Planned Versus Actual
		4.5	Submittals Monitoring	4.5.1	Subcontractors
				4.5.2	Material
				4.5.3	Shop Drawings
5	Cost Management	5.1	Cost Control	5.1.1	Work Performance
				5.1.2	S-Curve
				5.1.3	Forecasted Cost
		5.2	Change Orders		
		5.3	Progress Payment		
		5.4	Variation Orders		
6	Quality Management	6.1	Control Quality	6.1.1	Quality Metrics
				6.1.2	Quality Checklist
				6.1.3	Material Inspection
				6.1.4	Work Inspection
				6.1.5	Rework
				6.1.6	Testing
				6.1.7	Regulatory Compliance
7	Resource Management	7.1	Conflict Resolution		
		7.2	Performance Analysis		
		7.3	Material Management		
8	Communication Management	8.1	Meetings	8.1.1	Progress Meetings
				8.1.2	Coordination Meetings
				8.1.3	Safety Meetings
				8.1.4	Quality Meetings
		8.2	Submittal Control	8.2.1	Drawings
				8.2.2	Material
		8.3	Documents Control	8.3.1	Correspondence
9	Risk Management	9.1	Monitor and Control Risk	9.1.1	Scope Change Risk
				9.1.2	Schedule Change Risk
				9.1.3	Cost Change Risk
				9.1.4	Mitigate Risk
				9.1.5	Risk Audit

Serial Number	Management Processes	Activities	Elements
10	Contract Management	10.1 Inspection	
		10.2 Checklists	
		10.3 Handling of Claims, Disputes	
11	Health, Safety, and Environment	11.1 Prevention Measures	11.1.1 Accidents Avoidance/ Mitigation
			11.1.2 Firefighting System
			11.1.3 Loss Prevention Measures
		11.2 Application of Codes and Standards	
12	Financial Management	12.1 Financial Control	12.1.1 Payments to Project Team Members
			12.1.2 Payments to Contractor(s)/ Subcontractor(s)
			12.1.3 Material Purchases
			12.1.4 Variation Order Payment
			12.1.5 Insurance and Bonds
		12.2 Cash Flow	
13	Claim Management	13.1 Claim Prevention	13.1.1 Proper Design Review
			13.1.2 Unambiguous Contract Documents Language
			13.1.3 Practical Schedule
			13.1.4 Qualified Contractor(s)
			13.1.5 Competent Project Team Members
			13.1.6 RFI Review Procedure
			13.1.7 Negotiations
			13.1.8 Appropriate Project Delivery System

Source: Abdul Razzak Rumane. (2016). *Handbook of Construction Management: Scope, Schedule, and Cost Control.* Reprinted with permission of Taylor & Francis Group.

3.2.5 CLOSING PROCESS GROUP

Table 3.5 illustrates major construction activities relating to Project Closing Process Group.

3.3 MANAGEMENT PROCESSES

Management is a systematic way of management of processes in an efficient and effective manner toward accomplishment of organizational goals. Management is a systematic application of knowledge derived from general principles, concepts, theories, and techniques and embodied in the management functions which are

TABLE 3.5

Major Project Activities Relating to Closing Process Group

Serial Number	Management Processes	Activities	Elements
1	Integration Management	1.1 Close Project or Phase	1.1.1 Testing and Commissioning
			1.1.2 Authorities' Approvals
			1.1.3 Punch List/Snag List
			1.1.4 Handover of Project/Facility
			1.1.5 As-built Drawings
			1.1.6 Technical Manuals
			1.1.7 Spare Parts
			1.1.8 Lesson Learned
2	Resource Management	2.1 Close Project Team	2.1.1 Demobilization
			2.1.2 New Assignment
		2.2 Material and Equipment	2.2.1 Excess Material Removal/Disposal
			2.2.2 Equipment Removal
3	Contract Management	3.1 Close Contract	3.1.1 Project Acceptance/Takeover
			3.1.2 Issuance of Substantial Completion Certificate
			3.1.3 Occupancy
4	Financial Management	4.1 Financial Administration and Records	4.1.1 Payments to all Contractors, Subcontractors and Other Team Members
			4.1.2 Bank Guarantees/Warranties
5	Claim Management	5.1 Claim Resolution	5.1.1 Settlement of Claims

Source: Abdul Razzak Rumane. (2016). *Handbook of Construction Management: Scope, Schedule, and Cost Control.* Reprinted with permission of Taylor & Francis Group.

variables in terms of business practices as per organizational needs and requirements. Regardless of the type and scope of processes, there must be a plan that is organized, implemented, controlled, and maintained. There are five managerial functions which are known as the foundation for all management concepts. These are as follows.

1. Planning
2. Organizing
3. Staffing
4. Directing
5. Controlling

There are mainly three key attributes in a project, which have to be managed effectively and efficiently to achieve a successful project. These are

1. Scope
2. Time (Schedule)
3. Cost (Budget).

These are known as "Quality Trilogy" or "Triple Constraints".

From the project quality perspective, the phenomenon of these three components is called the "construction project quality trilogy" and is illustrated in Figure 3.5.

Triple Constraints is a framework to evaluate and balance these competing demands. It became a way to track and monitor projects. In pictorial form, Triple Constraints is a triangle in which one cannot adjust or alter one side of it without any effect on, or altering, the other side(s).

• If Scope is increased, the Cost will increase or the Time must be extended or both.
• It Time is reduced, then cost must increase or scope must decrease or both.
• If Cost is reduced, then the scope must be decreased or the time must increase or both.

In order to achieve a successful project, these key attributes, effectively and efficiently, track the progress of the work from the inception to completion of construction and handover of the project for successful completion. These three attributes

FIGURE 3.5 Triple constraints.

have functional relationship with many other processes, activities, and elements/sub-systems of the project. To achieve a successful project to the satisfaction of owner/end user, the construction/project manager has to manage the project in a systematic manner at every stage of the project and balance these attributes in conjunction with all the other activities which may affect the successful completion of the project. This can be done by implementing, amalgamating, and coordinating some or all the activities/elements of management functions, management processes, and project life cycle phases (technical processes). Thus, construction management process can be described as implementation and interaction of following functions and processes;

- Management functions
- Management processes
- Project phases (Technical Processes)

Figure 3.6 illustrates construction management process elements.

In practice, it is difficult to separate one element from others while executing a project. Interaction and/or combination among some or all of the elements/activities of these processes and their effective implementation and applications are essential throughout the life cycle of the project to conveniently manage the project. There are 47 Project Management Processes that are further grouped into 10 separate Knowledge Areas.

Figure 3.7 illustrates integration diagram of components/activities of three major elements of construction management process.

3.3.1 INTEGRATION MANAGEMENT

Integration Management is coordination and implementation of five Project Management Process Groups (Initiating, Planning, Executing, Monitoring, and Controlling) from the time the project is conceived right to closeout stage. Integration

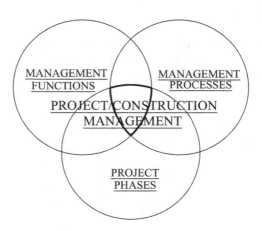

FIGURE 3.6   Project/construction management integration.

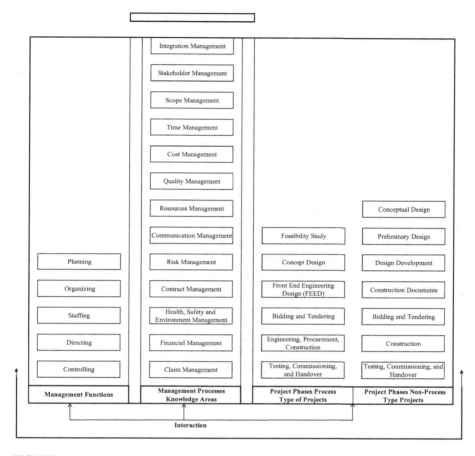

FIGURE 3.7   Project management process elements integration diagram.

Source: Abdul Razzak Rumane. (2016). Handbook of Construction Management: Scope, Schedule, and Cost Control.

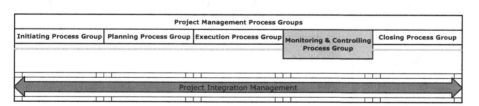

FIGURE 3.8   Project integration management.

Management involves putting all the process groups. Figure 3.8 illustrates Integration Management Process Cycle.

Integration Management of construction project includes all the activities performed to effectively control the final output of project.

3.3.2 STAKEHOLDER MANAGEMENT

A stakeholder is anyone who has involvement, interest, or impact in the project processes in a positive or negative way. Stakeholders play vital role in determining, formulation, and successful implementation of project processes. Stakeholders can mainly be classified as:

- Direct stakeholders
- Indirect stakeholders
- Positive stakeholders
- Negative stakeholders
- Legitimacy and Power

Figure 3.9 illustrates stakeholders having involvement or interest in the construction project.

These stakeholders have significant influence/impact on the outcome of the project. It is important to identify stakeholders who have interest and significant influence in the project. The stakeholders include members from within organization and people, agencies, and authorities outside the organization. A stakeholder register/log

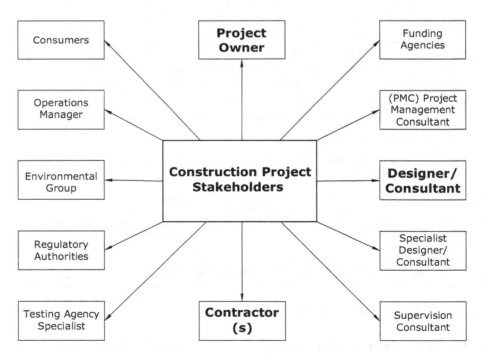

FIGURE 3.9 Construction project stakeholders.

Source: Abdul Razzak Rumane. (2016). *Handbook of Construction Management: Scope, Schedule, and Cost Control*. Reprinted with permission from Taylor & Francis Group.

is developed using different types of classification models for stakeholder's analysis. The stakeholder register/log is maintained and updated throughout the life cycle of project. It is essential that while analyzing the risk, it is important to address the needs of project stakeholders effectively predicting how the project will be affected and how the stakeholders will be affected, and the input from these stakeholders is taken into consideration in order to run a successful project.

Table 3.6 is an example Stakeholders Responsibility Matrix for Construction Project.

It is required to involve stakeholders through communication and working together to address the needs/expectations and issues of the stakeholders. Successful completion of the project depends on meeting the expectations of stakeholders. Stakeholder engagement in a project help to

- Reduce risk in the project
- Sharing experience and skills, thus resulting mitigate the threats and uncertainties
- Gain stakeholders' support
- To respond efficiently and effectively to the difficulties that may arise or issues that need to be resolved
- Reduce conflict
- To deal with changing needs of stakeholder
- To ensure that the project deliverables meet stakeholder expectations
- Successful completion of project within schedule, budget, and as per the approved scope

3.3.3   SCOPE MANAGEMENT

Project Scope Management consists of the processes to ensure that the project includes all the works required to complete the project successfully. Project scope management consists of six processes, which are as follows:

1. Plan Scope Management
2. Collect Requirements
3. Define Scope
4. Create WBS
5. Validate Scope
6. Control Scope

The scope management is the process which includes the activities to formulate and define the client's need/requirements by establishing project objectives and goals properly addressed in order for the project to have clear direction and controlling what is or is not involved in the project. The project scope documents explain the boundaries of the project, establishes project responsibilities for each team member, and sets up procedures for how completed works will be verified and approved. The scope describes the features and functions of the end product or the services to be provided by the project. During project, the scope documentation helps the project

TABLE 3.6
Stakeholders Responsibility Matrix

Serial Number	Activity	Owner/Client	Construction Manager/Project Manager	Designer/Consultant	Contractor	Supervisor	Regulatory Authority	Funding Agency	End User/Facility Manager	Notes/Comments
1	Project initiation	P	–	–	–	–	–	B	B	
2	Selection of construction manager	P	–	–	–	–	–	–	–	
3	Selection of designer	P	B	–	–	–	–	–	–	
4	Preparation of terms of reference (TOR)	A	P	–	–	–	–	–	–	
5	Preparation of design	A	B	P	–	–	R	–	–	
6	Value engineering	A	R	P	–	–	–	–	–	
7	Preparation of contract documents	A	B	P	–	–	–	–	–	
8	Project schedule	A	B	P	–	–	–	C	C	
9	Project budget	A	B	P	–	–	–	B	–	
10	Preparation of tendering documents	A	P	B	–	–	–	–	–	
11	Submission of bid	C	C	-	P	–	–	–	–	
12	Evaluation of bid	C	C	P	–	–	–	–	–	
13	Selection of contractor	A	P	B	–	–	–	C	C	
14	Approval of subcontractor	A	B	B	P	–	–	–	–	
15	Approval of contractor's staff	A	B	B	P	–	–	–	–	

#	Task								
16	Execution of works	C	C	R	P	R	–	–	–
17	Supervision of works	C	C	R	P	P	–	–	–
18	Approval of material	A	C	R	P	B	–	–	–
19	Approval of shop drawings	C	C	A	P	B	–	–	–
20	Construction schedule	A	C	R	P	B	–	–	–
21	Monitoring progress	P	C	P	P	B			
22	Monitoring cost	P	C	P	B	B			
23	Payments	A	R	R	P	B	–	–	–
24	Request for information	C	C	R	P	B			
25	Approval of change	A	B	R	P	B	–	–	–
26	Quality plan	C	B	R	P	B	–	–	–
27	Project quality	C	R	R	P	P			
28	Meetings	E	E	P	E	E	–	–	–
29	Safety plan	C	B	R	P	B	–	–	–
30	Site safety	C	C	B	P	P	–	–	–
31	Testing and commissioning	C	C	R	P	D	–	–	C
32	Authorities approval	C	C	B	P	B	A	–	–
33	Snag list	C	C	R	P	P	–	–	C
34	Substantial completion certificate	A	R	P	C	–	–	–	C

LEGEND: P = Prepare/Initiate/Responsible, R = Review/Comment, B = Advise/Assist, A = Approve, E = Attend, C = Inform team remain focused and on task. The scope statement also provides the project team with guidelines for making decisions about change requests during the project. It is essential that the scope statement should be unambiguous and clearly written to enable all the members of project team understand the project scope to achieve project objectives and goals.

Project development is a process spanning from the commencement of project initiation and ends with close out and finalizing project records after project construction. The project development process is initiated in response to an identified need by the owner/end user. It covers a range of time-framed activities extending from identification of a project need, development of contract documents, and construction of the project.

By applying the concept of Scope Management Processes methodology in development of construction project, following construction-related activities can be evolved:

1. Develop Scope Management Plan

   a. Project assumptions
   b. Constraints
   c. Key deliverables
   d. Project organization
   e. Roles and responsibilities
   f. Dependencies
   g. Milestones
   h. Project cost
   i. Quality
   j. Risks
   k. Safety regulations
   l. Environmental considerations
   m. Change control

2. Collect Requirements

   a. Need statement
   b. Project goals and objectives
   c. Terms of reference (TOR)

3. Develop Project Scope Documents

   a. Collect owner's preferred requirements
   b. Develop design

      I.   Concept design
      II. Schematic design
      III. Detail design

   c. Project specifications

      I. General specifications
      II. Particular specifications

     d. Contract documents

        I. General conditions
        II. Particular conditions
        III. Tender/bidding documents

4. WBS

    4.1 Project breakdown structure

       a. Project life cycle phases
       b. Project scope
       c. Bill of quantities
       d. Schedule milestones
       e. Cost estimates
       f. Quality management
       g. Resource management
       h. Risk management
       i. Documents

    4.2 Organizational breakdown structure
    4.3 WBS dictionary
    4.4 Responsibility assignment matrix
    4.5 Scope baseline

5. Validate scope

       a. Review of design documents
       b. Approval of design documents
       c. Review of contract documents
       d. Approval of contract documents
       e. Regulatory approvals
       f. Acceptance of project

6. Control Scope

       a. Scope change control
       b. Variation orders
       c. Change orders
       d. Performance measures

3.3.3.1 Develop Scope Management Plan

Project Scope Management Plan is a part of overall Project Management Plan that describes how the scope will be defined, developed, validated, and controlled. It explains how the project will be managed and how scope changes will be incorporated into the project management plan. Scope Management Plan establishes a structured process to ensure that the work performed by the project team is clearly within the established parameters and ensures that all project objectives are achieved.

    The key benefit of this process is that it provides guidelines and direction of how the scope will be managed throughout the project. Scope Management Plan documents

- Scope definition
- The scope management approach
- Roles and responsibilities of stakeholders pertaining to the project scope
- WBS (activity list)
- Organization breakdown structure
- Procedures to verify and approve the completed works
- Managing any changes in the project scope baseline
- Guidelines for making decision about change requests during execution of project and controlling project scope

3.3.3.2   Collect Requirements

Construction project development is initiated with the identification of need to develop a new facility or renovation/refurbishment of existing facility. It is essential to get a clear definition of the identified need or the problem to be solved by the new project. The owner's need must be well defined, indicating the minimum requirements of quality and performance, an approved budget, and required completion date. The need should be based on real (perceived) requirements. The identified need is then assessed and analyzed to develop need statement. Need assessment is conducted to determine the need. Need assessment is a systematic process for determining and addressing needs or "gaps" between current conditions and desired conditions "want". Need analysis is the process of identifying and evaluation need. The need statement is written based on the need analysis and is used to perform feasibility study to develop project goals and objectives and subsequently to prepare project scope documents.

The feasibility study takes its starting point from the output of project identification need. The need statement is the input to perform feasibility study. The main purpose of feasibility study is to evaluate the project need and decide whether to proceed with the project or stop. Depending on the circumstances, the feasibility study may be short or lengthy, simple or complex. In any case, it is the principal requirement in project development as it gives owner/client an early assessment of the viability of the project and the degree of risk involved.

Feasibility study can be categorized into following functions:

- Legal
- Marketing
- Technical and engineering
- Financial and economical
- Social
- Environmental
- Risk
- Scheduling of project

The project feasibility study is usually performed by the owner through his own team or by engaging a specialist agency or individual.

After completion and approval of feasibility study, it is possible to establish project goals and objectives. The goals and objectives must be

1. Specific
2. Measurable
3. Attainable/achievable
4. Realist
5. Time (cost) limited

Once project goals and objectives are established, a comprehensive scope statement (Terms of Reference) is prepared by the owner/client or by the project manager on behalf of the owner describing in details the project objectives and requirements to develop the project. Terms of Reference is a document that describes the intention of a project, the approach in which it will be constructed, and how it will be implemented. It can also be described as a specification of a team member's responsibilities and influence within a project. A Term of Reference (TOR) is an outline of a project, including its mission statement, its procedures and rules, and the different administrative aspects of the entire project.

3.3.3.3   Develop Project Scope Documents

Construction project scope documents are developed based on the requirements described in the Terms of Reference (TOR) prepared by the Owner/Client or Construction Manager/Project Manager on behalf of the Owner/Client. The TOR gives the Designer (Consultant) a clear understanding for the development of the project. The Designer (Consultant) utilizes TOR to develop contract documents to suit the project delivery system. The contract documents for Design-Bid-Build type of project delivery system mainly consist of

1. Scope of work
2. Design drawings
3. Bill of quantities
4. Technical specifications
5. Conditions of contract
6. Project schedule
7. Tender/bidding documents

In case of Design-Build type of project delivery system, tendering/bidding documents are prepared taking into consideration performance specifications for the project.

Figure 3.10 is an illustrative flow chart for development of terms of reference (construction project documents), and Figure 3.11 illustrates construction project development process.

3.3.3.4   Work Breakdown Structures

Work Breakdown Structure (WBS) is a hierarchical representation of system levels. WBS is a family tree, consists of a number of levels, starting with the complete scope of work at level 1 at the top and progressing downward through as many levels as necessary to obtain work elements (activities) that can be conveniently managed. WBS involves envisioning the project as a hierarchy of goal, objectives, activities, sub-activities, and work packages. WBS is constructed by

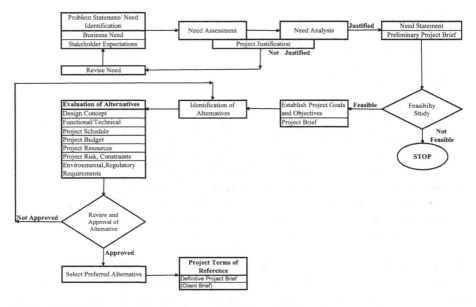

FIGURE 3.10 Flowchart for development of terms of reference.

Source: Abdul Razzak Rumane. (2016). *Handbook of Construction Management*. Reprinted with permission of Taylor & Francis Group.

dividing the project into major elements with each of these being divided into
sub-elements. This is done till a breakdown is done in terms of manageable units
of work for which responsibility can be defined. The hierarchical decomposition of activities continues until the entire project is displayed as a network of
separately identified and nonoverlapping activities. Each activity will be single
purposed, of specific time duration, and manageable, its time and cost estimates
easily derived, deliverables clearly understood, and responsibility for its completion clearly assigned.

In order to manage and control the project at different levels in most effective
manner, the project is broken down into a group of smaller subprojects/subsystems
and then to small well-defined activities. Each element (activity) should be

- Definable
- Manageable
- Measurable
- Estimable
- Independent
- Integratable
- Adaptable

Figure 3.12 illustrate an approach to development of work breakdown structure.

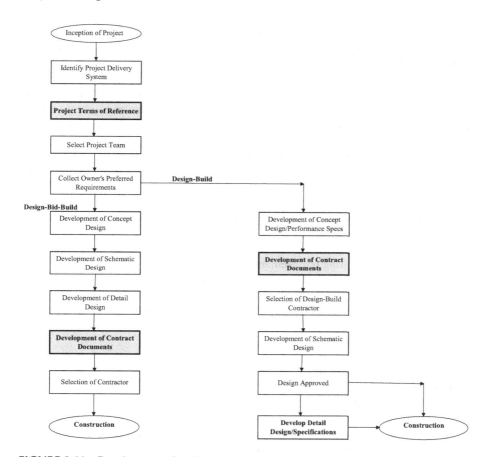

FIGURE 3.11 Development of project scope documents.

Source: Abdul Razzak Rumane. (2016). *Handbook of Construction Management*. Reprinted with permission of Taylor & Francis Group.

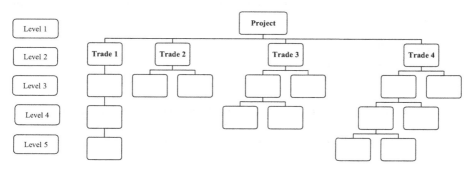

FIGURE 3.12 Approach to development of work breakdown structure.

3.3.3.4.1 Project Breakdown Structures

Construction projects are constantly increasing in technical complexity, and the relationships and contractual groupings of those who are involved are also more complex and contractually varied. Work Breakdown Structure (WBS) approach to construction projects help understand the entire process of project management and to manage and control its activities at different levels of various phases to ensure timely completion of the project with economical use of resources to make the construction project most qualitative, competitive, and economical. WBS is a deliverable-oriented grouping of project work elements shown in graphical display and the total scope of work organized and subdivided into small and manageable components (activities).

WBS development involves following major steps:

1. Identify final product(s) necessary to achieve total project scope
2. Identify major deliverables
3. Divide these major deliverables to a level of detail appropriate to meet managing and controlling requirements of the project
4. Divide each of these deliverables into its components.

While preparing WBS, following points need to be considered:

- Complexity of the project.
- Size of the project.
- Reporting requirements.
- Resource allocation.
- Works included in the scope of work only included in the WBS.
- Duration of task or activities should be as per industry common practice which is "8–80 rule". This rule recommends that the lowest level of work should be no less than 8 hours and no more than 80 hours.
- Each level is assigned unique identification number.

The lowest level of work breakdown structure is known as "work package". A work package can be divided into specific activities to be performed.

Figure 3.13 illustrate typical levels of WBS Process in construction project.

Traditional construction projects have the involvement of three main parties. These are:

1. Owner/Client
2. Designer/Consultant
3. Contractor

The involvement and interaction between owner/client, designer/consultant, and contractor(s) depends on the construction project procurement strategy followed by the owner. Based on the strategy, the WBS for construction project can be developed to suit the requirements of the project development at different stages (study, design, and construction) of the project.

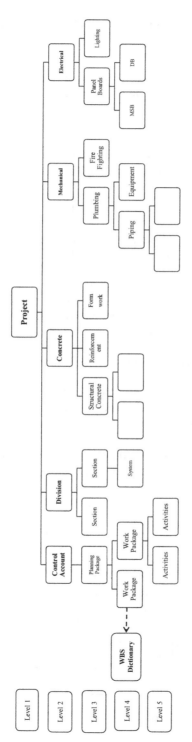

FIGURE 3.13   Typical levels of WBS process.

There are typically two types of Work Breakdown Structure. These are

1. Project WBS

   - I t is prepared by the contractor taking into considerations the requirements listed in contract documents. It is also known as Contractor's Construction Schedule

2. Project Summary WBS

   - It is prepared by the Designer (Consultant) summarizing the entire project requirements. It is part of contract documents signed between Owner/Client and Contractor.

There are different approaches followed for development of WBS. These are:

- Physical location which is further divided into floor levels and zones
- Project development stages such as study, design, and construction

Figure 3.14 illustrates WBS for project design.

In most construction projects, CSI MasterFormat® coding system is followed to develop WBS. Following figures are guidelines to develop WBS taking into consideration Divisions, Sections, and Titles from MasterFormat®. More detailed levels can be developed according to the project requirements.

Figure 3.15 illustrate WBS for concrete works in Construction Project.

3.3.3.4.2 Organizational Breakdown Structure (OBS)

Organizational Breakdown Structure (OBS) is a hierarchical organizational relationship of the project teams, including subcontractors, responsible for managing the designate scope of work described within the work breakdown structure (WBS). It is used as a framework for assigning work relationship. The WBS identifies what work to be done whereas OBS identifies the individual that will do the work. An organizational

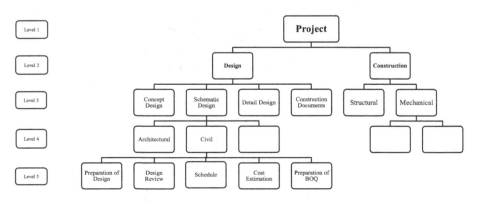

FIGURE 3.14 Typical levels of WBS process (project design).

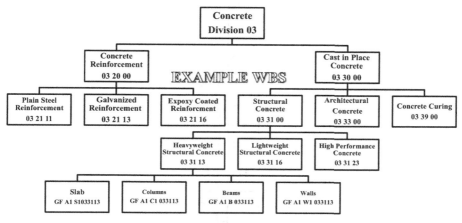

FIGURE 3.15   WBS for concrete works.

breakdown structure for the project can be simple or complex depending on the size and complexity of the project. WBS with appropriate type of organization (projectized, functional, matrix) assures that all the scope of work is accounted for, and each element of works is assigned to the level of responsibility for planning, tracking progress, costing, and reporting. The organizational and personal relationship can be established at any of several levels within the project and functional organization.

Figure 3.16 illustrates Construction Supervisor's site organization for construction project.

Figure 3.17 illustrates contractor's site organization for construction project.

3.3.3.4.3   WBS Dictionary

WBS Dictionary is a set of companion documents to WBS, which describes WBS elements in the work breakdown structure and includes

1. Code of account identifier for each WBS element
2. Scope description (statement of work)
3. Bill of quantities
4. Activities
5. Schedule milestone
6. Cost estimates
7. Resources requirements
8. Quality requirements
9. Technical and other references
10. Responsible organization/person
11. Deliverables
12. Contract information

Table 3.7 illustrates an example of WBS Dictionary for construction project

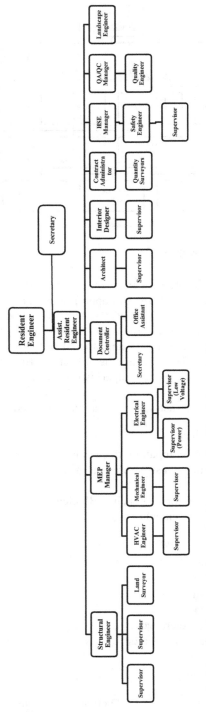

FIGURE 3.16 Consultant's (supervisor's) organizational breakdown structure.

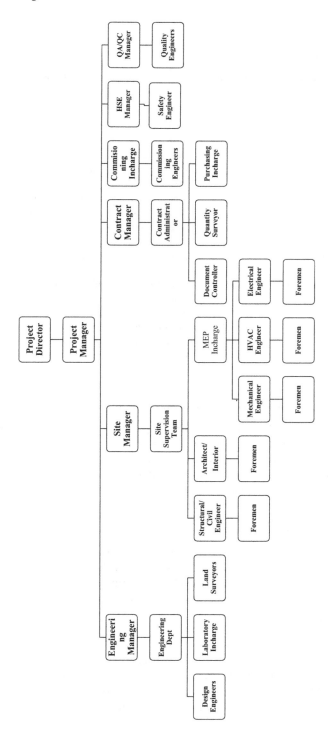

FIGURE 3.17 Contractor's organizational breakdown structure.

TABLE 3.7
WBS Dictionary

Serial Number	WBS Element	WBS Code	WBS Element Description
1	Demolition/ site preparation	024113	Clear and remove surface matters, windblown sand, heaps of soil, or debris of any kind on the proposed roads, shoulders, side slopes, and existing road and shoulder, if any, all as per specifications and drawings.
2	Cast in place concrete	033000	Plain concrete using sulphate resistant cement including formwork and additives.
3	Concrete unit masonry assembly	042200	Block work non-load bearing including control joints, filler and sealant at wall heads, mortar, anchors, reinforcement.
4	Bituminous damp proofing	071113	2 nos. layers of torch applied 4-mm thick water-proofing membrane including laps and bituminous primer.
5	Plumbing system	221400	Supply, install, commission, and handover water drainage system.
6	HVAC system	230593	Testing, adjusting, and balancing of HVAC.
7	Switchboards and panelboards	262413	Supply, install, and test the Main Low-Tension Board complete with all accessories as per the requirements of the drawings and specifications.

3.3.3.4.4   Responsibility Assignment Matrix (RAM)

Responsibility Assignment Matrix (RAM) depicts the intersection of WBS and OBS. It identifies the specific responsibility for specific project task and provides a realistic picture of the resources needed and can identify if the project has enough resources for successful completion of the project. RAM uses the WBS and OBS to link deliverables and/or activities to resources and relates the WBS element/task to the organization and the named individual who is responsible for the assigned scope of a control account. Following is the procedure to prepare RAM:

- Develop WBS and prepare the complete list of project deliverables
- Identify team members who will be assigned for the project
- Allocate responsibility to team members according to their field of competency, expertise, required experience that will support the work to produce different project deliverables
- Prepare the RAM

Figure 3.18 illustrates the relation between WBS and OBS for construction project.

Responsibility Assignment Matrix can be a simple tick box (check mark) or RACI type. Table 3.8 illustrates Example Responsibility Assignment Matrix for construction project, and Table 3.9 illustrates Example RACI Matrix for construction project.

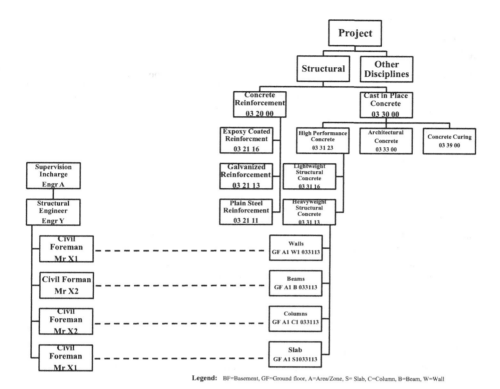

FIGURE 3.18 RAM for concrete works.

TABLE 3.8
RAM (Responsibility Assignment Matrix)

WBS Element	Office, Role, or Person				
	Owner	Resident Engineer	Planning and Control	Quality Manager	Quantity Surveyor
Notice to proceed	X				
Approval, permits	X				
Conduct meetings		X			
Action on Request for Information (RFI)		X			
Project monitoring			X		
Construction quality				X	
Review of progress (interim) payment					X

TABLE 3.9

RACI

WBS Element	Person			
	Owner	Project Manager	Consultant	Contractor
Bonds and guarantees	I	I	A	R
Subcontractor approval	A	C	C	R
Construction schedule	I	C	A	R
Meetings	C	C	R	I
Submittal logs	I	C	A	R
As-built drawings	I	I	A	R
Substantial completion certificate	A	C	R	I

Abbreviations: **R: Responsible, A: Accountable, C: Consulted, I: Informed**

R: Responsible: The person(s) responsible for performance of the concerned activity.

A: Accountable: The person(s) accountable for ensuring the activity is completed.

C: Consulted: The person(s) who must be consulted prior to or during the execution of the activity.

I: Informed: The person(s) who must be informed about the progress and outcome of the activity.

Work Breakdown Structure is a critical tool for organizing the work and preparing realistic schedule and cost estimates. It helps reporting, monitoring, and controlling the project. WBS supports integrating responsibilities for performing various works with various organizations and individuals having direct relationship between the WBS elements related to the identified individual through RAM.

3.3.3.4.5 Scope Baseline

As per PMI PMBOK, "The Scope Baseline is the approved version of a scope statement, work breakdown structure (WBS), and its associated WBS Dictionary, that can be changed only through formal change control procedures and is used as a basis for comparison".

For construction project, scope baseline is technical scope baseline that describes the performance capabilities that the project must provide at the end of construction phase to meet the owner's needs. The scope base line are the contract documents developed by the designer based on the requirements described in Terms of Reference (TOR). The scope baseline (contract documents) is developed based on following construction documents which are handed over to the successful bidder:

- Working drawings
- Technical specifications (particular specifications)
- Bill of quantities (BOQ)
- General specifications
- General conditions
- Particular condition

Contractor has to follow these documents for implementation/execution of project works.

3.3.3.5 Validate Scope

Validate scope is the process of formalizing acceptance of completed project deliverables. It is a method to ensure that

- Project design conforms to current applicable codes and standards
- Project design has taken care of all the requirements listed under Terms of Reference
- Assess whether Value Engineering analysis has been performed and the recommendations are incorporated in the project baseline
- All the material and equipment installed comply with specification requirements
- All the works at site are performed as per approved shop drawings by approved material
- Regulatory approvals are obtained
- Installed/executed works are checked, inspected at every stage to confirm that they have been installed/executed as specified, using specified and approved materials, installation method recommended by the manufacturer to meet intended use of the project
- Corrective actions or defect repairs are completed
- Inspection and tests are carried out to ascertain operational requirement
- The work is documented, and changes are recorded
- Records of inspection and tests are maintained to verify approved construction methods and materials were used
- Outstanding defects, works (punch list) are listed and documented
- As-built drawings, documents, manuals are ready for handover to the client
- Start-up test plans are established to demonstrate that all the systems installed in the project meet required operations and safety requirements
- Handover/takeover program is established
- All the requirements for facility management are documented with all the information and knowledge that is required to strategically and physically manage the new facility

In construction project, the project elements (intermediate deliverables) need to be verified, reviewed, approved, and accepted at different stages of the project life cycle to ensure that completed project deliverables meet owner's needs and expectations (goals and objectives). It is the assessment of readiness of construction or execution and to confirm the completeness and accuracy of the project as per agreed-upon scope baseline.

It is performed mainly during following phases of construction project:

1. Design phase
2. Construction phase
3. Testing, commissioning, and handover phase

The following are the four main steps needed to establish an assessment, review, approve, and accept the project:

1. Identify the elements, items, material, system, products to be reviewed and checked in each of the trade (architectural, structural, fire suppression, plumbing, HVAC, electrical, low voltage systems, landscape, external, etc.)
2. Identify the frequency of checking and inspection
3. Identify the agency/persons authorized to check and approve
4. Identify the indications/criteria for performance monitoring and control

Figure 3.19 illustrates scope validation process for construction project.

3.3.3.6 Control Scope

Control scope is the process of monitoring the project scope and managing any changes to the scope baseline. It is common that despite all the efforts devoted to develop the contract documents (scope baseline), the contract documents cannot provide complete information about every possible condition or circumstance that the construction team may encounter. Figure 3.20 illustrates scope control process in construction project.

During construction process, circumstances may come to light, which necessitate minor or major changes to the original contract. These changes may occur due to following causes:

1. Differences/errors in contract documents
2. Construction methodology
3. Non-availability of specified material
4. Regulatory changes to use certain type of material
5. Technological changes/introduction of new technology
6. Value engineering process
7. Additional work instructed
8. Omission of some works

Table 3.10 lists the causes of changes in construction projects.

These changes are identified as the construction proceeds. These changes or adjustments are beneficial and help build the facility to achieve project objective. Prompt identification of such requirements helps both the Owner and Contractor to avoid unnecessary disruption of work and its impact on cost and time. The impacts and consequences of changes in the construction project vary according to

- Type and nature of changes
- Time of occurrence or observance of error or omission
- Change needed for the benefit of the project

A critical change may have a negative impact on the project baseline(s). It is important to establish a reliable change control system to manage changes to the project

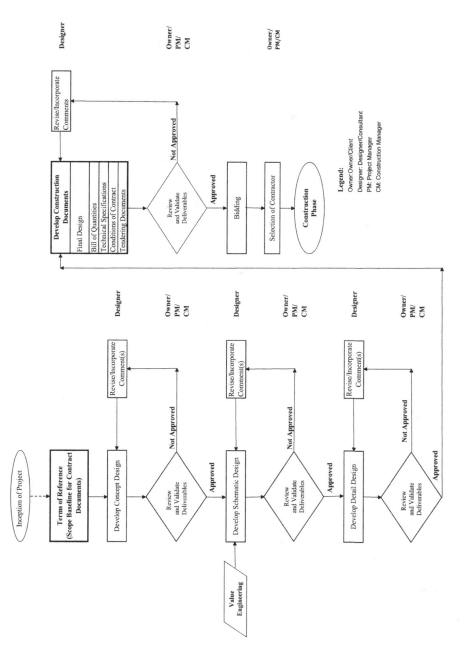

FIGURE 3.19 Scope validation process for construction project (design and bidding stage).

Source: Abdul Razzak Rumane. (2016). *Handbook of Construction Management.* Reprinted with permission of Taylor & Francis Group.

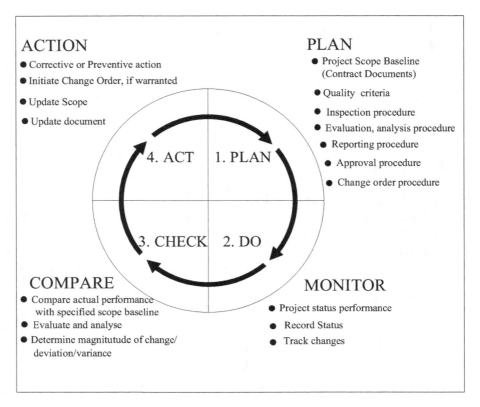

FIGURE 3.20   Scope control process.

scope. The changes should be managed to maximize the benefits to the project and minimize the negative impacts and effects on the project. The change control system should include:

- Submission procedure
- Evaluation and review procedure
- Impact on project
- Approval
- Reporting
- Baseline(s) updates
- Document update

The changes that arise during construction can be either initiated by the owner or by the contractor or even by any of the stakeholder to the construction project.

Identification of discrepancies/errors and changes in the specified scope are common in construction projects. Prompt identification of such requirements helps both the Owner and Contractor to avoid unnecessary disruption of work and its impact on

TABLE 3.10
Causes of Changes in Construction Projects

Serial Number		Causes
I	**Owner**	
	I-1	Delay in making the site available on time
	I-2	Change of plans
	I-3	Financial problems/payment delays
	I-4	Change of schedule
	I-5	Addition of work
	I-6	Omission of work
	I-7	Project objectives are not well defined
	I-8	Different site conditions
	I-9	Value Engineering
II	**Designer (Consultant)**	
	II-1	Inadequate specifications
		a) Design errors
		b) Omissions
	II-2	Scope of work not well defined
	II-3	Conflict between contract documents
	II-4	Coordination among different trades and services
	II-5	Design changes/modifications
	II-6	Introduction of latest technology
III	**Contractor**	
	III-1	Process/methodology
	III-2	Substitution of material
	III-3	Nonavailability of specified material
	III-4	Charges payable to outside party due to the cancellation of certain items/products
	III-5	Delay in approval
	III-6	Contractor's financial difficulties
	III-7	Unavailability of manpower
	III-8	Unavailability of equipment
	III-9	Material not meeting the specifications
	III-10	Workmanship not up to the mark
IV	**Miscellaneous**	
	IV-1	New regulations
	IV-2	Safety considerations
	IV-3	Weather conditions
	IV-4	Unforeseen circumstances
	IV-5	Inflation
	IV-6	Fluctuation in exchange rate
	IV-7	Government policies

Source: Abdul Razzak Rumane. (2013). *Quality Tools for Managing Construction Projects.* Reprinted with permission of Taylor & Francis Group.

cost and time. Contractor uses Request for Information (RFI) form to request technical information from the supervision team. These queries are normally resolved by the concerned supervision engineer. However, it is likely that the matter has to be referred to the designer as Request for Information (RFI) has many other considerations to be taken care of which may be beyond the capacity of supervision team member to resolve. Normally, there is a defined period to respond to RFI. Such queries may result in variation to the contract documents. It is in the interest of both Owner and Contractor to resolve RFI expeditiously to avoid its effect on construction schedule.

Figure 3.21 illustrates Request for Information (RFI) Form which contractor submits to the consultant to clarify differences/errors observed in the contract documents, change in construction methodology, change in the specified material, and so on.

Figure 3.22 illustrates process to resolve scope change (Contractor initiated).

Figure 3.23 illustrates a Variation Order Request Form the contractor submits to the owner/consultant for approval of change(s) in the contract.

Figure 3.24 illustrates process to resolve request for variation (contractor initiated).

Figure 3.25 illustrates a Site Works Instruction (SWI) Form. It gives instruction to the contractor to proceed with the change(s). All the necessary documents are sent along with the SWI to the contractor. SWI is also used to instruct contractor for owner-initiated changes.

Similarly, if the contractor requires any modification to the specified method, then the contractor submits a Request for Modification to the owner/consultant. Figure 3.26 illustrates the request for modification. Usually these modifications are carried out by the contractor, without any extra cost and time obligation toward the contract.

Figure 3.27 illustrates process to resolve scope change (owner initiated)

It is the normal practice that, for the benefit of project, the Engineer's Representative assesses the cost and time related to SWI or requests for change in the scope and obtains preliminary approval from the Owner, and the contractor is asked to proceed with such changes. The cost and time implementation is negotiated and formalized simultaneously/later to issue the formal variation order. In all the circumstances where a change in contract is necessary, owner approval has to be obtained. Figure 3.28 illustrates the form used by the Engineer's Representative to obtain Change Order Approval from owner.

Once cost and time implications are negotiated and finalized and both the owner and contractor approve the same, Variation Order is issued to the contractor and changes are adjusted with contract sum and schedule. Figure 3.29a illustrates Variation Order Form issued to formalize the change order, and Figure 3.29b illustrates the Attachment to Variation Order.

3.3.4 SCHEDULE MANAGEMENT

Planning and scheduling are often used synonymously for preparing construction program because both are performed interactively. Planning is the process of identifying the activities necessary to complete the project, while scheduling is the process of determining the sequential order of the planned activities and the time required

---

Project Name

Consultant Name

REQUEST FOR INFORMATION (TECHNICAL)

CONTRACT NO. : _____   R.F. I. NO. _____
CONTRACTOR. : _____   DATE : _____

To: Resident Engineer _____

REF:

SUBJECT:

REQUEST FOR INFORMATION (Technical)

This form is used by the contractor to request information and is normally
sent to the A/E who responds on the same form.

SAMPLE FORM

CONTRACTOR: _____

DISTRIBUTION: Employer ☐   Engineer ☐   R.E. ☐

RESPONSE BY R.E.:

Signature of R. E. _____   Date _____

RESPONSE RECEIVED:
FOR CONTRACTOR: _____   DATE: _____

DISTRIBUTION: Employer☐   Engineer☐   R.E. ☐

FIGURE 3.21   A sample request for information form.

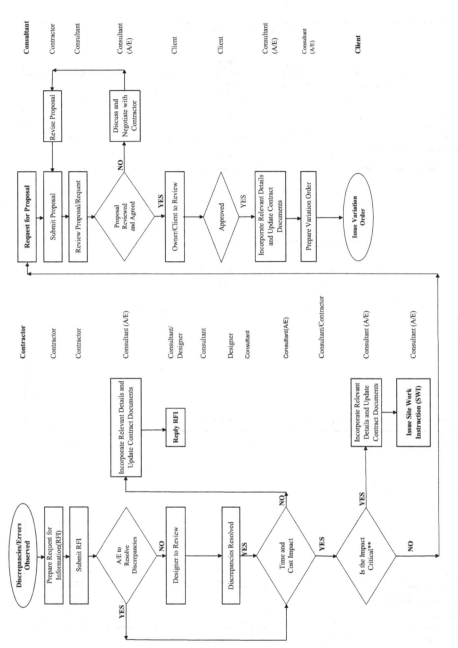

FIGURE 3.22 Process to resolve scope change (contractor initiated).

Source: Abdul Razzak Rumane. (2016). *Handbook of Construction Management.* Reprinted with permission of Taylor & Francis Group.

Project Name
Consultant Name

REQUEST FOR VARIATION

CONTRACT NO. : _____ NO. : _____
CONTRACTOR: _____ DATE : _____

TO : _____

SAMPLE FORM

PROPOSED VARIATION:

☐ PRODUCT ☐ METHOD OF FABRICATION ☐ METHOD OF INSTALLATION

SPECIFIED PRODUCT _____
PROPOSED PRODUCT _____
SPEC. SECTION # _____ PAGE # _____ ARTICLE # _____
DRG REF _____ DRG # _____ REV# _____
SPECIFIED MANUFACTURER _____
PROPOSED MANUFACTURER _____
BRIEF PRODUCT DESCRIPTION _____

REASON FOR PROPOSED VARIATION

| CHANGE
IN DESIGN | REQUIRED BY
AUTHORITIES | SITE
CONDITIONS | SWI |

COST AND TIME EFFECT

COST   NO ☐   YES ☐   AMOUNT --------(ADDITION)
TIME   NO ☐   YES ☐   DAYS ----------

ATTACHMENTS:
  1 Schedule of additions/ommissions
  2 Bill Summary
  3 Rate Analysis
  4 Measurements

Technical and cost comparison sheets must be attached with this request, other wise it will not be reviewed. Contractor shall fill and submit two formsto the OWNER.
Front sheet only shall be returned to Contractor with OWNER action.

WE (THE MAIN CONTRACTOR) CERTIFIES AND UNDERTAKES THAT :

CONTRACTOR'S REP _____ DATE/TIME _____
RECEIVED BY A/E _____ DATE/TIME _____

REVIEW AND ACTION BY OWNER

☐ Approved   ☐ Not Approved   ☐ Approved as Noted   ☐ Incomplete Data Resubmit

COMMENTS

APPROVED SUBJECT TO COMPLIANCE WITH CONTRACT DOCUMENTS

Authorised Signature _____ _____ DATE/TIME: _____

THE APPROVAL OF ANY VARIATION REQUEST SHALL BE SOLELY AT THE DIRECTION OF THE OWNER AND
SUCH APPROVAL SHALL IN NO WAY RELIEVE THE CONTRACTOR OF ANY OF HIS LIABILITIES AND OBLIGATIONS UNDER THE CONTRACT.

RECEIVED BY CONTRACTOR: _____ DATE/TIME: _____

cc: OWNER ☐   EMPLOYER ☐   R.E. ☐   ☐

FIGURE 3.23   A sample request for variation form.

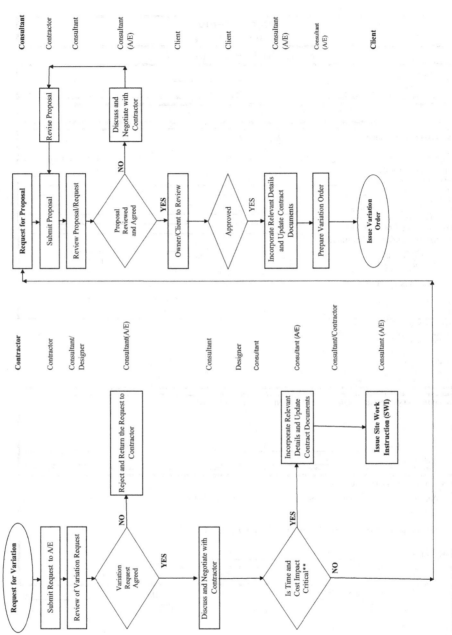

FIGURE 3.24 Process to resolve request for variation.

Source: Abdul Razzak Rumane. (2016). *Handbook of Construction Management.* Reprinted with permission of Taylor & Francis Group.

Project Name
Consultant Name

SITE WORK INSTRUCTION

CONTRACT NO. : _____   NO.   : _____
CONTRACTOR   : _____   DATE : _____

SUBJECT:

SITE WORK INSTRUCTION (SWI)

SAMPLE FORM

S.W.I. involves an anticipated change in the work. All S.W.I. must be authorized
and signed by the Owner Representative OR Authorised Signatory.

An S.W.I. is an instruction to the contractor to proceed prior to the issuing of a Variation
Order (V.O.). Whenever time allows, a V.O. will be issued instead of S.W.I.

Owner Rep. Signature   **Date**

THIS SITE WORKS INSTRUCTION (S.W.I.) IS A NOTICE TO PROCEED AND MAY INVOLVE CHANGE IN COST AND OR TIME.
YOU ARE REQUIRED TO ADVISE THE ENGINEER WITHIN 14 DAYS OF ANY ADDITIONAL COST AND
OR TIME REQUIRED TO COMPLY WITH THIS INSTRUCTION.

RECEIVED FOR
CONTRACTOR: _____   **DATE:** _____

DISTRIBUTION: Owner☐   Engineer☐   R.E.☐

FIGURE 3.25 A sample site work instruction form.

Project Name
Consultant Name

REQUEST FOR MODIFICATION

(AT NO EXTRA COST & OR TIME TO THE EMPLOYER)

CONTRACTOR : _____	DATE : _____
CONTRACT NO : _____	NO : _____

TO   :   Owner Name   CC: A/E   ☐

PROPOSED MODIFICATION TO:

☐ DESIGN DRAWING   ☐ METHOD OF FABRICATION   ☐ METHOD OF INSTALLATION

DESIGN DRAWING NO _____

SECTION # _____

CONTRACTOR'S PROPOSED DRAWING NO. _____

REASON FOR MODIFICATION _____

COST AND TIME SAVINGS   *At no extra cost and or time to the Employer*

COST	NO	☐	YES	☐	AMOUNT ------- (DEDUCTION) _____
TIME	NO	☐	YES	☐	DAYS

ATTACHMENTS:
Supplier confirmation letter   SAMPLE FORM

CONTRACTOR'S REP   _____   DATE/TIME   _____

RECEIVED BY A/E   _____   DATE/TIME   _____

REVIEW AND ACTION BY OWNER

☐ Approved   ☐ Not Approved   ☐ Approved as Noted   ☐ Incomplete Data Resubmit

COMMENTS

Authorised Signatories   _____   DATE/TIME: _____

RECEIVED BY CONTRACTOR: _____   DATE/TIME: _____

cc: Owner☐   Engineer ☐   R.E. ☐   ☐

FIGURE 3.26   A sample request for modification form.

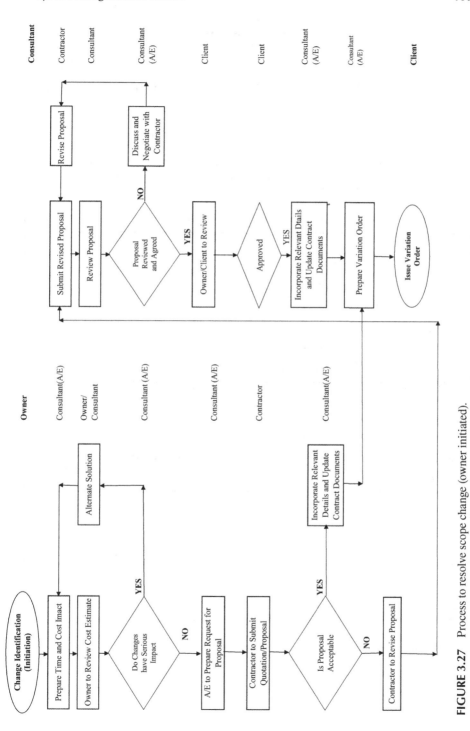

FIGURE 3.27 Process to resolve scope change (owner initiated).

Source: Abdul Razzak Rumane. (2016). *Handbook of Construction Management.* Reprinted with permission of Taylor & Francis Group.

| **Project Name** |
| **Consultant Name** |

VARIATION ORDER PROPOSAL (VOP)

CONTRACT NO: _____   VOP No. _____

CONTRACTOR: _____   Date: _____

The "Employer" decision is requested for approval / rejection of this Variation Order Proposal as described below.
Should the VOP be approved by the "Employer" an order to proceed shall be issued to the Contractor for his further action.

INITIATED / PROPOSED BY:

☐ ENGINEER   ☐ ENGINEER'S REP.   ☐ CONTRACTOR   ☐ OTHERS

REASONS:

   ☐ CHANGE IN DESIGN   ☐ SWI _____

   ☐ REQUIRED BY AUTHORITIES   _____

   ☐ REQUIRED BY SITE CONDITIONS

BRIEF DESCRIPTION & LOCATION:

SAMPLE FORM

BASIS OF V.O.EVALUATION:
PRORATA BOQ PRICES: ☐

PROPOSAL from CONTRACTOR: ☐

APPROXIMATE COST IMPACT _____   APPROXIMATE % _____

APPROXIMATE TIME IMPACT _____   ANY DELAY   _____

RELATED REFERENCES:

ENGINEER'S RECOMMENDATION:

RESIDENT ENGINEER'S SIGNATURE _____   DATE: _____

Distribution: ☐ Employer   ☐ Engineer   ☐ Resident Engineer   ☐

FIGURE 3.28   A sample variation order proposal form.

Project Name
Consultant Name

VARIATION ORDER (VO)

CONTRACT NO.   : _____   V.O. NO. : _____

CONTRACTOR   _____   DATE   : _____

In accordance with clause ----- of Document ---- Conditions of Contract, you hereby ordered to undertake work and/or amend the Contract as detailed below:

DESCRIPTION OF WORK

SAMPLE FORM

According to Clause ---------, Extension of Time for Completion and Valuation of VO are as follows:

Original Contract Price: _____   Contract Completion Date: _____

Previous VOs: _____   Previous Extension(s): _____

Value This VO: _____   Extension This VO: _____

Revised Contract Price: _____   Revised Completion Date: _____

We the undersigned Contractor hereby agree to carry out the works as ordered by this Variation Order and accept in full and final settlement of all related consequential costs and time

Recommended by Resident Engineer   Agreed by Contractor

Recommended by Engineer   Agreed by Employer

Distribution   ☐ Employer   ☐ Engineer   ☐ Resident Engineer   ☐ Contractor

FIGURE 3.29A   A sample variation order form.

Project Name
Consultant Name

ATTACHMENT TO VARIATION ORDER

CONTRACTOR: _____CONTRACT No.: _____

V. O. No. _____   Date: _____

1.   Site Works instructions incorporated into this Variation Order:

2.   Previous correspondence references (attached):

SAMPLE FORM

3.   Revised Drawings and Specifications (attached):

4.   Schedule of Omissions/Additions:

5.   Rate Analysis for New Items:

FIGURE 3.29B   A sample VO attachment form.

to complete the activity. Scheduling is the mechanical process of formalizing the planned functions, and assigning the starting and completion dates to each part or activity of the work will be carried out in such a way that the whole work proceeds in a logical sequence and in an orderly and systematic manner. Scheduling is

TABLE 3.11

Advantages of Project Planning and Scheduling

Serial Number	Advantages
1	It facilitates management by objectives.
2	It facilitates to execute the work in an organized and structured manner.
3	It eliminates or minimizes uncertainties.
4	It reduces risk to the minimum.
5	It helps in proper coordination.
6	It helps in integration of project activities for a smooth flow of project work.
7	It helps to reduce rework.
8	It improves the efficiency of the process and increase productivity.
9	It improves communication.
10	It establishes timely reporting system.
11	It establishes the duration of each activity.
12	It provides the basis for monitoring and controlling of project work.
13	It helps to establish a benchmark for tracking the quantity, cost, and timing of work required to complete the project.
14	It helps foresee problems at early stage.
15	It helps to know the responsibility and authority of people involved in the project.

a time-based graphical presentation of project activities/tasks utilizing information about available resources and time constraints. Table 3.11 illustrates advantages of project planning and scheduling.

Schedule/time management includes the processes required to manage the timely completion of the project. Project time management consists of seven processes, which are as follows:

1. Plan Schedule Management
2. Define Activities
3. Sequence Activities
4. Estimate Activity Resources
5. Estimate Activity Duration
6. Develop Schedule

As per Construction-Extension-PMBOK® Guide-Third Edition, there are three additional processes applicable for construction projects. These are

1. Activity Weights Definitions
2. Progress Curves Development
3. Progress Monitoring

In construction project, Project Time Management methodology can be termed Schedule Management having the following construction-related activities:

1. Identify Project Activities/Tasks

    1.1 Study Stage
    1.2 Design Stage
    1.3 Bidding and Tendering
    1.4 Construction Stage

2. Sequence Project Activities

    2.1 Study Stage
    2.2 Design Stage
    2.3 Bidding and Tendering
    2.4 Construction Stage

3. Develop Project Network Diagram

    3.1 Study Stage
    3.2 Design Stage
    3.3 Bidding and Tendering
    3.4 Construction Stage

4. Estimate Activity Resources

    4.1 Study Stage
    4.2 Design Stage
    4.3 Construction Stage

5. Estimate Activity Duration

    5.1 Study Stage
    5.2 Design Stage
    5.3 Bidding and Tendering
    5.4 Construction Stage

6. Develop Schedule

    6.1 Project Master Schedule (Project Life Cycle)
    6.2 Design Development Schedule
    6.3 Construction (Project) Schedule (Tendering Documents)
    6.4 Contractor's Construction Schedule

        6.4.1 Detailed Schedule
        6.4.2 Monthly Schedule
        6.4.3 Weekly/Biweekly Schedule
        6.4.4 Look Ahead (15 Days)

7. Analyze Schedule
8. Project Schedule (Schedule Baseline)
9. Monitor and Control Schedule

    9.1 Schedule Monitoring
    9.2 Progress Status
    9.3 Progress Reporting

9.4 Forecasting
9.5 Schedule Update
9.6 Schedule Control
9.7 Performance Reporting

Figure 3.30 illustrates Schedule Development Process.

3.3.4.1 Identify Project Activities

Project activities are the lowest level of Work Breakdown Structure evolved from decomposition of Work Package. Each project activity has a specific duration in which the activity is performed (started and completed). The project activity consumes time (duration), consumes resources, has a definable start and finish, is associated with cost, and is easy to monitor and manage. It is assignable, measurable, and quantifiable. In construction projects, the terms "project activities" and "project tasks" are used interchangeably. The definition of project activity is an important factor for identification and documentation. Many of the activities are repeated in construction project, but their use and performance requirements are different. For example, pipes can be water system pipes, firefighting system pipes, pipes for chilled water in HVAC system, therefore, proper identification and documentation of specific activity are important. Activities are normally categorized into three types. These are

1. Management/administrative
2. Procurement
3. Production

3.3.4.2 List Project Activities (Bill of Quantities)

Activity list is a tabulation of activities to be performed in the project and is to be included in the project schedule. Activity list should contain following information related to activity:

- Activity name
- Activity identification number
- Brief description of the activity

List of activities is known as Bill of Quantities (BOQ). BOQ is an itemized listing of project-specific activities identified by the drawings and specifications prepared by the designer/quantity surveyor (consultant) to meet the project objectives. The quantities may be measured in number, length, area, volume, weight, time depending upon the scale of unit required to define the activity. BOQ is basically listed under different trades the activity/task belongs to. BOQ is frequently used to

- Develop total estimate of the project
- Develop bidding documents
- Monitoring the progress of the project
- Making progress payments

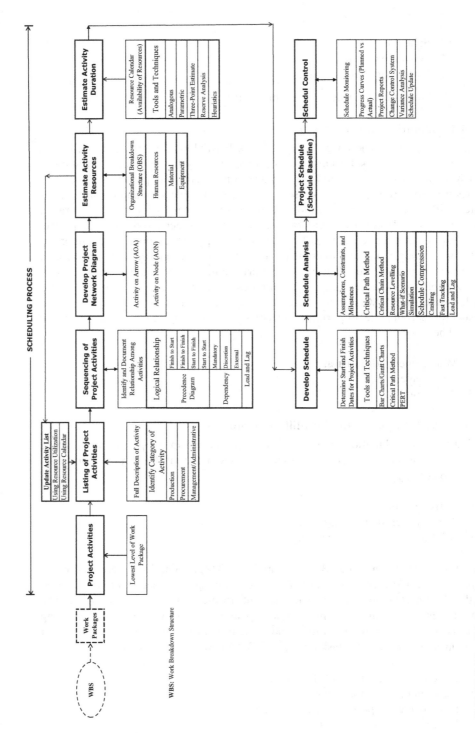

FIGURE 3.30 Schedule development process.

Source: Abdul Razzak Rumane. (2016). *Handbook of Construction Management.* Reprinted with permission of Taylor & Francis Group.

3.3.4.3   Sequence Project Activities

Once project activities/tasks have been defined and listed, it is required to identify the relationship and dependency among the project activities. The activities may have direct relationship or may be further constrained by indirect relationship. Following are direct logical relationships or dependencies among project-related activities:

- **Precedence**

    1. Finish-to-Start: Activity A must finish before activity B can begin.
    2. Start-to-Start: Activity A must start before activity B can start.
    3. Finish-to-Finish: Activity A must finish before activity B can finish.
    4. Start-to-Finish: Activity A must start before activity B can finish.

- **Dependency**

    a. **Mandatory**

        - Inherent in the nature of work being performed. It is also called hard logic. In building construction, superstructure can't begin unless the foundation work is complete.

    b. **Discretionary**

        - These are preferred or preferential logic. These are used at the discretion of project team. It is also called soft logic.

    c. **External**

        - These dependencies are outside of project's control. For example, approval from regulatory authority/agency.

- **Lead and Lag**

Lead may be added to start an activity before the predecessor activity is completed (a jump of the successor activity), while lag is inserted waiting time (time delay) between activities. Lead allows acceleration of successor activity, and lag delays in the start of successor activity. For example, concrete curing time is added before the start of tiling works.

- **Relationship**

    a. **Predecessor**

        - The relationship as to which activity must occur "before" the other

    b. **Successor**

        - The relationship as to which activity must occur "after" the other

    c. **Concurrent**

        - When one activity can occur "at the same time" as the other

Following diagram represents activity relationship.

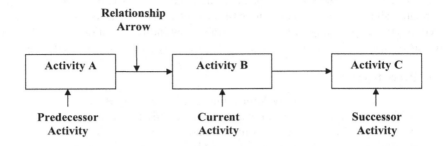

Figure 3.31 illustrates dependency relationship diagram.

7.3.4.4   Develop Project Network Diagram

Network diagram is a schematic display of project activities indicating logical relationship among the activities. It shows how the work progresses from the start till the completion of the project. There are two methods typically used to prepare the network diagram. These are:

1. Activity-on-Arrow (AOA)
2. Activity-on-Node (AON)

Activity-on-Arrow: Arrow diagrams or activity-on-arrows (A-O-A) is a diagramming method to represent the activities on arrows and connect them at nodes (circles) to show the dependencies. With A-O-A method, the detailed information about each activity is placed on arrow or as footnotes at the bottom.

Figure 3.32 illustrates activity-on-arrow diagramming method for concrete foundation work.

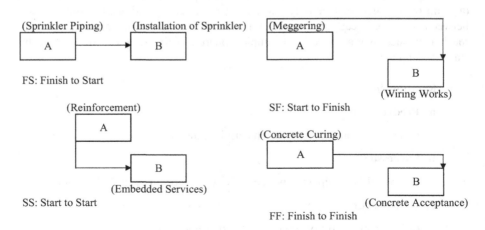

FIGURE 3.31   Dependency relationship diagram.

Activity-on-Node: "Activity on Node" network diagram has the activity informa-tion written in a small box that are the nodes of the diagram. Arrows connect the boxes to show the logical relationships between pairs of the activities.

Figure 3.33 illustrates activity-on-node diagramming method for concrete foun-dation work.

"Activity-on-Node" (A-O-N) is also referred to as the PDM (precedence diagram-ming method) technique because it shows the activities in a node (box) with arrows showing dependencies. Figure 3.34 illustrates precedence diagramming method (PDM). PDM is used to establish precedence relationship or dependencies (sequenc-ing of activities) and also in the Critical Path Methodology (CPM) for constructing the project schedule network diagram.

7.3.4.5 Estimate Activity Resources

It is the identification and description of types and quantities of resources required to perform the activity/task. Resource estimates are made by the project team which are based on the resource productivity, related experience, and availability of resources to carry a particular activity/task. Efficient utilization of resources is critical to suc-cessful project. Organizational Breakdown Structure is used to determine the related

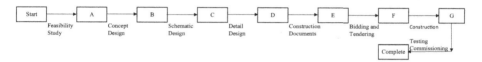

FIGURE 3.32 Arrow diagramming method for design phases.

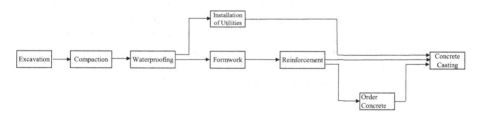

FIGURE 3.33 Activity-on-node diagram.

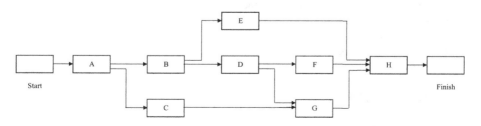

FIGURE 3.34 PDM diagramming method.

resource for a particular activity. There are different types of resources required in a certain quantity to perform and complete the activity. These are

- Human resources (manpower)
- Equipment
- Material

In construction projects, the availability of subcontractor (sub-consultant, subcontractor for designated works) is also to be considered while estimating the resources to perform the work.

While estimating the resources, following points should be considered:

- Availability of particular resource (resource histogram)
- Availability of exact number of resources
- Identification of the required skill (manpower) to perform the assigned activity/task
- Availability of specific type of equipment and the installation/operating crew
- Availability of space to perform certain type of activity simultaneously with other activity
- Availability of specified material
- Availability of fund to perform the activity

3.3.4.6 Estimate Activity Duration

Once the activities/tasks are sequenced and the types and quantities of resources are identified and determined, it is required to estimate the amount of time required to complete each activity in order to enable preparation of schedule.

Activity duration is the time between the start and finish of an activity/task. Estimation of activity duration is important to approximate the amount of time or period required to complete all project activities with the available resources. Following points are to be considered while estimating the activity duration:

- Work methodology
- Nature of work
- Resource (labor, equipment) type and availability
- Resource (labor, equipment) productivity
- Resource calendar
- Time contingency
- Quantum of work
- Quality of work
- Environmental factors
- Weather conditions
- Organizational factors
- Work restriction

There are several tools and techniques that are used to estimate activity duration. Following are the basic techniques most widely used by the project professionals to estimate activity estimation to prepare the schedule:

1. Expert judgment

   • Expert judgment normally comes from the project members having expertise in the subject matter. It is the mix of historical information from previous projects. It also uses published database about duration estimation.

2. Analogous estimating

   • This technique uses the information from previous projects of similar nature and size. It is also known as top-down estimation.

3. Parametric estimating

   • This technique uses mathematical, statistical, or quantitative relationship to estimate activity duration. This method has better accuracy than that of expert judgment and analogous estimation techniques.

4. Three-point estimating

   • This technique is used to estimate the duration for an activity having higher risk and not known very well. The following formula is used to calculate activity duration:

   Activity Duration = (P + 4M + O)/6
   Where P is Pessimistic (worst case-scenario),
   M is Most Likely (Realistic), and
   O is Optimistic (Best case-scenario).
   To calculate the installation of block masonry work per square meter by a crew of one mason and two labors, following assumptions can be made:
   Pessimistic: 35 square meters per crew per day
   Most likely: 20 square meter per crew per day
   Optimistic: 15 square meter per crew per day
   Therefore estimated duration per crew per day is;
   Te = (35 + 4 × 20 + 15)/6 = 21.67, say 22 square meters

If the activity is well known with little risk, then it is possible to estimate the duration of an activity by using the available information from organization's database. This technique is known as "one-time estimate".

3.3.3.7 Develop Schedule

Once activity resources and activity duration are estimated, it is possible to determine the start and finish dates for each activity. Thus, by sequencing all the activities, overall project schedule can be developed showing start and finish of the project. Project schedule is calendar-based graphical representation of all the activities that need to be performed in order to achieve project scope objectives. Schedule is one of the critical elements for the successful management of project. In order to develop schedule, the following information is required;

1. Project scope statement (assumptions, milestones, constraints)'
2. Activities' list
3. Network diagram
4. Resources needed
5. Duration for each activity
6. Resource calendar (availability of resources)
7. Working calendar

There are various methods used to develop project schedule. Following methods/ tools are used to develop project schedule:

- Gantt or Bar Charts

    - Bar charts are graphical presentation of project schedule information by listing the project activities on the vertical axis and showing the corresponding start and finish date on the horizontal axis as well as the expected duration in a calendar format. Bar chart method was developed by Henry Gant in 1917 by listing the activities to a time scale by drawing in the bar chart format.

- Critical Path Method (CPM)

    - Critical path is the longest path through a network; hence, it is the shortest project time. Critical path method was developed by DuPont and Remington Rand in 1958. It was developed for industrial projects (maintenance programs for chemical plant), where activity durations are generally known. CPM is a network diagramming technique used to predict total project time. CPM identifies the activities in the critical path that are likely to affect the completion of project duration as required. There could be more than one critical path if the lengths of two or more paths are same. The critical path can change as the project progresses. Critical Path Method is represented by following types of Network Diagrams:

        1. Activity-on-Arrow
        2. Activity-on-Node

- Project (Program) Evaluation and Review Technique (PERT)

    - PERT is a project management tool used to schedule, organize, and coordinate activities/tasks within the project. PERT was developed by U.S. Navy for Polaris missile project (program). It was developed for R&D project where activity times are generally uncertain. PERT is a flow chart diagram that depicts the sequence of activities needed to complete the project and the time or cost associated with each activity. It uses three-point estimation formula {Estimated time Te = (To + 4Tm + Tp)/6} to calculate expected activity duration. Where *To* is optimistic estimate, *Tm* is most likely estimate, and *Tp* is pessimistic estimate.

CPM and PERT tools are also known as Mathematical Techniques.

Bar Chart or Gantt Chart and Critical Path Method (Network Diagramming or Precedence Diagramming Network) scheduling techniques are most commonly used techniques to develop project schedule.

Following are the terminologies used to prepare Network Diagramming;

- Activity Block:

The Activity Block is a graphic display of activity information.

Figure 3.35 is an illustrative block showing the related information about an activity.

- **Early Start (ES):** The earliest date an activity can start.
- **Early Finish (EF):** The earliest date an activity can finish.
- **Late Start (LS):** The latest date an activity can start.
- **Late Finish (LF):** The latest date an activity can finish.
- **Duration:** Number of working days or hours needed to perform the specific activity.
- **Mile stone:** It is a specific point(s) having significant importance shown in the project schedule having no duration.
- **Forward Pass:** Forward pass is used for calculating Early Start and Early Finish of each activity.
- **Backward Pass:** Backward pass is used for calculation of Late Start and Late Finish of each activity.
- **Float:** Float (Slack) is the measure of time the work can be delayed without having effect on the project completion date.

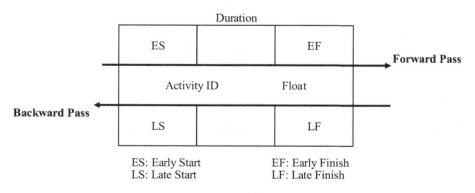

FIGURE 3.35   Activity block.

- **Advantages of CPM:**

1. It shows logical relationship between activities and their interdependence.
2. It is easy to understand the activities relationship.
3. It identifies the activities that are critical to timely completion of the project.
4. It shows the minimum completion time for the completion of the project.
5. It clearly demonstrates how a change in one activity impacts on other activities in the schedule.
6. It enables the easy calculation of shortest time to complete the project.
7. It provides accurate and detailed picture of forecasted project completion dates.
8. It is easy to apply on resources and determine resources or time trade-off.
9. It shows how much float other activities have.
10. It is easy to assign lead and lag between two activities to solve a dead time problem.

7.3.4.8   Analyze Schedule

Once the basic schedule is developed, it needs to be reviewed and analyzed to make sure the timing of each activity is aligned with resources and to ensure schedule accuracy with the relevant assumptions, constraints, and milestones. Scheduling is an iterative process. If the results obtained with the creation of schedule matches with project requirements, the iteration must stop. Otherwise the schedule has to be reviewed and analyzed by adjusting the dependencies and resources. Schedule analysis is important to determine schedule accuracy and create project schedule (schedule baseline). Following points should be considered to develop accurate and realistic baseline schedule:

1. All the activities are properly identified and listed to meet the scope of work.
2. Project logics dependencies and relationships are properly considered while sequencing of the activities to be performed.
3. All the resources and durations are correctly estimated.
4. Sufficient resources have been identified and allocated.
5. Assumptions and constraints are sound and true.
6. Schedule reserves/contingencies have been identified.
7. Realistic and feasible project duration is considered.
8. Regulatory requirements.
9. Contractual requirements.

Following tools and techniques are used to improve project schedule:

- **Schedule Compression**

In order to shorten the project schedule to meet the imposed dates, schedule constraints or other objectives without changing the scope of work. This is achieved by following methods:

- Crashing
  - To decrease total project duration by analyzing how to get maximum compression with least cost effect.
- Fast Tracking
  - Compression of project duration by changing the sequences of activities to do in parallel or allowing some activities to overlap.

- **Critical Chain Method**

It is the technique that modifies the project schedule to account for limited resources.

- **Resource Leveling**

This technique is used when shared or critical required resources are only available at certain times, are only available in limited quantities, or to keep resource usage at a constant level.

- **Simulation**

It involves calculating project durations with different sets of activity assumptions to assess the feasibility of project schedule under adverse conditions. Following are techniques used for simulation:

- Monte Carlo Analysis
- What-if Analysis

There are many computer-based programs available for preparing the network and critical path of activities for construction projects. These programs can be used to analyze the use of resources, review project progress, and forecast the effects of changes in the schedule of works or other resources. Most computer programs automate preparation and presentation of various planning tools such as Bar chart, PERT, and CPM analysis. The programs are capable of storing huge data, and help process and update the program quickly. It manipulates data for multiple usages from the planning and scheduling perspectives. It is useful for updating and tracking, sorting, filtering, and resource leveling.

3.3.4.9   Project Schedule (Baseline Schedule)

Baseline schedule is the project plan developed to meet the project objectives and the scope of work and accepted by the stakeholders as a benchmark for tracking and measuring the project performance and progress. An accurate baseline schedule is necessary to assess actual performance (status) of the project, determine the significance of variance, and forecast to complete the project. The baseline schedule is saved (Freeze) once the schedule is approved and accepted by all the stakeholders (project team members) at the start of the project. Any change to the baseline schedule must be approved by the change control board (authorized project team members).

The Baseline Schedule is

- Commitment by all stakeholders (project team members) to accomplish project objectives within the agreed time.
- Representation of project assumptions, constraints, and milestones.
- A benchmark to perform the activities.
- A datum for measuring project performance and progress and to forecast project completion date.
- Used as a framework for identification of impacts on the project.
- Saved (Freeze) as original plan upon acceptance by all the stakeholders (authorized project team members).
- Updated with formal approval by change control board (authorized project team members).

Construction project development has mainly four stages. These are

1. Study
2. Design
3. Bidding and Tendering
4. Construction

There are many activities/task that need to be performed during each of these stages. In order for a smooth flow and performance of activities to happen during these stages, it is necessary to prepare a schedule at every stage of construction project life cycle to perform the work in an organized and structured manner. Following are the stages/phases at which schedules are developed by various project participants;

- Pre-design phase (feasibility stage): By owner
- Design phases: By designer (consultant)

  - Concept design
  - Preliminary design
  - Detail design
  - Construction documents preparation

- Bidding and tendering and contract award: By owner and designer
- Construction phase: By contractor

  - Construction
  - Testing, commissioning, and handover

The schedule developed at these stages (phases) serve as a baseline schedule for the specific stage and help to accomplish all the key activities/tasks and to achieve overall project objectives.

3.3.4.9.1 Schedule Levels

There are different levels at which the project schedule is presented. There are five levels of schedules typically developed in construction projects. The objective is to establish appropriate levels of schedule detail for planning, scheduling, monitoring, controlling, and reporting on the overall project.

Level 1: Project Master Schedule (Executive Summary)
Level 2: Summary Master Schedule (Management Summary)
Level 3: Project Coordination Schedule (Publication Schedule)
Level 4: Project Working Level Schedule (Execution Schedule)
Level 5: Detail Schedule

The amount of information desired at each level depends on the requirements of stakeholders and the project stage (phase). Table 3.12 illustrates schedule level deliverables.

In order to improve the understanding and the communication among stakeholders involved with preparing, evaluating, and using project schedule, AACE International has published the guideline to classify schedule in five levels. Figure 3.36 illustrates classification versus levels so that schedule can be developed and/or presented.

Table 3.13 illustrates Generic Schedule Classification Matrix, and Table 3.14 illustrates Characteristics of Schedule Classifications.

TABLE 3.12
Schedule Levels

Schedule Level 1

Schedule Title	**Project Master Schedule (Executive Summary)**
Description	Level 1 schedule is a high-level schedule that reflects key milestones and high-level project activities by major phase, stage, or project being executed. This schedule level may represent summary activities of an execution stage, specifically engineering, procurement, construction, and start-up activities. Level 1 schedules provide high-level information that assists in the decision-making process (go/no go prioritization and criticality of projects).
	Level 1 schedule can be used to integrate multiple contractors/multiple schedules into an overall program management process.
	Level 1 audience include, but are not limited to client, senior executives and general managers.
	Level 1 schedule may be used in the proposal stage of a potential project/contract.
Scheduling Method	Graphical representation in the bar chart format.
Usage in Construction Projects	Study stage, Pre-design phase, feasibility study

(Continued)

TABLE 3.12
(Continued)

Schedule Level 2

Schedule Title	**Summary Master Schedule (Management Summary)**
Description	Level 2 schedules are generally prepared to communicate the integration of work throughout the life cycle of a project. Level 2 schedules may reflect, at a high level, interfaces between key deliverables and project participants (contractors) required to complete the identified deliverables. Level 2 schedules provide high-level information that assists in the project decision-making process (re-prioritization and criticality of project deliverables). Level 2 schedules assist in identifying project areas and deliverables that require actions and/or course correction. Audiences for this type of schedule include, but are not limited to general managers, sponsors, and program or project managers.
Scheduling Method	Typically presented in the Gantt (bar chart) format.
Usage in Construction Projects	Overall design schedule, overall construction schedule

Schedule Level 3

Schedule Title	**Project Coordination Schedule (Publication Schedule)**
Description	Level 3 schedules are generally prepared to communicate the execution of the deliverables for each of the contracting parties. Level 3 reflects the interfaces between key workgroups, disciplines, or crafts involved in the execution of the stage. Level 3 schedule includes all major milestones, major elements of design, engineering, procurement, construction, testing, commissioning, and handover. Level 3 schedules provide enough detail to identify critical activities. Level 3 schedules assist the team in identifying activities that could potentially affect the outcome of a stage or phase of work, allowing for mitigation and course correction in short course. Level 3 audiences include, but are not limited to, program or project managers, CMs or owner's representatives, superintendents, and general foremen.
Scheduling Method	Typically presented in Gantt or CPM network format and is generally the output of CPM scheduling software.
Usage in Construction Projects	Detail design, detail construction

Schedule Level 4

Schedule Title	**Project Working Level Schedule (Execution Schedule)**
Description	Level 4 schedules are prepared to communicate the production/execution of work packages at the deliverable level. Level 4 schedule reflects interfaces between key elements that drive completion of activities. Level 4 schedules usually provide enough detail to plan and coordinate contractor or multidiscipline/craft activities. Level 4 schedule displays the activities to be accomplished by identifying all the required resources. Level 4 audiences include but are not limited to project managers, superintendents, and general foremen.

Scheduling Method	Typically presented in Gantt or CPM network format
Usage in Construction Projects	Monthly Schedule, Look Ahead Schedules

Schedule Level 5

Schedule Title	**Detail Schedule**
Description	Level 5 schedules are prepared to communicate task requirements for completing activities identified in a detailed schedule.
	Level 5 schedules are usually considered working schedules that reflect hourly, daily, or weekly work requirements.
	Level 5 schedules are used to plan and schedule the utilization of resources (labor, equipment, and materials) in hourly, daily, or weekly units for each task.
	Level 5 audiences include but are not limited to superintendents, general foremen, and foremen.
Scheduling Method	Typically presented in an activity listing format without time-scaled graphical representation of work to accomplish.
Usage in Construction Projects	Daily Work Schedule, Daily Progress Report

Source: Adapted from AACE International Recommended Practice 27R-03 Schedule Classification System and AACE International Recommended Practice 37R-08, Schedule Levels of Detail-As Applied in Engineering, Procurement, and Construction. "Copyright © 2010 by AACE International; all rights reserved". Reprinted with permission of AACE International.

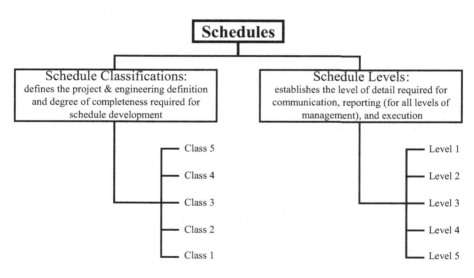

FIGURE 3.36   Schedule: classifications versus levels.

Source: Adapted from Figure 1 from the AACE International Recommended Practice 27R-03 Schedule Classification System, latest version of November 12, 2010. Reprinted with permission of AACE International 726 East Park Avenue, #180 Fairmont, WV 26554.

TABLE 3.13
Generic Schedule Classification Matrix

	Primary Characteristic	Secondary Characteristic	
Schedule Class	Degree of Project Definition (Expressed as % of Complete Definition)	End Usage	Scheduling Methods Used
Class 5	0% to 2%	Concept Screening	Top-down planning using high-level milestones and key project events
Class 4	1% to 15%	Feasibility study	Top-down planning using high-level milestones and key project events. Semi-detailed
Class 3	10% to 40%	Budget, authorization, or control	"Package" Top-down planning using key events. Semi-detailed
Class 2	30% to 75%	Control or bid/tender	Bottom-up planning. Detailed
Class 1	65% to 100%	Bid/tender	Bottom-up planning. Detailed

Source: Adapted from AACE International Recommended Practice 18R-97. Recommended Practice 18R-97 provides range of percentages for each class. "Copyright © 2020 by AACE International; all rights reserved". Reprinted with permission of AACE International.

TABLE 3.14
Characteristics of Schedule Classification

Class 5 Schedule Characteristics

Description	Class 5 schedules are generally prepared on the basis of very limited information and subsequently have a wide accuracy range.
	The Class 5 schedule is considered a preliminary document, usually presented in either Gantt (bar chart) or table form. The class 5 schedule should have, as a minimum, a single summary per stage with major project milestones identified
Degree of Project Definition Required	0% to 2%
End Usage	Class 5 schedules are prepared for any number of strategic business planning purposes, such as but not limited to: market studies, assessment of initial viability, evaluation of alternative schemes, project screening, project location studies, evaluation of resource need and budgeting, long range capital planning, etc.

Scheduling Methods Used	Gantt, bar chart, milestone/activity table. Top-down planning using high-level milestones and key project events.

Class 4 Schedule Characteristics

Description	Class 4 schedules are generally prepared based on limited information and subsequently have fairly wide accuracy ranges. They are typically used for project screening, determination of "do-ability", concept evaluation, and to support preliminary budget approval. The Class 4 schedule is usually presented in either Gantt (bar chart) or table form. The Class 4 schedule should define the high-level deliverables for each specific stage going forward (since the previous stage has passed). This document should also provide an understanding regarding the timing of key events, such as independent project reviews, committee approvals, as well as determining the timing of funding approvals. A high-level WBS may be established at this time.
Degree of Project Definition Required	1 % to 15 %
End Usage	Class 4 schedules are prepared for a number of purposes, such as but not limited to: detailed strategic planning, business development, project screening at more developed stages, alternate scheme analysis, confirmation of "do-ability", economic and/or technical feasibility, and to support preliminary budget approval or approval to proceed to next stage. It is recommended that the Class 4 schedule be reconciled to the Class 5 schedule to reflect the changes or variations identified as a result of more project definition and design. This will provide an understanding of the changes from one schedule to the next.
Scheduling Methods Used	Gantt, bar chart, milestone/activity table, top-down planning using high-level milestones and key project events, semi-detailed.

Class 3 Schedule Characteristics

Description	Class 3 schedules are generally prepared to form the basis of execution for budget authorization, appropriation, and/or funding. As such, they typically form the initial control schedule against which all actual dates and resources will be monitored.
	The Class 3 schedule should be a resource-loaded, logic-driven schedule developed using the precedence diagramming method (PDM). The schedule should be developed using relationships that support the overall true representation of the execution of the project (with respect to start to start and finish to finish relationships with lags). The amount of detail should define, as a minimum, the work package (WP) level (or similar deliverable) per process type/unit and any intermediate key steps necessary to determine the execution path. (The WP rolls up into the predefined WBS.) In some circumstances, where there is a high degree of parallel activities, the critical nature of the project, or the extreme complexity and/or size of the project, it may be warranted to provide further detail of the schedule to assist in the control of the project.
Degree of Project Definition Required	10% to 40%

(Continued)

TABLE 3.14
(Continued)

End Usage	Class 3 schedules are typically prepared to support full project funding requests and become the first of the project phase "control schedules" against which all start and completion dates and resources will be monitored for variations to the schedule. They are used as the project schedule until replaced by more detailed schedules.
	It is recommended that the Class 3 schedule be reconciled to the Class 4 schedule to reflect the changes or variations identified as a result of more project definition and design.
Scheduling Methods Used	PDM, PERT, Gantt/bar charts "Package" top-down planning using key events. Semi-detailed.

Class 2 Schedule Characteristics

Description	Class 2 schedules are generally prepared to form a detailed control baseline against which all project work is monitored in terms of task starts and completions and progress control.
	The Class 2 schedule is a detailed resource-loaded, logic-driven schedule that should be developed using the critical path method (CPM) process. The amount of detail should define as a minimum, the required deliverables per contract per work package (WP). The schedule should further define any additional steps necessary to determine the critical path of the project necessary for the appropriate degree of control.
Degree of Project Definition Required	30% to 70%
End Usage	Class 2 schedules are typically prepared as the detailed control baseline against which all actual start and completion dates and resources will now be monitored for variations to the schedule and form a part of the change/variation control program.
	It is recommended that the Class 2 schedule be reconciled to the Class 3 schedule to reflect the changes or variations identified as a result of more project definition and design.
Scheduling Methods Used	Gantt/bar charts, PDM, PERT Bottom-up planning. Detailed.

Class 1 Schedule Characteristics

Description	Class 1 schedules are generally prepared for discrete parts or sections of the total project rather than generating this amount of detail for the entire project. The updated schedule is often referred to as the current control schedule and becomes the new baseline for the cost/schedule control of the project. The Class 1 schedule may be a detailed, resource-loaded, logic-driven schedule and is considered a "production schedule" used for establishing daily or weekly work requirements.
Degree of Project Definition Required	70% to 100%

| End Usage | Class 1 schedules are typically prepared to form the current control schedule to be used as the final control baseline against which all actual start and completion dates and resources will now be monitored for variations to the schedule, and form a part of the change/variation control program. They may be used to evaluate bid-schedule checking to support vendor/contractor negotiations or claim evaluations and dispute resolution. It is recommended that the Class 1 schedule be reconciled to the Class 2 schedule to reflect the changes or variations identified as a result of more project definition and design. |
| Scheduling Methods Used | Gantt/bar charts, PDM, PERT Bottom-up planning. Detailed. |

Source: Adapted from AACE International Recommended Practice 27R-03. "Copyright © 2010 by AACE International; all rights reserved". Reprinted with permission of AACE International.

3.3.4.10 Monitor and Control Schedule

Monitor and Control Schedule is the process to determine the current status of the schedule, identify the influencing factors that cause the schedule changes, determine that the schedule has changed, and manage the changes in the approved project schedule baseline by updating and taking appropriate actions, if necessary, to minimize deviation from the approved schedule.

Monitoring is collecting, recording, and reporting information concerning project performance. Monitoring involves the measurement of current status of the project accomplishment and performance.

Current schedule (As-built schedule) is a reflection of current situation of all activities/tasks, milestones, sequencing, resources, duration, constraints, and project update.

Control process is established for managing the current schedule. Controlling is using the actual data collected through monitoring and comparing the same to the planned performance to bring actual performance to the level of planned performance by correcting the variances or implementing approved changes. Analysis of variance between the baseline and current schedule dates and duration provides necessary information for management and stakeholders' decision.

Monitoring in construction projects is normally done by compiling status of various activities in the form of progress reports. These are prepared by the contractor, supervision team (consultant), and construction/project management team. The objectives of project monitoring and control are

1. To report the necessary information in details and in appropriate form which can be interpreted by management and other concerned personnel to provide them with the information about how the resources are being used to achieve project objectives.

2. To provide an organized and efficient means of measuring, collecting, verifying, and quantifying data reflecting the progress and status of execution of project activities, with respect to schedule, cost, resources, procurement, and quality.
3. To provide an organized, efficient, and accurate means of converting the data from the execution process into information.
4. To identify and isolate the most important and critical information about the project activities to enable decision-making personnel to take corrective action for the benefit of the project.
5. Forecast and predict the future progress of activities to be performed.

Following information is required to prepare current (As-built) schedule and compare with the baseline schedule:

- Percentage completion of each activity based on approved checklist
- Actual start/finish dates for the completed activities
- Activities scheduled to start but not yet started
- Activities scheduled to complete but under progress
- Remaining duration to complete each activity
- Percentage of completed activities
- Percentage of partially completed activities
- Percentage of activities not yet started
- Material, equipment yet to be received
- Available resources
- Regulatory approvals
- Milestones not yet reached
- Logic and duration revision to keep the schedule unchanged
- Problems and issues
- Risks
- Change orders

After analyzing the current status with the actual, the schedule performance report is prepared consisting of the following information:

- Project status: where the schedule stands at current situation
- Project progress: plan versus actual
- Forecasting: prediction of future status and progress trend

Figure 3.37 illustrates Schedule Monitoring and Controlling Process.

3.3.5 COST MANAGEMENT

Cost management is the process involving planning, cost estimating, budgeting, and cost controlling to ensure that the project is successfully completed within the approved budget.

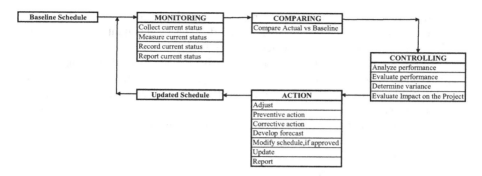

FIGURE 3.37 Schedule monitoring and controlling process.

Source: Abdul Razzak Rumane. (2016). *Handbook of Construction Management*. Reprinted with permission of Taylor & Francis Group.

Construction projects are mainly capital investment projects. They are customized and non-repetitive in nature.

Cost management in construction project is planning and managing the cost of facility throughout the project life cycle. The capital cost for a construction project include following expenses:

1. Land acquisition

   • This cost depends on the location and market price of the land. The owner may purchase the land on ownership or may lease the land to construct the facility.

2. Project construction

   • Cost expended by the owner for:
     • Project study stage
     • Project design development (concept, schematic, detail, construction documents, bidding and tendering)
     • Construction supervision (construction management, project management)
     • Site preparation
     • Construction of project (material, equipment, systems, furnishing, and human resources)
     • Inspection and testing (may be included as part of costs for different stages/phases)

Apart from the aforementioned expenses such as license fee, permits, regulatory taxes, insurance, and project-related owner's overhead cost are incurred by the owner.

In addition to the aforementioned costs, operation and maintenance costs also are required till the functional life of the facility.

Total costs related to construction are also known as Hard Costs. Hard costs are those arising mainly by the decisions of the designer (consultant) engaged in the preparation of design for the construction project.

The construction project initiation starts with the identification of need and a business case. The owner is interested to have best facility meeting the owner's schedule and within the budget to have the best facility to meet investment plan and objectives.

Cost management in construction project includes following processes:

1. Estimate cost
2. Prepare project budget
3. Control cost

3.3.5.1 Estimate Cost

It is important to estimate the total cost involved in construction of the project in order to evolve the project budget and manage cost variance on the project. Cost estimation is a prediction of most likely total cost of the identified need and business case. Construction project estimates are based on identifying, quantifying, and estimating the costs of all the resources (material, equipment, systems, furnishing, and people) required to complete all the activities within the required schedule and specifications. Cost estimates in construction projects vary as they progress through feasibility to the bidding/tendering stages. Construction project costs are generally estimated at following stages;

- Project feasibility study stage
- Client brief or terms of reference
- Concept design
- Schematic design
- Detail design
- Contract document
- Contract sum agreed during bidding and tendering stage is considered as cost baseline (contract cost) and is used to manage cost variance in the project.

3.3.5.1.1 Cost Estimation Tools

Following are the basic tools and techniques generally used for cost estimation in construction projects;

- Analogous or top-down estimates
  - Analogous estimating method relates to using the actual cost of a previous, similar nature of project as the basis for estimating the cost of current project. Analogous estimation is used when there is limited amount of detailed information about the project is available.
  - Top-down estimating approach recognizes all the activities, systems, equipment required to complete the project and are included in the total cost of the project.

- Bottom-up estimates

  - Bottom-up estimates involve the estimating cost of all the individual activities, material, equipment, resources, and aggregate all the costs to get total project cost.
  - Bottom-up estimates assure that all known activities, components, systems are accounted to get the total cost.

- Parametric estimate

  - It uses statistical relationship between the historical data and other variables to estimate the total cost of the project.
  - For example, in construction of buildings, following factors are used;

    - Buildings (offices, factories, mix use): Square meter of floor area
    - Schools: Number of students
    - Hospitals: Number of patents
    - Hotels: Number of bed/rooms
    - Theatre/cinema: Number of seats
    - Parking facility: Per parking bay

In construction projects, the cost estimates vary as the project design progresses. At the inception of project, the cost estimate is based on Rough Order of Magnitude. When detail design is available, the cost estimates become definitive.

The cost estimated by the designer (consultant) is based on assumptions and historical data available from experience on similar projects.

The contractor's estimation is more realistic and is based on the actual costs the contractor will incur to execute the project.

Table 3.15 illustrates cost estimation levels for construction projects.

TABLE 3.15
Cost Estimation Levels for Construction Projects

Project Stage/ Phase	Tools/ Methodology	Accuracy	Purpose
Inception	Analogous	−50% to +100%	Project initiation (Rough Order of Magnitude)
Feasibility	Analogous	−25% to +75%	Justification to proceed (screening estimate)
Concept Design	Parametric	−10% to +25%	Budgetary (conceptual estimate)
Preliminary Design	Elemental parametric	−10% to +25%	Budgetary (preliminary estimate)
Detail Design	Elemental parametric/ detailed costing	−5% to +10%	Detailed estimate
Bidding and Tendering	Detailed costing	−5% to +5%	Bid estimate/definitive estimate
Construction	Detailed costing	Project cost baseline (contracted value)	Contract cost (control estimate)

Contract awarded (negotiated) total project cost is considered as most accurate cost as it is derived from the detailed information provided in the bidding/tendering documents. In case of Design-Bid-Build type of project delivery system, detail Bill of Quantity (BOQ) is worked out by the Designer (Consultant) and included in the bidding/tendering documents. The BOQ is used to prepare estimate.

In order to estimate accurate cost of the project, the contractor has to consider following points;

- Location of the project
- Site conditions
- Project specifications
- Codes and standard to be followed
- Project schedule
- Risks and constraints in the project
- Inflation
- Project delivery system
- Taxes
- Currency exchange rate, if applicable
- Contingency
- Cost toward subcontracted items

In case of Design-Build type of project delivery system, the project cost is developed based on the performance specifications and conceptual BOQ.

Figure 3.38 is an example Analysis of Estimation of Prices Form to work out activity unit price and total project cost estimate.

3.3.5.2 Establish Budget

Establish Budget is the process of aggregating in a time-phased manner the estimated costs to perform all the known activities to establish an authorized baseline.

In construction project, Budget is established at various stages as the project life cycle progresses. It helps project owner to control, adjust the project cost, and achieve the intended project objectives within the investment limit. The degree of accuracy of establishing project budget varies at different stages of the project.

At early stage of project, that is, the feasibility stage, the budget is developed on specific project information using order of magnitude estimation tool. The budget is further refined in the succeeding phases as more information and details are available. The budget prepared by the designer (consultant) at the bidding/tendering stage is of definitive nature as at this stage the project details are clearly defined and detailed drawings, BOQ, technical specifications and project schedule are available for cost estimation. The project budget is the approved funding amount required to complete the project within the approved schedule. The project budget is based on the total anticipated cost to complete the project and also includes contingency and management reserves. Figure 3.39 illustrates process of establishing construction budget (Hard Cost).

Project Name
Analysis of Estimation of Prices

Serial Number	BOQ Reference	Brief Description	Unit	Material Cost	Labor Cost	Equipment Cost	Sub Total -1 Activity Cost Estimate	Overhead & Profit (% of e)	Sub Total-2 (e+f)	Main Contractor's Overhead (% of (g))	Sub Total-3 (g+h)	Main Contractor's Profit (Mark up) (% of (j))	Total (j+k) Unit Price
			(a)	(b)	(c)	(d)	(e)=(b+c+d)	(f)	(g)	(h)	(j)	(k)	(l)

SAMPLE FORM

FIGURE 3.38 A sample analysis of estimation of prices form.

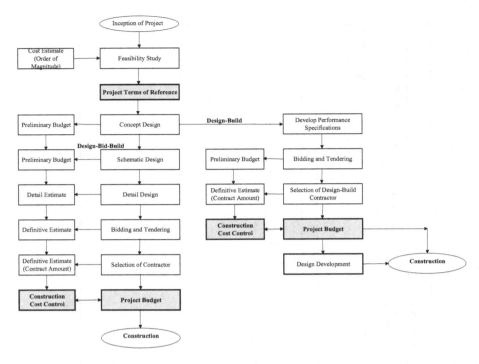

FIGURE 3.39   Process of establishing construction budget.

3.3.5.2.1   Cost Baseline

Construction project's Cost Baseline is developed based on construction budget. Cost baseline are time-phased budgets that are used as a basis against which to measure, monitor, and control overall cost performance of the project. The Cost Baseline is usually shown in an S-Curve graph. The S-Curve predicts how much amount will be spent over the established project schedule (time). The S-Curve is used to measure project performance and predict the expenses over project duration. The Cost Baseline helps owner/client to know the project funding requirements. Funding requirements, total and periodic, are derived from the cost baseline. Figure 3.40 illustrates example S-Curve (Budgeted).

Any changes to the baseline need approval of Change Control Board. Figure 3.41 illustrates Baseline Change Request Form.

3.3.5.3   Control Cost

Cost Control is the process of monitoring the status of the project to update the project cost and managing changes to baseline. This process provides the means to recognize variance from the approved plan, evaluate possible alternative, and take corrective action to minimize the risk. In order to have successful cost control, it is essential to have the necessary information and data to take appropriate action. If the necessary information and regular updates are not available or if the action is inefficiently executed, then the risk to cost control on a project is raised considerably.

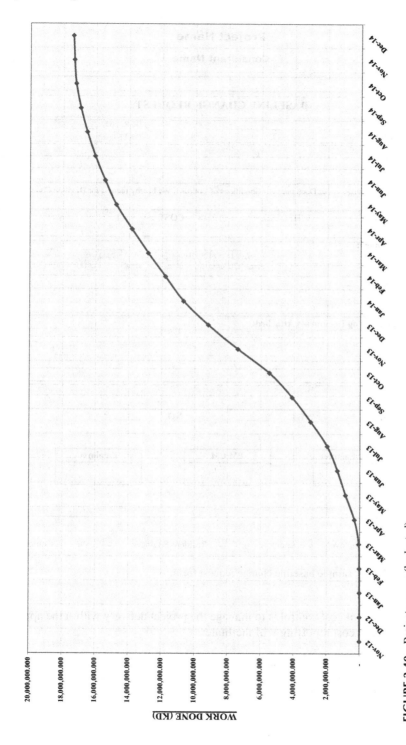

FIGURE 3.40 Projects-curve (budgeted).

Project Name
Consultant Name

BASELINE CHANGE REQUEST

CONTRACT NO.   :_____   BCR. NO. :_____
CONTRACTOR   _____   DATE   :_____

BCR ORIGINATOR _____

In accordance with clause ----- of Document ---- Conditions of Contract, We hereby request for Baseline Change as follows;

SCOPE   ☐   SCHEDULE   ☐   COST   ☐

Reason for Change:

SAMPLE FORM

Change Impact:

Following Supporting Documents Attached:

1) Cost Estimate

2) Schedule

3) Other

APPRIVALS: YES   _____   NO   _____

Signature:

Resident Engineer	PM/CM	Employer

Date:

Distributio ☐   Employer ☐   Engineer   ☐   Resident Engineer ☐   Contractor

FIGURE 3.41   A sample baseline change request form.

The purpose of cost control is to manage the project delivery within the approved budget. Regular cost reporting will facilitate:

- Establish project cost to date
- Anticipate final budget of the project
- Cash flow requirements
- Understanding potential risk to the project

Cost Control process focuses on:

- Identifying the factors that influence the changes to the cost baseline
- Determining the cost baseline has changed
- Ensuring that the changes are beneficial for the project
- Establishing the cost control structure and policy
- Managing the actual changes when and as they occur
- Monitoring cost performance to detect cost variance from the actual budget
- Recording all appropriate changes
- Informing/reporting concerned stakeholders about the approved changes
- Preventing unauthorized changes to the cost baseline
- Working to bring cost overruns within acceptable limits

Construction project development has mainly four stages. These are:

1. Study
2. Design
3. Bidding and tendering
4. Construction

It is essential to perform cost controlling process at these stages in order to ensure that total project cost does not increase the investment plan set by the owner and to achieve project value proposition.

In construction projects, the designer plays an important role to influence the final cost of project to ensure the owner that there will not be budget overrun. The designer has to not only control the project cost but also ensure that design:

- Has minimum errors
- Minimum omissions
- Properly written specifications
- Realistic construction schedule
- Includes all the requirements to achieve owner's need and objectives

In the case of the Design-Bid-Build type of project delivery system, cost control during the study stage, design stage, and bidding/tendering stage is carried out by the owner/designer. The contractor is responsible for controlling the costs during construction phase.

In the case of the Design-Build type of project delivery system, the contractor carries out most of the cost control. During design phase, if the Designer (Consultant) finds that the approved budget is inadequate, then they should seek revision/adjustment to the budget from any contingency amount that the owner might have.

During the construction phase, it is the contractor who is interested to complete the construction within the contracted amount and the Construction Manager/Project Manager controls the cost to ensure that there is no overrun and changes to the cost baseline.

The contractor has to mainly control the following costs:

- Material
- Labor
- Resource (labor) productivity
- Equipment/plant productivity
- Rework
- Delays
- Subcontractor's work
- Currency exchange

The Construction Manager/Project Manager, on behalf of the owner, should

- Compare and monitor project progress
- Identify and control approved variations
- Identify cash flow forecasts
- Compare current budget forecast for remaining works

3.3.5.3.1 Earned Value Management

Earned Value Management is a methodology used to measure and evaluate project performance against cost, schedule, and scope baseline. It compares the amount of planned work with what is actually accomplished to determine whether the project is progressing as planned or not. Earned Value Analysis is used to:

- Measure progress of the project budget, schedule, scope to know how much percentage of:
  - Budget is spent
  - Time has elapsed
  - Work is done
- Forecast its completion date and cost
- Provide budget and schedule variances

Following are the basic terminologies used in Earned Value Management.

1. BCWS → Budgeted Cost of Work Scheduled or Planned Value (PV)

   It is planned cost of total amount of work scheduled to be performed by the milestone date.

2. BCWP → Budgeted Cost of Work Performed or Earned Value (EV)

   It is the actual cost incurred to accomplish the work that has been done to date.

3. ACWP → Actual Cost of Work Performed or Actual Cost (AC)

   It is the planned cost to complete the work that has been done.

These three key values are used in various combinations to determine cost and schedule performance and provide an estimated cost of the project at its completion.

Table 3.16 illustrates the terms used in Earned Value Management (EVM) and their interpretation whereas Table 3.17 illustrates the formulas used in EVM and their Interpretations.

TABLE 3.16
Earned Value Management Terms

Serial Number	Term	Description	Interpretation
1	PV (BCWS)	Planned Value (Budgeted Cost of Work Scheduled)	Planned cost of total amount of work scheduled to be performed by the milestone date.
2	EV(BCWP)	Earned Value (Budgeted Cost of Work Performed)	Planned/budgeted cost to complete the work that has been done.
3	AC(ACWP)	Actual Cost (Actual Cost of Work Performed)	Actual cost incurred to accomplish the work that has been done to date.
4	BAC	Budget At Completion	Estimated total cost of project when completed.
5	EAC	Estimate At Completion	Expected total cost of project when completed.
6	ETC	Estimate To Completion	Expected additional cost needed to complete the project.
7	VAC	Variance At Completion	Amount over budget or under budget we expect at the end of the project.

TABLE 3.17
Parameters of Earned Value Management

Serial Number	Name	Formula	Interpretation
1	Cost Variance (CV)	BCWP-ACWP (EV-AC)	A comparison of the budgeted cost of work performed with actual cost. POSITIVE result means under budget. NEGATIVE result means over budget.
2	Schedule Variance	BCWP-BCWS (EV-PV)	A comparison of amount of work performed during a given period of time with that of what was scheduled to be performed. POSITIVE result means ahead of schedule. NEGATIVE result means behind schedule
3	Cost Performance Index (CPI)	EV/AC BCWP/ACWP	Greater than 1 means work is being produced for less than planned. Less than 1 means the work is costing more than planned.

(Continued)

TABLE 3.17
(Continued)

Serial Number	Name	Formula	Interpretation
4	Schedule Performance Index (SPI)	EV/PV BCWP/BCWS	Greater than 1 means project is ahead of schedule Less than 1 means project has accomplished less than planned and behind schedule
5	Estimate At Completion (EAC)	AC+ETC	Expected total cost of project when completed.
6	Estimate To Complete (ETC)	EAC-AC	Expected additional cost needed to complete the project.
7	Variance At Completion (VAC)	BAC-EAC	How much over budget or under budget we expect at the end of the project?

3.3.6   QUALITY MANAGEMENT

Quality management is an organization-wide approach to understand customer needs and deliver the solutions to fulfill and satisfy the customer. Quality management is managing and implementation of quality system to achieve customer satisfaction at the lowest overall cost to the organization while continuing to improve the process. Quality system is a framework for quality management. It embraces the organization structure, policies, procedures, and processes needed to implement quality management system.

Quality management in construction projects is different from that of manufacturing. Quality in construction projects is not only the quality of products and equipment used in the construction, it is also the total management approach to complete the facility as per the scope of works to customer/owner satisfaction within the budget and to be completed within specified schedule to meet owner's defined purpose. Quality management in construction addresses both the management of project and the product of the project and all the components of the product. It also involves incorporation of changes or improvements, if needed. Construction Project quality is fulfillment of owner's needs as per defined scope of works within a budget and specified schedule to satisfy owner's/user's requirements.

Quality management system in construction projects mainly consists of:

• Quality management planning
• Quality assurance
• Quality control

Each of these processes and activities is to be performed during the following main stages of construction project:

1. Study
2. Design
3. Bidding and tendering
4. Construction

3.3.6.1   Develop Quality Management Plan

The quality management plan for construction projects is part of the overall project documentation addressing and describing the procedures to manage construction quality and project deliverables. The quality management plan identifies following key components:

- Details of the quality standards and codes to be complied
- Project objectives, project scope of work
- Stakeholders' quality requirements
- Regulatory requirements
- Quality matrix for different stages
- Design criteria
- Design procedures
- Detailed construction drawings
- Detailed work procedure
- Well-defined specification for all the materials, products, components, and equipment to be used to construct the facility
- Manpower and other resources to be used for the project
- Inspection and testing procedures
- Quality assurance activities
- Quality control activities
- Defect prevention, corrective action, and rework procedure
- Project completion schedule
- Cost of the project
- Documentation and reporting procedure

3.3.6.1.1   Designer's Quality Management Plan

Construction projects have the involvement of owner, designer (consultant), and contractor. In order to achieve project objectives, both designer as well as contractor has to develop project quality management plan. The designer's quality management plan shall be based on owner's project objectives whereas the contractor's plan shall take into consideration requirements of contract documents.

Figure 3.42 illustrates Project Quality Management Plan for Design Stage.

3.3.6.1.2   Contractor's Quality Control Plan

During construction stage, the contractor prepares Contractor's Quality Control Plan (CQCP) on the basis of project-specific requirements. The contractor's quality control plan (CQCP) is the contractor's everyday tool to ensure meeting the performance standards specified in the contract documents. Its contents are drawn from the company's quality system, the contract and related documents. It is a framework for the contractor's process for achieving quality construction. It is a document setting out the specific quality activities and resources pertaining to a particular contract or project. It is the documentation of contractor's process for delivering the level of construction quality required by the contract. A quality plan is virtually a quality manual tailor-made for the project and is based on contract requirements. Figure 3.43 illustrates process for development of contractor's Quality Control Plan (CQCP), and Table 3.18 illustrates table of contents of contractor's Quality Control Plan.

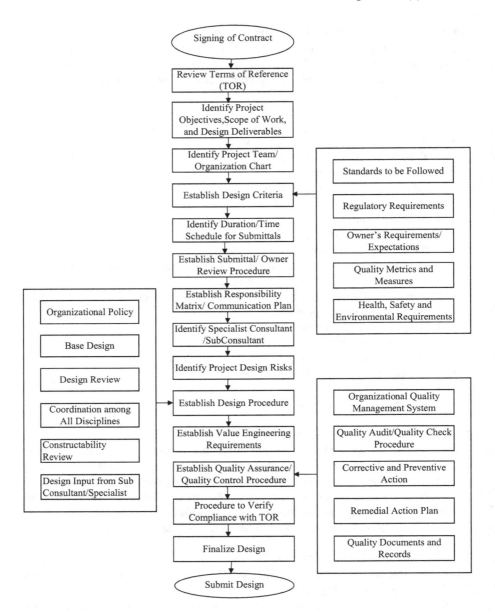

FIGURE 3.42 Project quality management plan for design stage.

Source: Abdul Razzak Rumane. (2013). Quality Tools for Managing Construction Projects. Reprinted with permission of Taylor & Francis Group.

3.3.6.1.3 Quality Matrix

In case of construction projects, an organizational framework is established and implemented mainly by three parties: owner, designer/consultant, and contractor.

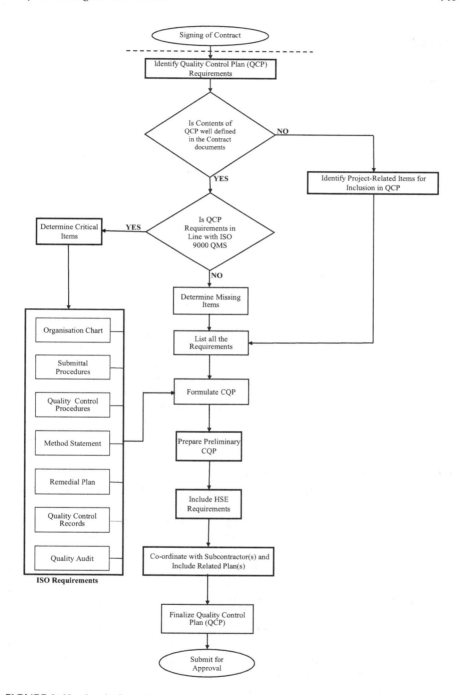

FIGURE 3.43 Logic flow diagram for development of contractor's quality control plan.

Source: Abdul Razzak Rumane. (2016). *Handbook of Construction Management*. Reprinted with permission of Taylor & Francis Group.

TABLE 3.18
Contents of Contractor's Quality Control Plan

Section		Topic
1		Introduction
2		Description of project
3		Quality Control Organization
4		Qualification of QC staff
5		Responsibilities of QC personnel
6		Procedure for submittals
7		Quality Control Procedure
	7.1	Procurement
	7.2	Inspection of site activities (checklists)
	7.3	Inspection and testing procedure for systems
	7.4	Off-site manufacturing, inspection, and testing
	7.5	Procedure for laboratory testing of material
	7.6	Inspection of material received at site
	7.7	Protection of works
8		Method statement for various installation activities
9		Project-specific procedures
10		Quality control records
11		Company's quality manual and procedure
12		Periodical testing
13		Quality updating program
14		Quality auditing program
15		Testing, commissioning, and handover
16		Health, safety, and environment

3.3.6.2 Perform Quality Assurance

Quality assurance in construction projects covers all activities performed by design team, contractor, and quality controller/auditor (supervision staff) to meet owner's objectives as specified and to ensure and guarantee that the project/facility is fully functional to the satisfaction of owner/end user. Auditing is part of the quality assurance function.

Quality assurance is the activity for providing evidence to establish confidence among all concerned that quality-related activities are being performed effectively. All these planned or systematic actions are necessary to provide adequate confidence that a product or service will satisfy given requirements for quality.

Quality assurance covers all activities from design, development, production/construction, installation, servicing to documentation and also includes regulations of the quality of raw materials; assemblies, products and components, services related to production; and management, production and inspection processes. Following are major activities to be performed for quality assurance of the construction project;

- Confirm that owners needs and requirements are included in the scope of works Terms of Reference (TOR)

TABLE 3.19

The Responsibilities Matrix for Quality Control Related Personnel

Linear Responsibility Cart

Serial Number	Description	Owner Owner/ Project Manager Construction Manager	Consultant Consultant/ Designer	Contractor Manager	Quality In-Charge	Contractor Quality Engineers	Site Engineers	Safety Officer	Head Office
1	Specify quality standards	□	■						
2	Prepare quality control plan			□	■	□			□
3	Control distribution of plans and specifications			□	■	□			□
4	Submittals		■	■	□		■		
5	Prepare procurement documents		□	□			■		■
6	Prepare construction method procedures		□	□	□	□	■		
7	Inspect work in progress		■	■		□	■		
8	Accept work in progress		■						
9	Stop work in progress	■	□						
10	Inspect materials upon receipt	□	■	□	■	□	■		
11	Monitor and evaluate quality of works		■	□	■	■	■		
12	Maintain quality records			□	■				
13	Determine disposition of nonconforming items	□	□	■					
14	Investigate failures	□	□	■	■	□	■		
15	Site safety			□				■	
16	Testing and commissioning	□	■	□			■		
17	Acceptance of completed works	■	□						

■ Primary responsibility
□ Advise/Assist

Source: Abdul Razzak Rumane (2010), *Quality Management in Construction Projects.* Reprinted with permission from Taylor & Francis Group.

- Review and confirm design compliance to TOR
- Executed works comply with the specified standards and codes
- Conformance to regulatory requirements
- Works executed as per approved shop drawings
- Installation of approved material, equipment on the project
- Method of installation as per approved method statement or manufacturer's recommendation
- Coordination among all the trades
- Continuous inspection during construction/installation process
- Identify and correct the deficiencies
- Timely submission and review of transmittals

3.3.6.3 Control Quality

Quality control in construction projects is performed at every stage through use of various control charts, diagrams, checklists, etc., and can be defined as:

- Checking and review of project design
- Checking and review of bidding/tendering documents
- Analysis of contractor's bids
- Checking of executed/installed works to confirm that works have been performed/executed as specified, using specified/approved materials, installation methods and specified references, codes, standards to meet intended use
- Controlling budget
- Planning, monitoring, and controlling project schedule

The construction project quality control process is a part of contract documents which provide details about specific quality practices, resources, and activities relevant to the project. The purpose of quality control during construction is to ensure that the work is accomplished in accordance with the requirements specified in the contract. Inspection of construction works is carried out throughout the construction period either by the construction supervision team (consultant) or appointed inspector agency. Quality is an important aspect of construction project. The quality of construction project must meet the requirements specified in the contact documents. Normally contractor provides onsite inspection and testing facilities at construction site. On a construction site, inspection and testing are carried out at three stages during the construction period to ensure quality compliance.

1. During construction process. This is carried with the checklist request submitted by the contractor for testing of ongoing works before proceeding to next step.
2. Receipt of subcontractor or purchased material or services. This is performed by a material Inspection Request submitted by the contractor to the consultant upon the receipt of material.
3. Before final delivery or commissioning and handover.

3.3.7   Resource Management

Resource management in construction is mainly related to management of following processes:

1. Human resources (project teams)

   - Project owner team (project manager and construction manager)
   - Designer (consultant)

2. Construction resources (contractor)

   - Manpower
   - Equipment
   - Material

Construction projects are of non-repetitive nature and have definite beginning and definite end. The important factor in construction project is to complete the facility as per defined scope within specified schedule and budget. In order to meet the project quality, the construction manager firm, designer firm, and contractor have to ensure that right human resources and right type of materials and equipment are available at the right time. Since construction projects are of temporary nature, every time a new project starts, the human resource configuration changes, depending on the company policies and practices to engage human resources for a particular project.

Similarly, it has to be ensured that specified materials and equipment are available to meet the installation/execution schedule.

3.3.7.1   Manage Human Resource

Construction project human resource management process includes following:

1. Plan human resources (project teams)
2. Acquire project team
3. Develop project team
4. Manage project team
5. Release/demobilize project team

3.3.7.1.1   Plan Human Resources (Project Teams)

Every project needs people to complete the project. Traditional construction projects involve three main groups. These are:

- Owner
- Designer (consultant)
- Contractor (s)

Depending on the type of project delivery system, each of these groups needs human resources to manage the project. Human resource planning is executed in the initial stage for overall requirement of the project and is performed iteratively and interactively during each of the project stage with other aspects of project such as scope, schedule, and cost of the project and adjusted as per project requirements.

For example, the construction/project management firms plan their human resources as per the scope of work and project schedule. The designer (consultant) plans their resources depending upon the size of the project and design deliverables. Contractor is responsible for all type of human resources to complete the project.

Human resource planning process includes organizational planning taking into consideration the requirements of the project and the stakeholders involved, detailing the project roles, responsibilities, required skills and reporting relationships to the appropriate people or group of people. Human resource planning includes:

- Documenting staffing requirements (roles, responsibilities, skills)
- Project organization chart
- Staff deployment charts
- Team acquiring process
- Team development, training needs
- Release/demobilization plan

3.3.7.1.2   Acquire Project Team

Acquire project team is the process of obtaining project team members for completing the project. There are mainly three project teams associated with three groups involved in a construction project. These are:

1. Construction Management or Project Management Firm (depending on the type of project delivery system selected by the owner)
2. Designer (consultant)
3. Contractor's core staff

Acquiring project team is a process of getting right people having right knowledge, skills, and experience that is required to work on the project and to perform the assignment within a given time frame. A project team is a group of professionals committed to achieving common goal to complete the project successfully. Acquire project team is a continuous process which is performed throughout the project as and when a particular category of team member is required as per the project deployment chart. For example, a resident engineer for a project is acquired at the beginning of the project and is released after completion of the project. Other team members are acquired as per the agreed-upon deployment chart depending on the project needs and project organizational structure. Acquiring project team includes:

- Negotiation
- Internal or externally hired member
- Deployment chart

While engaging a team member, it is necessary to consider following criteria:

- Required level of experience
- Organizational Breakdown Structure
- Project-specific requirements/needs
- Professional qualification
- Member's experience in similar type of project
- Project knowledge
- Technical skills and capabilities in relevant areas
- Management skills
- Collaborative skills
- Communication capabilities
- Availability of the candidate to meet work schedule
- Professional membership and training certification
- Cost
- Staff approval requirements by the project owner/client
- Regulatory constraints

Figure 3.44 illustrates project team acquiring process.

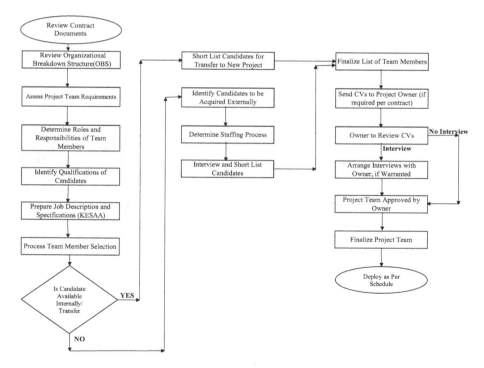

FIGURE 3.44 Project team acquisition process.

Source: Abdul Razzak Rumane. (2016). *Handbook of Construction Management*. Reprinted with permission of Taylor & Francis Group.

3.3.7.1.3 Develop Project Team

Project team is a combination of professionals acquired to do project-specific activities to complete the project successfully. Each of the project team members is committed to achieve common goal to complete the project as per defined scope within specified time and budget. Most construction projects have projectized or matrix type of organizational structure. In either case, the team members comprise different background and disciplines.

There are five stages of team development. These are:

1. Forming
2. Storming
3. Norming
4. Performing
5. Adjourning

In construction projects, there are mainly three project teams. Once the team members are assigned and a team is formed, a meeting is to be arranged where every member is introduced and given clear understanding about:

- Project goals
- Project overall mission
- Roles and responsibilities of each team member
- Coordination process
- Working in collaborative climate
- Communication method
- Member's relationship with each other
- Commitment of each member toward the project

Normally, project team members are selected having the required competency in the specific field for which they are assigned and are responsible to perform their efficiently. However, it is necessary to improve the knowledge and skills of team members in order to increase their ability and competency to complete project work to meet or exceed the specified quality of the project. In order to enhance the competency and ability of the project member, appropriate training can be arranged by ascertaining the gap for which training is to be arranged for the team member. Performance evaluation process helps determine the training requirements for the project team member and providing training identified by gap analysis improves technical competencies of project team member.

3.3.7.1.4 Manage Project Team

It is a process to keep track team member performance, provide feedback, resolve issues, and manage changes to ensure project performance optimization.

Construction projects are of temporary nature, and project team members are collected from different background and disciplines; therefore it is inevitable that issues/

conflicts may arise among the members. There are mainly three project teams in the construction. Each project manager has to resolve the issue as they arise and has to manage the team effectively by maintaining cohesion among all the team members. This can be achieved by:

- Establishing ground rules
- Coordinating with all team members to understand their issues
- Creating shared vision among team members
- Tracking team performance
- Training, recognition, and rewards
- Conducting meetings, exchanging relevant information, and resolving issues
- Problem identification and providing quick solution
- Conflict management

3.3.1.1.4.1 Manage Conflict There exist several types of conflict. Each conflict can assume a different intensity at different stages of the project. The causes of disagreement vary in different phases of the project as different members are involved at different phases of the project. Following are typical source of conflict that may arise during the project:

- Priorities
- Project schedule
- Resources
- Cost
- Scope change
- Technical opinions
- Personality conflict
- Communication problem
- Lack of coordination by team members
- Administrative procedures

The project manager has to resolve conflict by searching an alternate solution. Following methods are normally used to resolve conflicts:

1. Withdrawing/voiding

   It means both parties retreat from the conflict issue.

2. Smoothing/accommodating

   It is emphasizing upon friendly relationship and agreement rather than differences of opinions.

3. Collaborating

   In this method, parties try to incorporate multiple viewpoints in order to lead to consensus.

4. Confronting/Problem Solving

It is a fact-based approach where both parties solve their problems by focusing on the issues, looking at alternatives, and selecting the best alternative.

5. Compromising

Compromising is finding solution that brings some degree of satisfaction to both parties.

6. Forcing

Forcing is the use of authority and power in resolving the conflict by exerting one's viewpoint over another at the expense of another party.

Figure 3.45 illustrates conflict management flow chart.

3.3.7.1.5   Release/Demobilize Project Team

Construction projects are of temporary endeavor. Every time a project is initiated, project team members are acquired to complete the project. The organization structure depends on the strategic policy of the organization. Once the project is completed, the project team members are released and sent back to their original functional department or engaged in other projects or terminated from the work.

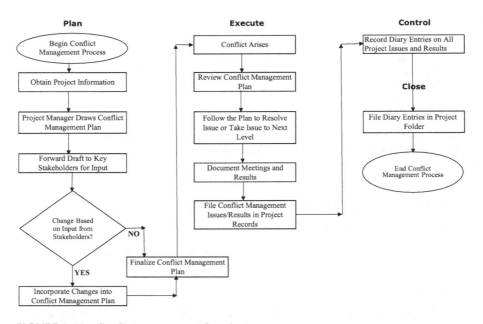

FIGURE 3.45   Conflict management flowchart.

3.3.7.2   Construction Resources Management

In most construction projects, the contractor is responsible to manage its resources. It includes

1. Manpower
2. Material
3. Equipment
4. Subcontractor(s)

3.3.7.2.1   Manage Manpower

It is contractor's responsibility to arrange necessary manpower to execute the project. This includes;

- Contractor's site staff to supervise the construction (core staff)
- Field workers to execute the works

Contract documents normally specify a list of minimum number of core staff to be available at site during construction period. Absence of any of these staff may result in penalty to be imposed on contractor by the owner.

Contractor has to consider the qualification and related experience mentioned in the contract documents while acquiring the team members to work on the project.

3.3.7.2.2   Manage Equipment

Contractor has to provide the required equipment to execute the construction works. Normally, contract documents specify that a minimum number of equipment are to be available at site during construction process to ensure smooth operation of all the construction activities. The typical equipment list is as follows;

- Tower Crane
- Mobile Crane
- Normal Mixture
- Concrete Mixing Plant
- Dump Trucks
- Compressor
- Vibrators
- Water Pumps
- Compactors
- Concrete Pumps
- Trucks
- Concrete Trucks
- Diesel Generator Set(s)

Contractor has to maintain the equipment in good conditions to ensure efficient utilization of the equipment.

3.3.7.2.3   Manage Material

In most construction projects, the contractor is responsible for procurement of material, equipment, and systems to be installed on the project. The contractors have their own procurement strategies. While submitting the bid, contractor obtains the quotations from various suppliers/subcontractors. The contractor has to consider following, as a minimum while finalizing the material procurement:

- Contractual commitment
- Specification compliance
- Statutory obligations
- Time
- Cost
- Performance

Figure 3.46 illustrates material management process for construction projects.

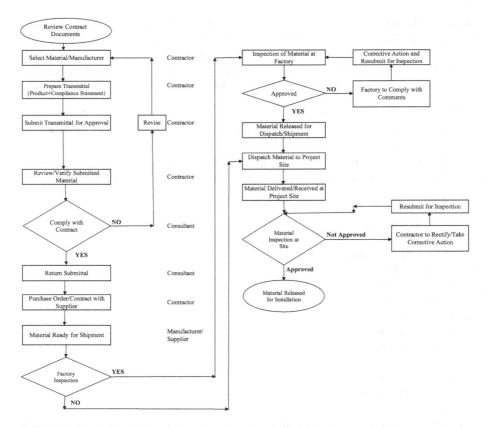

FIGURE 3.46   Material management process for construction project.

Source: Abdul Razzak Rumane. (2016). *Handbook of Construction Management*. Reprinted with permission of Taylor & Francis Group.

3.3.7.2.4 Manage Subcontractor

In construction projects, the main contractor (general contractor) is responsible to manage subcontractors' work. Much of the works required to be carried out on major projects are performed by subcontractors or specialist contractors. The main contractor has to manage their works by carefully planning, scheduling, and coordinating to complete the project successfully. Areas of subcontracting are generally listed in the particular conditions of the contract document. The main contractor is responsible to manage subcontractors' work.

3.3.8 COMMUNICATION MANAGEMENT

Communication is the process through which information is transmitted by a sender to a receiver via a channel/medium. The receiver encodes the message and gives feedback to the sender. Figure 3.47 illustrates a communication model.

The standard methods of communication methods widely used are either written or oral. Written means letters, notices, email, or messaging (electronic media). Oral means meeting, telephonic conversation. Apart from these two mechanisms, no-verbal communication is another method used to assess communication within organization. Communication is either formal or informal.

Formal communication is planned and delivered as part of standard operating policies and procedures of the organization.

Informal communication is not mandated or otherwise required but occurs as part of people functioning collectively.

Communication can be internal and external.

Direction of communication can be top to bottom vertical communication (downward) or bottom to top (upward) or horizontal.

Communication Management includes the processes that are required to ensure timely and appropriate planning, collection, creation, storage, retrieval, management, control, monitoring, and ultimate disposition of project information. Project communication management includes the following processes:

1. Plan communication
2. Manage communication
3. Control communication

Construction project has the involvement of many stakeholders. The project team must provide timely and accurate information to identified stakeholders who will

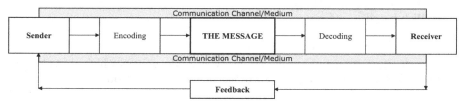

FIGURE 3.47 Communication model.

receive communications. Effective communication is one of the most important factors contributing to the success of project. Project communication is the responsibility of everyone in the project team. Communication management process in construction project consists of following activities:

1. Develop communication plan
2. Manage communication
3. Control documents

3.3.8.1 Develop Communication Plan

Communication plan helps project team members to identify internal and external stakeholders involved in the project. Figure 3.48 illustrates Communication Plan Development process.

A comprehensive communication plan for construction project can be developed by analyzing the questions listed in Table 3.20.

3.3.8.1.1 Establish Communication Matrix

For smooth flow of communication in construction project, proper communication matrix among all the stakeholders needs to be established at the start of each stage of the project. Communication matrix is used as a guideline of what information to communicate, who is the team member to initiate, who will receive and take appropriate action, when to communicate, and method of communication. Table 3.21 illustrates example guidelines to prepare communication matrix for site administration during construction phase.

3.3.8.1.2 Determine Communication Methods

The method of communication in construction projects depends on the project stage and needs of the stakeholders. There are a number of methods for determining requirements of stakeholders; however, it is imperative that they are completely understood in order to effectively manage their interest, expectations, and influence and ensure a successful project. Following are the common methods used in construction projects:

- Letters and other hard copies
- Specific type of transmittal forms
- Meeting
- E-mails
- Telephone calls
- Conferencing (voicemail)
- Web-based

3.3.8.1.3 Establish Submittal Procedure

In a construction project, there are various types of documents which are to be sent to different stakeholders. Proper correspondence and reporting method is important to distribute this information.

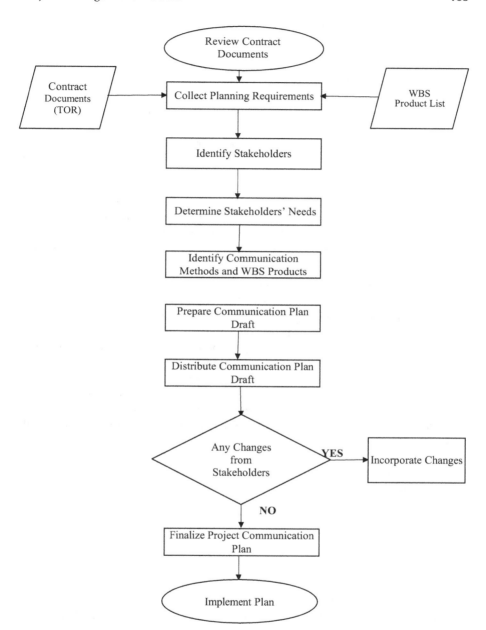

FIGURE 3.48 Communication plan development process.

TABLE 3.20
Analysis for Communication Matrix

Serial Number	W+H	Related Analyzing Question
1	What	What is the purpose of communication
2	What	What type of information needs to be communicated
3	Who	Who will initiate (send) the communication
4	Who	Who are the stakeholders to receive the information
5	When	When is the information to be sent (frequency)
6	What	What method of communication is to be used
7	How many	How many copies to be distributed
8	How much	How much time to wait to receive the feedback
9	How	How to archive the documents

Construction projects have involvement of many stakeholders. Large numbers of documents move forward and backward between these stakeholders for information or action. During the design stage, the communication is mainly between the Owner (Project/Construction Manager) and Designer (Consultant). However, once the contractor is selected depending on type of project delivery system (Design-Build Contractor or Construction Contractor or CM @ Risk), then the Contractor is actively involved in the project communication system. Apart from these three parties, some other stakeholders, who have interest, expectations, and influence in the project, are also sent the copy of information/documents for their information or action.

During the design stage, following are the major documents the Designer sends to the Owner:

- Design drawings
- Reports
- Minutes of meetings
- Review comments
- Contract specifications
- Bidding/tendering documents

3.3.8.2 Manage Communication

Contract documents specify number of original (paper print) and copies to be transmitted to A/E (Consultant) for review and approval. Figure 3.49a illustrates an example submittal process (paper based) and Figure 3.49b illustrates an example submittal process (electronic).

With the advent of electronic submittal transmittal system, electronic documents such as Portable Document Format (PDF) and Building Information Model (BIM) are used for submittal purpose. Contractor, A/E (consultant), or owner may use Internet-based project management software having transmission, tracking, and

TABLE 3.21
Sample Form Communication Matrix

Project Name

Name of Construction/Project Manager:

Contractor Name:

Name of Consultant:

Project Number:

Serial Number	Type of Document	Originator	Receiver(s)	Purpose	Frequency	Method	Responsible Person for Action	Comments

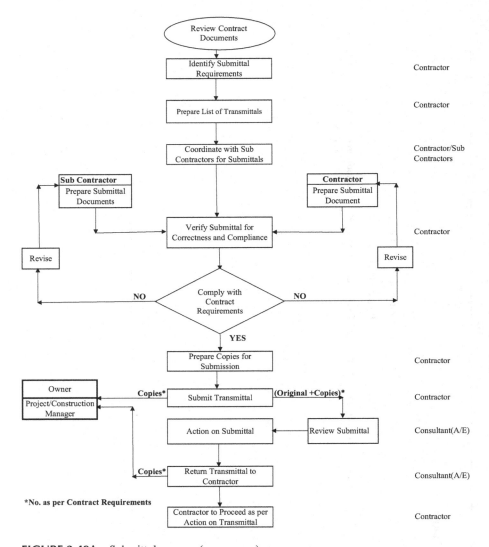

FIGURE 3.49A   Submittal process (paper copy).

email features; however, utilization of electronic documentation system should be specified in the contract documents.

3.3.8.2.1   Manage Submittals

There are different types of logs used in construction projects to monitor the submission of material, shop drawings, and other submittals.

3.3.8.2.2   Conduct Meetings

Project meeting refers to as a face-to-face communication method to exchange information among team members and stakeholders.

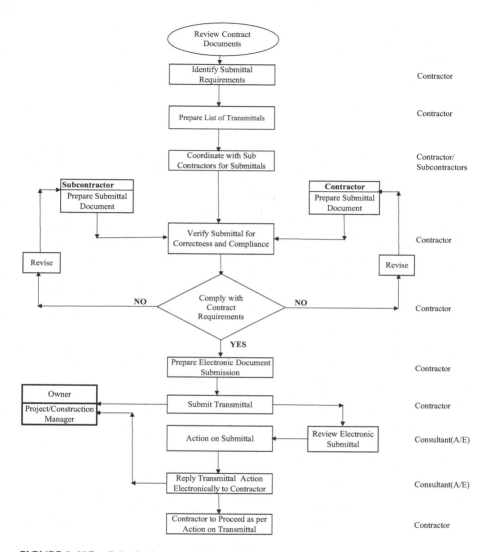

FIGURE 3.49B   Submittal process (electronic).

In construction project, there are many types of meetings held during all the stages/phases of the project. Each meeting has its own purpose and structure. The meetings are used to distribute information, discuss issues, make proposals, suggest solutions, share information, and contribute the ideas for improvement and successful completion of project. Following types of meetings are normally held in construction projects.

1. Kick-off meeting
2. Planning meeting

3. Pre-tender meeting
4. Pre-construction meeting
5. Coordination meeting
6. Progress meeting
7. Change control meeting
8. Quality meeting
9. Safety meeting

The frequency of meetings depends on the project stage/phase. The frequency of meetings in the construction phase is normally specified in the contract documents. To conduct a project meeting, it is advisable to prepare and circulate the agenda well in advance. In case of progress meeting during construction phase, the Resident Engineer prepares the agenda and circulates to the participants expected to attend the meeting. The contractor informs the Resident Engineer in advance about the points the contractor would like to discuss. These points are also included in the agenda. Figure 3.50 illustrates a typical agenda format for meeting. Figure 3.51 is a sample form to list meeting attendees, Figure 3.52 illustrates a format for the preparation of minutes of meeting, and Figure 3.53 is a sample transmittal form for the distribution of minutes of meeting.

3.3.8.2.3 Manage Documents

In construction projects, all the related documents are sent and received (exchanged) through the use of transmittal form. These forms are generally issued to the contractor along with other contract documents. Project team members are required to follow the procedures specified in the contract.

For any communication with external stakeholders such as regulatory bodies, company letterhead is used to communicate the matter.

3.3.8.3 Control Documents

Contract conditions specify the time allowed to process the transmittal and other contract-related communication. Contractor, consultant, and construction/project manager maintain logs for all incoming and outgoing documents. Follow-up is also done among internal project team members to expedite the required action to be taken against the transmittal.

3.3.9 RISK MANAGEMENT

Please refer Chapter 1 for risk management process.

3.3.10 CONTRACT MANAGEMENT

Project Procurement/Contract Management in construction projects is an organizational method, process, and procedure to obtain the required construction products. It includes the process to acquire construction facility complete with all the related product/material, equipment, and services from outside contractors/companies to the satisfaction of the owner/client/end user.

PROJECT NAME

Contract Number:			
Type of Meeting:		Date of Meeting:	
Place of Meeting:		Time of Meeting:	
Owner:			
Project/Construction Manager			
A/E (Consultant)			

AGENDA

SAMPLE FORM

1.Points to be discussed:

   1.1 ....

   1.2 ....

   1.3 ....

   1.4 ....

     ....

     ....

2. Any other Issues:

Signed by: ....   Position: ....

Deate: ....

FIGURE 3.50   Agenda format for meeting.

Conventional notions of the procurement/purchasing cycle, which is normally applied in batch production, mass production, or in merchandising are less appropriate to the realm of construction projects. The procurement of construction projects also involves commissioning professional services and creating a specific solution. The process is complex, involving the interaction of owner/client, project/construction manager, designer/consultant, contractor (s), suppliers, and various regulatory bodies. Generally, a construction project comprises building materials (civil), electromechanical items, finishing items, instrumentation, and equipment. Construction

	PROJECT NAME Construction/Project Manager Name CONSULTANT(A/E) NAME			
	ATTENDEES			

Meeting Type: _____   Meeting No.: _____
                                    Location: _____
Project Phase: _____   Date: _____
                                    Time: _____

S No.	NAME	POSITION	COMPANY	SIGNATURE
1				
2				
3				
4				
5				
6				
7				
8				
9				
10				
11				
12				
13				
14				
15				

SAMPLE FORM

FIGURE 3.51 A sample form to list meeting attendees.

involves installation and integration of various types of materials/products, equipment, systems, or other components to complete the project/facility to ensure that the facility is fully functional to the satisfaction of the owner/end user. Contract management involves:

- identification of;
  - What are the services in-house available
  - What services to be procured from outside agencies/organizations

PROJECT NAME

Minutes of Meeting

Contract NO.:	
Owner Name:	
Project/Construction Manager:	
Contractor Name:	

Meeting Type:		Minutes Number:	
		Date:	
Meeting Location:		Time:	

Attendees SAMPLE FORM

Number	Name	Position	Company

ITEM	DESCRIPTION OF DISCUSSION	STATUS	PRIORITY	ACTION			
				By	Due	Started	Completed

Distribution	Owner	PM/CM	Engineer's Representative	Contractor	Other
Original Copies	☐☐	☐☐	☐☐	☐☐	☐☐

FIGURE 3.52 Minutes of meeting format.

- How to procure (direct contract, competitive bidding)
- How much to procure
- How to select a supplier/contractor
- How to arrive at appropriate price, terms, and conditions

PROJECT NAME

Transmittal for Minutes of Meeting

Transmittal Number:		Date:	
Contract Number:		Contractor:	
Owner:			
Project/Construction Manager		SAMPLE FORM	
A/E (Consultant)			

Meeting Type:		Minutes Number:	
		Date:	
Meeting		Time:	

ATTENDEES			
NO	NAME	POSITION	COMPANY

Your Use ☐		For Review and Comment ☐	
Your Approval ☐		As Requested ☐	

Engineer's Representative

_____   _____   _____
Name   Signature   Date

The attached minutes constitutes our understanding of the points discussed and conclusions reached. All participants are requested to review these minutes and inform Engineer's Representative of their comments, if any, latest in 72 hours of the date of receiving these minutes.

FIGURE 3.53 Transmittal for minutes of meeting.

- Signing of contract
- Timely delivery
- Receiving right type of material/system
- Timely execution of work

- Inspection of work to maintain the quality of the project
- Completion of project within an agreed-upon schedule
- Completion of project within an agreed-upon budget
- Documenting reports and plans

3.3.10.1 Plan Procurement Management Process

In construction projects, the involvement of outside companies/parties starts at the early stage of project development process. The owner/client has to decide which work is to be procured, constructed by others. Every organization has their procurement system to procure services, contracts, and product from others. Figure 3.54 illustrates the stages at which outside agency (contractor) is selected as per the procurement strategy for particular type of project delivery system. At each of these stages, the procurement management process (Bidding and Tendering Process) takes place. Figure 3.55 illustrates procurement management process life cycle to select the contractor (outside agency).

Also, the owner has to determine the type of contract/pricing methods to select the contractor (outside agency).

3.3.10.2 Select Project Delivery System

Construction project delivery system is organizational relationship of three elemental parties, Owner, Designer (Consultant), Contractor, the roles and responsibilities of the parties and the general sequence of activities to be performed to deliver the project. The roles and responsibilities of these parties vary considerably under different project delivery system. Following are the main categories of project delivery system;

1. Traditional System
2. Integrated System
3. Management Oriented System
4. Integrated Project Delivery System

Each category is further classified and subclassified to various project delivery systems.

The selection of project delivery system mainly depends on the project size, complexity of the project, innovation, uncertainty, urgency, and the degree of involvement of the Owner.

3.3.10.3 Select Project Team (Contractor)

In construction projects, the owner engages outside agencies such as construction/project manager, designer, consultant, and a contractor to do following works:

1. Feasibility studies, if in-house facility is not available or inadequate
2. Project management, construction management
3. Project design
4. Construction

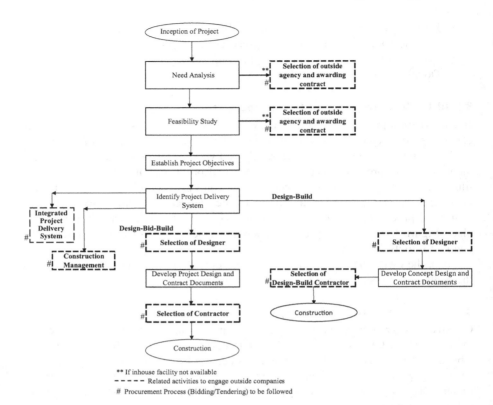

FIGURE 3.54   Procurement management process stages for construction projects.

Source: Abdul Razzak Rumane. (2013). Quality Tools for Managing Construction Projects. Reprinted with permission of Taylor & Francis Group.

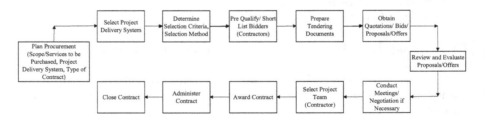

FIGURE 3.55   Contract management process.

Source: Abdul Razzak Rumane. (2016). *Handbook of Construction Management*. Reprinted with permission of Taylor & Francis Group.

The selection of project team (contractor) is mainly done as follows;

1. Screening of Qualified Contractors (Pre-Qualification of Contractor)
2. Selecting Contractor using Contracting Method such as:

     a. Competitive bidding
     b. Competitive negotiations
     c. Direct negotiation

3. Awarding contract

While engaging an outside agency (contractor), following selection criteria to be considered as a minimum to pr qualify the agency (contractor);

- Available skill level
- Relevant/past performance on similar type of work
- Reputation about their works
- Number of projects (works) successfully completed
- Technical competence
- Knowledge about the type of projects (works) for which likely to be engaged
- Available resources
- Commitment to creating best value
- Commitment to containing sustainability
- Rapport/behavior
- Communication

The aforementioned information is gathered through a Request for Information, Request for Pre-Qualification, or Pre-Qualification Questionnaires from the prospective agency, consultant, and contractor.

After short listing of contractors, bid documents are distributed to submit the proposal. Following are the most common contracting methods for the selection of project team:

1. Low bid
2. Best value

     a. Total cost
     b. Fees

3. Qualification-based selection (QBS)

3.3.10.3.1 Perform Short Listing (Pre-Qualification) Process

The first stage in selection of project team (contractor) is by short listing the prospective contractors. The information is gathered through:

- Request for Information
- Request for Qualification
- Pre-Qualification Questionnaires

3.3.3.3.1.1 Pre-Qualification of Feasibility Study Consultant

Table 3.22 illustrates the required qualifications to select a consultant to conduct feasibility study. Normally, the consultants are selected on quality-based selection method.

TABLE 3.22
Consultant's Qualification for Feasibility Study

Serial Number	Description
1	Experience in conducting feasibility study
2	Experience in conducting feasibility study in similar type and nature of projects
3	Fair and neutral with no prior opinion about what decision should be made
4	Experience in strategic and analytical analysis
5	Knowledge of analytical approach and background
6	Ability to collect large number of important and necessary data via work sessions, interviews, surveys, and other methods.
7	Market knowledge
8	Ability to review and analyze market information
9	Knowledge of market trend in similar type of projects/facility
10	Multidisciplinary-experienced team having proven record in following fields: a) Financial analyst b) Engineering/technical expertise c) Policy experts d) Project scheduling
11	Experience in review of demographic and economic data

Source: Abdul Razzak Rumane. (2013). *Quality Tools for Managing Construction Projects.* Reprinted with permission of Taylor & Francis Group.

3.3.3.3.1.2 Pre-Qualification of Construction Manager Construction Manager is a professional firm or an individual having an expertise in the management of construction processes. There are two types of Construction Managers (Construction Management Systems). These are:

1. Agency CM
2. CM at-Risk

While selecting the construction manager, the owner has to ensure that CM has professional expertise in managing the following main areas of the project;

- Scope
- Schedule
- Cost
- Monitoring and control
- Quality
- Safety

The owner has to consider regulatory requirements (licensing) while selecting the CM. Table 3.23 lists Pre-Qualification Questionnaires (PQQ) to select Construction Manager.

TABLE 3.23

Pre-Qualification Questionnaires (PQQ) for Selecting Construction Manager

Serial Number	Question	Answer
1	Name of the Organization and Address	
2	Organization's Registration and License Number	
3	ISO Certification	
4	Total Experience (Years)	
	4.1 Agency construction manager	
	4.2 CM at-risk	
5	Size of Project (Maximum Amount Single Project)	
	5.1 Agency construction manager	
	5.2 CM at-risk	
6	List Similar Type (Type to be Mentioned) of Projects Completed	
	6.1 Agency construction manager	
	6.2 CM at-risk	
7	Type of Services Provided as Agency CM/CM for Aforementioned Projects	
	7.1 During project study	
	7.2 During design	
	7.3 During bidding/tendering	
	7.4 During construction	
	7.5 During startup	
8	Total Experience in Green Building Construction	
9	Total Management Experience	
	9.1 Project scope	
	9.2 Project planning and scheduling	
	9.3 Project costs	
	9.4 Project quality	
	9.5 Technical and financial risk	
	9.6 HSE	
	9.7 Stakeholder management	
	9.8 Project monitoring and control	
	9.9 Conflict management	
	9.10 Negotiations	
	9.11 Information technology	
10	Experience in Conducting Value Engineering	
11	Resources	
	11.1 Management	
	11.2 Engineering	
	11.3 Technician	
	11.4 Construction equipment	
12	Total Turnover for Last 5 years	
13	Submit Audited Financial Reports for Last 3 Years	
14	Experience in Training of Owner's Personnel	
15	Knowledge about Regulatory Procedures	
16	Litigation (Dispute, Claims) on Earlier Projects	

3.3.3.3.1.3 Pre-Qualification of Designer (A/E)   Designer (A/E) consists of architects or engineers or consultant. They are the owner's appointed entity accountable to convert owner's conception and need into specific facility with detailed directions through drawings and specifications within the economic objectives. They are responsible for the design of the project and in certain cases supervision of construction process. Table 3.24 lists Pre-Qualification Questionnaires (PQQ) to select the Designer (A/E).

3.3.3.3.1.4 Pre-Qualification of Contractor (Design-Build)   Table 3.25 lists Pre-Qualification Questionnaire (PQQ) to select contractor (Design-Build).

3.3.3.3.1.5 Pre-Qualification of Contractor (Design-Bid-Build)   Table 3.26 lists Pre-Qualification Questionnaire (PQQ) to select contractor

TABLE 3.24
Pre-Qualification Questionnaires (PQQ) for Selecting Designer (A/E)

Serial Number	Question	Answer
1	Name of the Organization and Address	
2	Organization's Registration and License Number	
3	ISO Certification	
4	LEED or Similar Certification	
5	Total Experience (years) in Designing Following Type of Projects	
	5.1 Residential	
	5.2 Commercial (Mix use)	
	5.3 Institutional (Governmental)	
	5.4 Industrial	
	5.5 Infrastructure	
	5.6 Design-Build (Specify type)	
6	Size of Project (Maximum Amount Single Project)	
	6.1 Residential	
	6.2 Commercial (Mix use)	
	6.3 Institutional (Governmental)	
	6.4 Industrial	
	6.5 Infrastructure	
	6.6 Design-Build (Specify type)	
7	List Successfully Completed Projects	
	7.1 Residential	
	7.2 Commercial (Mix use)	
	7.3 Institutional (Governmental)	
	7.4 Industrial	
	7.5 Infrastructure	
	7.6 Design-Build	

Serial Number	Question	Answer
8	List Similar Type (Type to be Mentioned) of Projects Completed	
	8.1 Project Name and Contracted Amount	
	8.2 Project Name and Contracted Amount	
	8.3 Project Name and Contracted Amount	
	8.4 Project Name and Contracted Amount	
	8.5 Project Name and Contracted Amount	
9	Total Experience in Green Building Design	
10	Joint Venture with any International Organization	
11	Resources	
	11.1 Management	
	11.2 Engineering	
	11.3 Technical	
	11.4 Design Equipment	
	11.5 Latest Software	
12	Design Production Capacity	
13	Design Standards	
14	Present Workload	
15	Experience in Value Engineering (List Projects)	
16	Financial Capability (Turnover for Last 5 years)	
17	Financial Audited Report for Last 3 Years	
18	Insurance and Bonding Capacity	
19	Organization Details	
	19.1 Responsibility Matrix	
	19.2 CVs of Design Team Members	
20	Design Review System (Quality Management during Design)	
21	Experience in Preparation of Contract Documents	
22	Knowledge about Regulatory Procedures and Requirements	
23	Experience in Training of Owner's Personnel	
24	List of professional Awards	
25	Litigation (Dispute, Claims) on Earlier Projects	

TABLE 3.25
Pre-Qualification Questionnaires (PQQ) for Selecting Design-Build Contractor

Serial Number	Question	Answer
1	Name of the Organization and address	
2	Organization's Registration and License Number	
3	ISO Certification	
4	Registration/Classification Status of the Organization	
5	Joint Venture with any International Contractor	
6	Total Turnover Last 5 years	

(Continued)

TABLE 3.25
(Continued)

Serial Number	Question	Answer
7	Audited Financial Report for Last 3 years	
8	Insurance and Bonding Capacity	
9	Total Experience (years) as Design-Build Contractor	
10	Total Experience (years) as Contractor	
	Design-Build Contracts Information	
11	Total Experience (years) in Construction of Following Type of Projects	
	11.1 Residential	
	11.2 Commercial (Mix use)	
	11.3 Institutional (Governmental)	
	11.4 Industrial	
	11.5 Infrastructure	
12	Size of Project (Maximum Amount Single Project)	
	12.1 Residential	
	12.2 Commercial (Mix use)	
	12.3 Institutional (Governmental)	
	12.4 Industrial	
	12.5 Infrastructure	
13	List Successfully Completed Projects	
	13.1 Residential	
	13.2 Commercial (Mix use)	
	13.3 Institutional (Governmental)	
	13.4 Industrial	
	13.5 Infrastructure	
14	List Similar Type (Type to be Mentioned) of Projects Completed	
	14.1 Project Name and Contracted Amount	
	14.2 Project Name and Contracted Amount	
	14.3 Project Name and Contracted Amount	
	14.4 Project Name and Contracted Amount	
	14.5 Project Name and Contracted Amount	
15	Resources	
	15.1 Management	
	15.2 Engineering	
	15.3 Technical	
	15.4 Foreman/Supervisor	
	15.5 Skilled Manpower	
	15.6 Unskilled Manpower	
	15.6 Plant and Equipment	
16	Quality Management Policy	
17	Health, Safety, and Environment Policy	
	17.1 Number of Accidents During Last 3 Years	
	17.2 Number of Fires at Site	
18	Current Projects	
19	Staff Development Policy	

Serial Number	Question	Answer
20	List of Delayed Projects	
21	List of Failed Contract	
	Designer's Information	
22	Total Years of Experience in Design-Build Type of Projects	
23	Size of Project (Maximum Value)	
24	List Similar Type of Successfully Completed Projects	
25	List Successfully Design-Build projects	
26	Resources	
	26.1 Architect	
	26.2 Structural Engineer	
	26.3 Civil Engineer	
	26.4 HVAC Engineer	
	26.5 Mechanical Engineer	
	26.6 Electrical Engineer	
	26.7 Low Voltage Engineer	
	26.8 Interior Designer	
	26.9 Landscape Engineer	
	26.10 CAD Technicians	
	26.11 Quantity Surveyor	
	26.12 Equipment	
	26.13 Design software	
27	LEED or Similar Certification	
28	Total Experience in Green Building Design	
29	Design Philosophy/Methodology	
30	Quality Management System	
31	HSE Consideration in Design	
32	List of Professional Awards	
33	Litigation (Dispute, Claims) on Earlier Projects	

TABLE 3.26
Pre-Qualification Questionnaires (PQQ) for Selecting Contractor

Serial Number	Question	Answer
1	Name of the Organization and Address	
2	Organization's Registration and License Number	
3	ISO Certification	
4	Registration/Classification Status of the Organization	
5	Joint Venture with any International Contractor	
6	Total Turnover Last 5 years	
7	Audited Financial Report for Last 3 years	
8	Insurance and Bonding Capacity	

(Continued)

TABLE 3.26
(Continued)

Serial Number	Question	Answer
9	Total Experience (years) in Construction of Following Type of Projects	
	9.1 Residential	
	9.2 Commercial (Mix use)	
	9.3 Institutional (Governmental)	
	9.4 Industrial	
	9.5 Infrastructure	
10	Size of Project (Maximum Amount Single Project)	
	10.1 Residential	
	10.2 Commercial (Mix use)	
	10.3 Institutional (Governmental)	
	10.4 Industrial	
	10.5 Infrastructure	
11	List Successfully Completed Projects	
	11.1 Residential	
	11.2 Commercial (Mix use)	
	11.3 Institutional (Governmental)	
	11.4 Industrial	
	11.5 Infrastructure	
12	List Similar Type (Type to be Mentioned) of Projects Completed	
	12.1 Project Name and Contracted Amount	
	12.2 Project Name and Contracted Amount	
	12.3 Project Name and Contracted Amount	
	12.4 Project Name and Contracted Amount	
	12.5 Project Name and Contracted Amount	
13	List of Subcontractors	
14	Resources	
	14.1 Management	
	14.2 Engineering	
	14.3 Technical	
	14.4 Foreman/Supervisor	
	14.5 Skilled Manpower	
	14.6 Unskilled Manpower	
	14.7 Plant and Equipment	
15	Current Projects	
16	Quality Management Policy	
17	Health, Safety and Environment Policy	
	17.1 Number of Accidents During Last 3 Years	
	17.2 Number of Fires at Site	
18	Staff Development Policy	
19	List of Delayed Projects	
21	List of Failed Contract	
22	List of professional Awards	
23	Litigation (Dispute, Claims) on Earlier Projects	

TABLE 3.27
Designer's (A/E) Selection Criteria

Serial Number	Evaluation Criteria	Weightage	Notes
1	**General Information**		
	a) Company Information		
2	**Business**	**10%**	
	a) LEED or Similar Certification	5%	
	b) ISO Certification	5%	
3	**Financial**	**20%**	
	a) Turnover	5%	
	b) Financial Standing	5%	
	c) Insurance and Bonding Limit	10%	
4	**Experience**	**30%**	
	a) Design Experience	10%	
	b) Similar Type of Projects	10%	
	c) Current Projects	10%	
5	**Design capability**	**10%**	
	a) Design Approach	5%	
	b) Design Capacity	5%	
6	**Resources**	**20%**	
	a) Design Team Qualification	10%	
	b) Design Team Composition	5%	
	c) Professional Certification	5%	
7	**Design Quality**	**5%**	
8	**Safety consideration in Design**	**5%**	

Note: The weightage mentioned in the TABLE are indicative only. The % can be determined as per the owner's strategy.

3.3.3.3.1.6 Evaluation of Pre-Qualification Documents Upon receiving the information relating to the qualification data, the documents are evaluated as per the selection criteria determined earlier by the project owner. Table 3.27 is evaluation criteria for the designer to qualify to participate in the tender.

3.3.10.4 Manage Construction Documents

The construction documents reviewed and approved by the owner are organized by the consultant for bidding purpose. Necessary information is inserted on these documents prior to the release for distribution to quailed bidders.

3.3.10.5 Conduct Bidding and Tendering

Regardless of the type of project delivery system, the contract arrangement between the owner and the contractor has to be established. Following are the most common types of contract/compensation methods are used in construction projects:

1. Firm fixed price or lump sum contract
2. Cost reimbursement contract
3. Re-measurement contract
4. Target price contract
5. Time and cost contract
6. Guaranteed Maximum Price (GMP) contract

Tendering and bidding documents are prepared based on the strategy to procure the contract. Bid/tendering documents are distributed to short listed contractors. Figure 3.56 illustrates the bidding process to select the contractor as per the

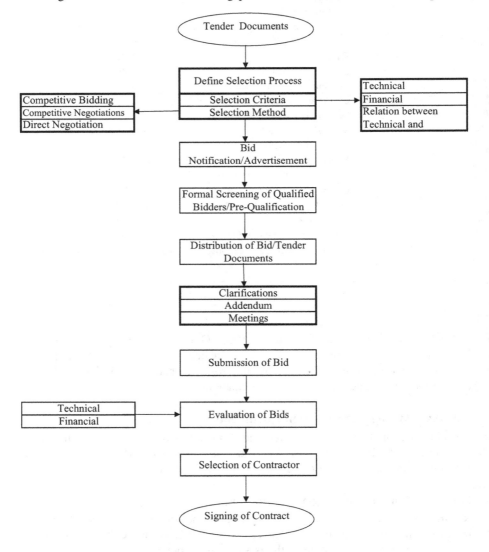

FIGURE 3.56 Bidding and tendering (procurement) process.

procurement method in accordance with the contractor selection strategy for the selection of contractor.

3.3.10.5.1 Evaluate Bids

The submitted bids are reviewed and evaluated by the contractor selection committee members for full compliance to the tender requirements.

The main purpose of bid evaluation is to determine that the bid responses are complete in all respects in accordance with the evaluation and selection methodology as specified in the tender documents.

Most government, public sector projects follow Low Bid selection method. There are three international bidding procedures which may be selected by the project owner to suit the nature of project procurement. These are:

1. Single stage—one envelop: In this procedure, the bidders submit bids in one envelope containing both technical and financial proposals. The bids are evaluated by the selection committee and send their recommendation to the Owner. Following the review and concurrence by the Owner, the contract is awarded to the lowest bidder.
2. Single stage—two envelopes: In this procedure, the bidders submit two envelopes, one containing technical proposal and the other the financial proposal. Initially, technical proposals are evaluated without referring to the price. Bidders whose proposal does not conform to the requirements may be rejected/not accepted. Following the technical proposal evaluation, the financial proposals of technically responsive bidders are reviewed. Upon review and concurrence by the Owner, the contract is awarded to the lowest bidder.
3. Two stages: In this procedure, during the first stage, the bidders submit their technical offers on the basis of operating and performance requirements but without price. The technical offers are evaluated by the selection committee. Any deviations to the specified performance requirements are discussed with the bidders who are allowed to revise or adjust the technical offer and resubmit the same.

   During the second stage, the bidders, whose technical offers are accepted, are invited to submit the final technical proposal and financial proposal. Both the proposals are reviewed by the selection committee, and, following review and concurrence by the Owner, the contract is awarded to the lowest bidder. This procedure is mainly applicable for turnkey projects, complex plants.

Table 3.28 lists the items to be reviewed prior to evaluation of bid documents.

3.3.10.5.2 Select Contractor

After the bids have been reviewed and evaluated by the consultant, they are sent to the owner for the approval and further action. In certain cases, the bids are reviewed by tender-approving agency. The comments raised by the agency are to be taken care before finalization of bid. In most cases, the bidder with lowest bid amount wins the contract. In case the bid amount is more than the approved project budget, the owner has to update the budget amount or negotiate with the contractor to meet owner's

TABLE 3.28
Checklist for Bid Evaluation

Serial Number	Description	Yes	No	Notes
A: Documents				
A.1	Bid submitted before closing time on the date specified in the bid documents			
A.2	Bidders' identification is verified			
A.3	Bid is properly signed by the authorized person			
A.4	Bid bond is included			
A.5	Required certificates are included			
A.6	Bidders' confirmation to the validity period of bid			
A.7	Confirmation to abide by the specified project schedule			
A.8	Bid documents have no reservation or conditions (limitation or liability)			
A.9	Preliminary method statement			
A.10	List of equipment and machinery			
A.11	List of proposed core staff as listed in the tender documents			
A.12	Complete responsiveness to the commercial terms and conditions			
A.13	All the required information is provided (completeness of information)			
A.14	All the supporting documents required to determine technical responsiveness is submitted			
B: Financial				
B.1	All the items are priced			
B.2	Bid amount clearly mentioned			
B.3	Prices of provision items			

expectations. Upon approval from the relevant agency, the owner awards the contract to one bidder

3.3.10.6 Administer Contract

Contract administration is the process of formal governance of contract and changes to the contract document. It is concerned with managing contractual relationship between various participants to successfully complete the facility to meet owner's objectives. It includes tasks such as:

- Administration of project requirement
- Administration of project team members
- Communication and management reporting
- Execution of contract
- Monitoring contract performance (scope, cost, schedule, quality, risk)

- Inspection and quality
- Variation order process
- Making changes to the contract documents by taking corrective action as needed
- Payment procedures

It is required that the contract administration procedure is clearly defined for the success of the contract and that the parties to the contract understand who does what, when, and how. Following are some typical procedures that should be in place for the management of the contract management activities:

1. Contract document maintenance and variation
2. Performance review system
3. Resource management and planning
4. Management reporting
5. Change control procedure
6. Variation order procedure
7. Payment procedure

Table 3.29 lists the contents of contract management plan.

3.3.10.7   Close Contract

It is a process of completing the contract, verification of completed activities, acceptance of all the deliverables, takeover of project, and issuance of substantial completion certificate. Normally, the contract documents stipulate all the requirements to be completed to close the contract. Table 3.30 illustrates the lists all the activities required to be recorded to close the contract.

TABLE 3.29
Contents of Contract Administration Plan

Serial Number	Topics
1	Contract summary, deliverables, and scope of work
2	Type of contract
3	Contract schedule
4	Contract cost
5	Project team members with roles and responsibilities
6	Core staff approval procedure
7	Contract communication matrix/management reporting
8	Coordination process
9	Liaison with regulatory authorities
10	Material/product/system review/approval process
11	Shop drawing review/approval process

(Continued)

TABLE 3.29
(Continued)

Serial Number	Topics
12	Project monitoring and control process
13	Contract change control process
	a) Scope
	b) Material
	c) Method
	d) Schedule
	e) Cost
14	Review of variation/change requests
15	Project holdup areas
16	Quality of performance
17	Inspection and acceptance criteria
18	Risk identification and management
19	Progress payment process
20	Claims, disputes, conflict, and litigation resolution
21	Contract documents and records
22	Post-contract liabilities
23	Contract closeout and independent audit

Source: Abdul Razzak Rumane. (2013). *Quality Tools for Managing Construction Projects.* Reprinted with permission of Taylor & Francis Group.

TABLE 3.30
Contract Closeout Checklist

Serial Number	Description	Yes/ No
Project Execution		
1	Contracted works completed	
2	Site Work Instructions completed	
3	Job Site Instructions completed	
4	Remedial Notes completed	
5	Non-Compliance Reports completed	
6	All services connected	
7	All the contracted works inspected and approved	
8	Testing and commissioning carried out and approved	
9	Any snags	
10	Is project fully functional?	
11	All other deliverable completed	
12	Spare parts delivered	
13	Is waste material disposed?	
14	Whether safety measures for use of hazardous material are established?	
15	Whether the project is safe for use/occupation?	

Serial Number	Description	Yes/ No
16	Whether all the deliverables are accepted?	
Project Documentation		
17	Authorities' approval obtained	
18	Record drawings submitted	
19	Record documents submitted	
20	As-built drawings submitted	
21	Technical manuals submitted	
22	Operation and maintenance manuals submitted	
23	Equipment/material warrantees/guarantees submitted	
24	Test results/test certificates submitted	
Training		
25	Training to owner/end user's personnel imparted	
Negotiated Settlements		
26	Whether all the claims and dispute negotiated and settled	
Payments		
27	All payments to subcontractors/specialist suppliers released	
28	Bank guarantees received	
29	Final payment released to main contractor	
Handing over/taking over		
30	Project handed over/taken over	
31	Operation/maintenance team taken over	
32	Excess project material handed over/taken over	
33	Facility manager in action	

Source: Abdul Razzak Rumane. (2010). *Quality Management in Construction Projects.* Reprinted with permission of Taylor & Francis Group.

TABLE 3.31
Contents of Contractor's Financial Plan

Section	Topic
1	Description of Project
2	Finance Management Organization
3	Estimated Cost of Project
4	Project Schedule
5	Type of Contracting Method
6	Source of Funds
7	Payment (Receivable) Schedule Breakdown
	7.1 Advance payment
	7.2 Monthly progress payment
8	Accounts Payable Schedule Breakdown
	8.1 Insurance, Bonds
	8.2 Staff salaries

(Continued)

TABLE 3.31
(Continued)

3.3.11 HEALTH, SAFETY, AND ENVIRONMENT MANAGEMENT (HSE)

Safety is a cornerstone of a successful project. Accidents cause needless loss of human and physical capital. Logically these losses can only be detrimental to the project schedule and cost. Safety, long neglected, is increasingly being recognized for its value as contractors and owners strive to avoid the increasing costs of injuries and fatalities.

The trend toward improved safety seems gradual but recognizable. As in other fields, technology advances continually provide better materials, methods, and equipment. Increasingly prefabricated materials and modular construction reduce worker exposure to more hazardous construction activities found with older conventional methods and materials. In recent years, researchers have found that incorporating safety in the design phase has a huge potential to impact exposures to hazardous situations.

3.3.11.1 Major Causes of Site Accidents

The results of site accidents range from minor injuries such as cuts and bruises to major ones such as broken limbs and paralysis. The worst accidents may result in fatalities. Thankfully, injury severity and the frequency of that injury tend to be inversely related. This discussion focuses primarily on the causes of fatalities, but in reality, the causes of accidents resulting in minor and major injuries are often remarkably similar. There is an element of randomness in injury outcomes. A few inches may be the only difference between a near miss and a fatality.

Almost 60% of construction fatalities are caused by four primary means. These hazards are known as the "Fatal Four". They consist of falls, electrocutions, struck by incidents, and caught in between incidents. Preventing these is the core of any construction safety program.

Electrocutions, another leading fatality cause, are responsible for about 10% of construction deaths. Primary causes of electrocution are:

- Contact with overhead power lines
- Contact with energized sources
- Improper use of extension and flexible cords

Safety countermeasures are:

- Maintaining a safe distance from overhead lines*
- Using ground fault circuit interrupters (GFCI)
- Inspecting portable tools and extension cords for damage

*Ten feet is considered the minimum safe clearance for power lines up to 50 kV. For a 200 kV line, the minimum clearance doubles to 20 feet.

When impact alone creates the injury, this is considered a struck-by event. Workers can be struck by flying, falling, swinging, or rolling objects. These events may be minimized by staying clear of lifted loads; maintaining awareness and respect for the swing radius of cranes, backhoes, etc.; and providing equipment operators protection from falling debris.

Struck by and caught in between events are somewhat similar. When crushing between two objects or being pulled into or getting caught in running machinery occurs, this is considered caught in between. Some common causes of this type are structural collapse, trench cave-in, or becoming trapped between equipment and a stationary object. A safe worksite will have bracing for unstable structures as well as sloping, benching, or shielding for trenching operations. Workers should be trained to maintain awareness of moving equipment, stay clear when possible, and avoid pinch points.

Thus far, the focus has been on accidents that cause fatalities and serious injuries. Any safety discussion should also include common causes of less serious injuries. Most construction work is physically challenging and often performed in harsh environments. Strains, sprains, bumps, bruises, and cuts are common. The inevitable bumps and scrapes can be minimized with Personal Protective Equipment (PPE). Using hard hats, gloves, safety glasses, and hearing protection must become second nature. To avoid strains, two people should be used to lift heavy loads and avoid awkward working positions.

Nature's elements can create a variety of problems. Rain, wind, cold, and heat each presents special challenges. Heat illness is a major concern in summer months. When temperatures are high, workers should take special care to keep hydrated, work in the shade when possible, and complete higher-effort tasks in the cooler morning hours.

Less obvious and often overlooked, a cluttered and dirty worksite is one common and easily preventable cause of accidents. Scattered tools, materials, and construction debris invite slips, trips, and falls. Loose nails create cuts and punctures. Site cleanliness and orderliness cannot be overemphasized. There is no doubt that a strong correlation exists between cleanliness and safety.

Another less obvious cause of accidents is the inability to recognize potential hazards on the jobsite. Young workers are particularly vulnerable because they often lack experience. Immigrant workers comprise another vulnerable group. Poor language skills make recognition training and awareness more difficult. Since the lack of ability to recognize potential hazards is frequently found in both groups, they may require specialized training to build recognition and teach avoidance. Mentoring and close supervision are also recommended to reduce their risk of injury.

3.3.11.2   Safety in Construction Projects and Process Industries

Loss Prevention Philosophy makes process industries, construction projects, and other industries believe that protection of its resources, including employees, and physical assets against human distress and financial loss resulting from accidental occurrences can be controlled. Disasters have taught us an improved loss control through a professional management system. The major elements in professional management system include Policy Declaration, Employee Training, Permit to Work Procedures, Safety Inspections, Predictive Maintenance, Safety Meetings, Safety Talks, Safety Suggestions, Loss Prevention Compliance Reviews, Executive Management Safety Reviews, Safe Operations Committee, Contractor Safety Program, Nonoperating Personnel Safety Orientation, Emergency Drills, Disaster Control Plan, and Employee Incentive Schemes.

Fundamental concepts and methodology of the program are to identify all loss exposures, evaluate the risk of each exposure, plan how to handle each risk, and manage according to plan. People are the first source of losses. These could be managers, engineers, or workers who plan, design, build, operate, and maintain the plants. Also, general public who may be subjected to the hazards! Second in the line are equipment—whether fixed plant, machines, tools, protective gear, or vehicles. Third are materials such as process substances, supplies, and products that have physical and chemical hazards affecting people, equipment and environment. Fourth is surrounding that includes buildings, surfaces and sub-surfaces, atmosphere, lighting, noise, radiation, hot or cold weather and social or economic conditions, which can affect safe performance of people, equipment, and materials.

Human sufferings and economic aspects of costs lost and work delays or property damage varies from minor to major. As we all know, a loss is the effect of an undesired event that could result in injury, property damage, and production upset. Root causes of accidents in process industry are 30% due to design failure, operational error, equipment failure; 20% attributed to maintenance and inspection deficiencies; 45% to inadequate supervision and training; and 5% due to natural phenomena true external influences.

3.3.11.2.1   *Policy Declaration*

It is important that the highest authority of the corporation/company and his management considers no phase of operation or administration as being of greater importance that accident prevention. It should be the policy of management to provide and maintain safe, healthful working conditions to follow operating practices that will safeguard employees and result in safe working conditions and efficient operations. A policy declaration to this effect should be developed and displayed for the knowledge of employees and its implementation should be ensured by the management.

3.3.11.2.2   Training

Safety training should be conducted for all employees that include the operations, maintenance, and engineering and other administration personnel.

New employees assigned to work including clerical staff are given an introduction to general activities of the industry with emphasis on loss prevention. As a part of this training program, new employees should receive training on first aid/CPR, firefighting, usage of personal protective equipment, hazardous chemical handling, and work permit procedures. Over and above the normal work-related training, regular employees should also be given refresher safety training on a set frequent scheduled basis. The employees' supervisor usually determines the training requirements.

3.3.11.2.3   Work Permit Procedures

Management should have the certification program for the employees to issue and receive work permits for a good control and safe execution of tasks to be performed by other than normal operating personnel within the defined hazardous areas. Professionals from the Loss Prevention/Safety Department should assist work permit procedure certification. Validity of this certification should be decided by the management on the basis of the nature of activities, number of employees, and other relevant criteria such as employee's knowledge, ability, attitude, and behavior. During this certification period, employees should undergo a refresher program to update and upgrade themselves on work permit procedures that may be organized by in-house Operations/Maintenance training groups.

Generally, process industries have work permits for hot and cold works and a permit for entry into confined space. Other work permits include precautionary permit for potential for accidental release of hydrocarbons, steam condensate, or any other toxic or reactive injurious materials during the course of any activity, excavation permits, permit to work on higher elevation or under water, and so on. The functioning of the work permit system and its strict adherence to the procedural requirements should be monitored by performing field surveys by dedicated management representative which should be backed by spot check reviews by loss prevention/safety professionals.

3.3.11.2.4   Safety Inspections

Over and above all daily, weekly, and monthly regular work-level inspections, process industry should make a commitment that all facilities, such as operating plants, maintenance shops, offices, and other workplaces be inspected as spot checks by the Management at least once every quarter to identify and eliminate hazards and to provide a safe working condition.

A typical operating plant safety inspection team should consist of members from Operations, Maintenance, Process Engineering, Loss Prevention/Safety, and Fire

Protection. An Operations' Superintendent should lead the team from an unaffected division along with the Process and Maintenance Engineers of the division to be surveyed and a representative each from Loss Prevention and Fire Protection. The team should review procedures and related software and conduct walks through inspection of the operating plants to not only verify satisfactory completion of previous items but also make a note of other additional observations. The findings should be discussed at a subsequent critique meeting and should be consolidated in a report. The operations Management should follow up on the corrective actions, taken on the findings.

Similar quarterly safety inspections should also be conducted on maintenance facilities such as machine shops, welding shops, sheet metal shops, relief valve shops, instrumentation shops, offices, and material warehouses.

3.3.11.2.5 Other Predictive Maintenance Inspections

Some other inspections, certifications, and predictive maintenance programs that have proved beneficial in process industries are as given in subsequent sections.

3.3.2.2.5.1 On-Stream Inspection Program   This records thickness measurements for all pressure vessels, equipment, and selected process line circuits. The OSI program involves various survey points in the plants on a pre-planned schedule.

3.3.2.2.5.2 Relief Valve Inspection Program   This program should normally monitor relief valves through a relief valve coordinator or a specialist from the Inspection group. The program is to forecast valves to be removed, to revise test intervals, and to evaluate specific problems.

3.3.2.2.5.3 Rotating Equipment Monitoring Program   Rotating equipment should be monitored on a biweekly/monthly/quarterly cycle based on the critical nature of the equipment by dedicated maintenance specialists using state-of-the-art sophisticated vibration analysis equipment.

3.3.2.2.5.4 Crane/Heavy Equipment Inspection   All cranes (mobile and fixed) used in the process industry should be inspected and certified by a crane inspection specialist. The crane/heavy equipment operators and riggers should also be certified for a particular crane/load during lifting operations.

3.3.2.2.5.5 Safety Meetings   Safety meetings and talks are considered vital in communicating Loss Prevention/Safety topics, motivating employees, and acknowledging contributions in the Loss Prevention/Safety Program.

3.3.2.2.5.6 Operations Safety Meetings   Every operations unit should organize a safety meeting at least once in two months. These meetings should be chaired by the Plant Foreman and attended by Plant Operation Supervisors, Lead Operators, and a representative from Loss Prevention/Safety if available. Various issues related to operations safety, observations during quarterly plant safety inspections, and near misses/incidents having lesson learning potential should be discussed during these meetings.

3.3.2.2.5.7 Maintenance Safety Meetings   Maintenance Department should organize safety meetings more frequently, preferably at least once two weeks. Due to the nature of their work, often the maintenance employees are exposed to unsafe conditions or commit unsafe acts. These meetings, usually chaired by Maintenance Foreman, are organized based on crafts and held at the work places. During these meetings, a Loss Prevention/Safety representative should also deliver a safety talk on specialized topics. Safety films, as visual training aid, should be shown during these meetings.

3.3.2.2.5.8 Safety Talks   Frequent weekly/monthly/quarterly safety talks should be conducted by the industry line organizations. The appropriate management should recognize the best safety talk. Discussions on related Operating Instructions as well as recently occurred on or off the job incidents/accidents involving losses should be included as topics of safety talks. The inmate discussion concept is the best and effective way to educate and update the employees with relevant information to help increase and improve their safety awareness both at work and at home.

3.3.2.2.5.9 Safety Suggestion Program   A Safety Suggestion Program should be introduced which encourages employee involvement in Loss Prevention. Employees whose suggestions are accepted by the management should be recognized for their contributions. Implementation of accepted safety suggestions should receive a greater prioritized attention for early benefits. Process industry history shows that the implementation of many of these suggestions have resulted in significant improvement in plant and personal safety.

3.3.2.2.5.10 Loss Prevention Compliance Reviews   Loss Prevention Compliance Reviews of the facility should be performed on a selective basis to monitor adherence with the Industry/Company Loss Prevention/Safety policy. These independent reviews provide objective feedback on the effectiveness of the Loss Prevention/Safety effort and should be performed by specialists such as Operations/Maintenance/ Engineering representatives from the proponent organization and other expertise from Electrical/Instrumentation, Rotating Equipment, On-stream Inspector, Crane/ Heavy Equipment Inspector, Loss & Fire Prevention and Industrial Hygiene/Health/ Environmental. Guidelines for performing reviews should be tailored to the size and type of review undertaken. Management should follow the team's recommendations for timely corrective actions taken.

3.3.2.2.5.11 Executive Management Safety Reviews   It is good industry practice to conduct an overall safety review of the facility by Corporate Executive Management at least once a year. Participants should review the status of implementation to all items reported during the previous tour. They should also review new constructions and modifications to equipment in respect of loss prevention factors. These executives should allow time for presentations on items requiring capital budgetary approval, which could improve process safety and personnel safety.

3.3.2.2.5.12 Safe Operations Committee   The Safe Operations Committee should be chartered to evaluate the Company Organization Loss Prevention Program and recommend ways and means for improvement. The Management level Company/Organization representative should chair the Committee. Generally, members of the Committee should include Managers/Superintendents of Operations, Maintenance and Engineering, and representatives from Training, Loss Prevention and Fire Protection. Key issues which should be addressed in such committees are analysis of deficiencies, effectiveness of programs, compliance with corporate general and engineering standards, review and follow-up of corrective actions taken on safety problems identified by safety inspection or other reviews, examine levels of priority given to various safety problems, and discuss employee involvement in various on-the-job or off-the-job activities. Reviewing of the incidents/accidents/near-misses should be given a greater importance for lesson learning potentials and make recommendations.

3.3.2.2.5.13 Contractor Safety Program   Special care and attention should be given to activities performed by the contractors. Safe execution of any tasks within the process facility is the sole responsibility of the proponent organization management regardless of contractor involvement. The proponents must ensure that the contractors have a good workable safety program that is well communicated and is fully implemented. The program should assure that contractors meet their responsibilities to protect personnel, equipment, and plant facilities. Contractual obligations are reviewed at pre-contract meetings and during regular site inspections while work is in progress. At the end, an evaluation report should be prepared for future reference to determine whether the contractor is worthy of being considered again.

3.3.2.2.5.14 Contractor Safety Orientation Program   It is important that company Management recognizes the hazards and the needs of nonoperating personnel who are not familiar with the process plant facilities and the procedures to follow in case of emergency. They should organize a safety program for all such employees, especially contractor employees working inside their facility. During the program, a safety awareness presentation should be made on various hazards in the facility, work permit procedural requirements, safety policy, and general functional procedures for their operations and emergency/evacuation procedures.

3.3.2.2.5.15 Fire/Emergency Drills   In industry, announced and unannounced Fire/Emergency Drills have been found of great value to tune employees to face any real in-plant emergencies. A written scenario should be prepared for each drill that sufficiently describes the role of operating personnel, location and magnitude of fire/emergency, equipment affected, source of a release, direction of flow, etc. A team consisting of representatives from Operations, Maintenance and Process

Engineering, Fire Protection, and Loss Prevention/Safety should review these drills by making observations. A meeting should be conducted after such drills to evaluate the opportunities for improvement.

3.3.2.2.5.16 Disaster Control Plan   Every process industry/organization should develop a Disaster Control Plan that should define the procedures for obtaining assistance from inside or outside the company support organizations during disasters and detail the functions of essential personnel assigned to control the disaster. At least two drills per annum, one of which is preferably unannounced, are recommended to evaluate the effectiveness of the plan and readiness of the people, and corrective actions are taken on any deficiencies noted.

3.3.2.2.5.17 Incentive Programs   Company/organizations should develop good safety promotional incentive programs such as safety competitions of employees and also for the contracting firms who may be involved in major project construction and/ or maintenance/repair jobs. Employee's individual and team performance should be recognized by the company to encourage and motivate employees to think and execute safety at all times and reduce losses that could occur due to injury or property damage or operations/production upsets.

3.3.11.3   Safety Hazards during Summer

During summer, people suffer from fatigue, exhaustion, heat-related illnesses, tornadoes, lightning storms, stinging insects, allergies, wildfires, poisonous plants, poisonous animals, and, kids drowning are reported each week in hot areas. Vehicle safety is also the most important as a killer. Children should not be left unattended inside a vehicle . . . not even for a split second! Temperature inside a car reaches well over 120 degrees F in less than one hour. Also, perfumes, cologne, hair sprays, or aerosol cans of any kind should not be kept inside a car to avoid car explosions!

As the temperature rises, the body stress also rises. There are two critical actions that can help to battle the heat. Acclimatization to the heat, that is, to get accustomed and used to heat and consume lot of water. Human body is a good regulator of heat. It reacts to heat by circulating blood and raising skin temperature. Excess heat is then released by sweating. Sweating maintains a stable body temperature if the humidity level is low enough to permit evaporation and also if the body fluids and salts you lose are adequately replaced. Many factors can cause the body to unbalance its ability to handle heat for example, age, weight, fitness, medical condition, and diet.

When the heat is combined with other stresses such as physical work, loss of body fluids, fatigue, or some medical conditions, it may lead to heat-related illnesses, disability, and even to death. Therefore, everyone including workers and children should be cautioned that this can happen to anybody—even to young and fit people.

Heat stress is a serious hazard. When body temperature rises even a few degrees above normal, that is, 98.6 degrees F, we can experience tiredness, irritability, inattention, muscle cramps and may become weak, disoriented, and dangerously ill. Heat stress reduces our work capacity and efficiency. People who are overweight, physically unfit, suffer from heart conditions, drink too much alcohol, and are not used to summer temperatures are at a greater risk and should seek medical advice. First aid for heat stress is nutritious food and getting acclimatized with heat and drinking lots of water.

Heat Rash: When people are constantly exposed to hot and humid air, heat rash can substantially reduce the ability to sweat and the subsequent body tolerance to heat. First aid for heat rash is to clean the affected areas thoroughly and dry them completely. Calamine or other soothing lotion may help relieve the discomfort.

Heat cramps are the final warning for heat stress. These may occur after prolonged exposure to heat. They are the painful intermittent spasms of the abdomen and other muscles. First aid for heat cramps varies. The best care is rest and moving the victim to a cool environment, giving him plenty of water and not giving pops, sparkling water, or alcohol.

Heat exhaustion may result from physical exertion in hot environments. Heat exhaustion develops when a person fails to replace body fluids and salt that is lost through sweating. You experience extreme weakness, fatigue, nausea, or a headache as heat exhaustion progresses. First aid for heat exhaustion is to rest in a shade or cool place, drink plenty of water, loosen clothing to allow for the body to cool and use cool wet rags to aid cooling.

Heat stroke is a serious medical condition that urgently requires medical attention in which sweating stops, which makes the skin hot and dry. Body temperature is very high (105 F. and rising). Symptoms of heat stroke are mental confusion, delirium, that is disordered speech, chills, dizziness, strong fast pulse, loss of consciousness, convulsions or coma, a body temperature of 105 degrees F or higher, and hot, dry skin that may be red or bluish. Remember, heat stroke is a medical emergency since brain damage and death are possible. Until medical help arrives, move the victim to a cool place. Call Emergency telephone number for help such as 9–9–9 or 1–1–0 or 9–1–1. You must use extreme caution when soaking clothing or applying water to a victim. Shock may occur if done too quickly with very cold water. Use a fan or ice packs. Douse or sprinkle the body continuously with a cool liquid and summon medical help.

Other sun-related health problems: Exposure to UV radiation can lead to most common skin cancers. Getting one or two sunburn blisters before the age of 18 doubles the risk for developing melanoma. This is important for children.

Long exposure to sunlight can lead to eye problems later in life, such as cataracts. Another one is a "burning" of the eye surface, called "snow blindness" from sunlight, which may lead to complications later in life.

Repeated exposure to the sun can cause premature aging effects. Sun-induced skin damage causes wrinkles, easy bruising, and brown spots on the skin.

Furthermore, sunburns can alter the distribution and function of disease-fighting white blood cells for up to 24 hours after exposure to the sun. Repeated over-exposure to UV radiation can cause more damage to the body's immune system. Mild sunburns can directly suppress the immune functions of human skin where the sunburn occurred, even in people with dark or brown skin. This means that it applies to all of us with Indian origin.

The sting of the summer: With increased temperatures, insects become very active. Insects are nuisance and can cause many health-related problems. Common stinging insects are bees, wasps, hornets, yellow jackets, fire ants, etc. Over 2 million people are allergic to stinging insects. An allergic reaction to an insect sting can occur within minutes, or even hours after the sting. Stinging insects are especially attracted to sweet fragrances of perfumes, colognes, hair sprays, picnic food, open soda or beer cans, and garbage areas. Avoid these attractants in order to lessen a person's chance of being stung.

Symptoms of an allergic reaction are itching and swelling in areas other than the sting site; tightness in the chest; and difficulty in breathing, hoarse voice, dizziness, and unconsciousness or cardiac arrest. The treatment to an allergic reaction is the use of epinephrine and other treatments. Epinephrine can be self-injected if directions are followed or administered by a doctor. Intravenous fluids, oxygen, and other treatments may also be needed. It is very important to call for medical assistance immediately, even if a person says, "I am okay" after administering epinephrine.

To remove the stinger, scrap a credit card or other object. Do not pinch and pull out the stinger, this will inject more venom. If breathing difficulties develop, *dial* Emergency telephone, for example, 999 or 110 or 911 for prompt medical care. Wash stung area with soap and water. If stung on a finger or hand, remove jewelry in case swelling may occur. Apply a cold compress.

Snakes, scorpions, and disease-carrying mosquitoes, desert flies, and ticks are also prevalent during the summer months.

Ticks can carry a wide variety of diseases since they contract the diseases from the host they attach to. Other than medical help, there are over 150 *natural repellents* to protect from ticks. The most common ones are Citronella, Eucalyptus, Lemon Leaves, Peppermint, Lavender, Cedar Oil, Canola, Rosemary, Pennyroyal, and Cajuput. Generally, these are considered safe to use in low dosage, but their effectiveness is limited to 30 minutes.

How can we have fun in summer? Like anything else, moderation is best. Avoid those beehives and hornet nests. Keep wastes, beverages, and food in enclosed containers. Wear clothing that provides protection from ticks and mosquitoes. Wear proper sunscreens. Wear sunglasses to protect your eyes from UV light. If you are sensitive to sunburns, avoid being in the sun when the sun is at its peak. Consume lots of water to stay hydrated. Cool down in air-cooled rooms or near fans. Wear light-colored, natural fiber clothing to help your body to repel heat absorption and cool easier. Avoid strenuous activities.

Control the heat at its source using insulating and reflective barriers on walls and windows of your home and keep blinds or curtains drawn. Keep doors and windows closed. Switch off lights behind you when done; also switch off your car engines radiating heat. Lower the shower temperatures. Reduce physical demands of work using mechanical assistance, for example, instead of lifting manually, use a wheeled trolley. Increase the frequency and length of rest breaks if you are doing hard work. Schedule jobs to cooler times of either early mornings or late evenings. Drink one cup of water every 20 minutes or so. Salt your food well, particularly while acclimatizing to a hot job. Low salt diet people having blood pressure problems should consult with their doctor. Properly get acclimatized to heat. Recognize symptoms of heat stress utilizing a "buddy" system.

3.3.11.4 Loss Prevention during Construction

Eliminate the hazard when possible. This can be extremely simple. For example, using modular components assembled at ground level can eliminate exposure to falls. Other hazards associated with placing conventional reinforcing steel and cast-in-place concrete can be avoided with the use of precast components manufactured in a controlled environment. No other means of control will be as effective as elimination.

Elimination is not always feasible. The next options are substitution and engineering controls. For example, a contractor might substitute scaffolding for ladders to reduce the danger of falls. After elimination and substitution, seek engineering controls to control the hazard at its source. Isolation and guarding such as limiting access to hazardous areas and removing defective equipment from service are examples.

Next in the hierarchy, administrative controls such as a written safety and health plan do not eliminate hazards but build the ability to recognize and avoid them. Though less effective than the previous controls, this aspect of loss prevention should not be neglected. To complement the written plan, employees should receive initial and refresher training in fall protection, electrical safety, lockout/tag out procedures, proper use of PPE, etc. Small firms often utilize a professional safety consultant to develop and implement training. Larger firms might invest in a full-time safety officer for training and troubleshooting. General knowledge training is vital for administrative controls to be effective.

In addition to general knowledge training, employees should receive site-specific training since each site has its own peculiarities and special concerns. Site conditions are constantly changing. These changing conditions and other areas of special emphasis should be addressed in weekly jobsite safety meetings sometimes referred to as toolbox talks.

When other more effective controls are infeasible, the last line of defense is personal protective equipment or PPE. Types of PPE include hard hats, safety glasses, gloves, hearing protection, respirators, etc. These items should be furnished by the employer with strict attention to compliant utilization by all. PPE will go far in eliminating many common injuries.

3.3.12 Financial Management

Construction projects are mainly capital investment projects which have the involvement of three main parties.

1. Owner
2. Designer
3. Contractor

The owner is responsible for financing the project to procure the services of Designer (A/E), Contractor, and other parties and has to compensate for their services for which the Owner has contracted with these parties to construct the facility/project. Other parties may include Project Manager, Construction Manager, Specialists, Material/Equipment suppliers, and many other players as per the procurement strategy adopted by the Owner. The relationship between Owner and these parties is Buyer–Seller relationship and is bound by the contract signed between Owner and these parties. The capital funding required for a construction project mainly include:

1. Land acquisition cost

   • There is not much to manage the finances for land acquisition cost which are normally known as per market conditions.

2. Cost related to license fee, permits, regulatory taxes, insurance, and project-related owner's overhead cost.

   • The cost toward each of these items is fixed and known and does not have major effect or variations to be managed and controlled.

3. Construction Project Cost

   • The overall cost expended during the life cycle of the construction project consists of:
   • Designer (A/E) fee
   • Construction supervisor (Consultant, Project Manager, Construction Manager) fee
   • Construction cost, which normally include supply, installation/execution, testing and commissioning, and maintenance for a period of one or two years (depends on the contract)

The construction cost is generally 80–85% of the construction project cost (construction phase). Balance is the fee for designer and supervision.

Construction project cannot proceed unless adequate funds are arranged by the Owner. These funds are arranged by either public financing or private financing.

Finance management is a process bringing together project planning, scheduling, budget, procurement, accounting, disbursement, control, auditing, reporting, and physical performance (progress) of the project with the aim of managing project resources and achieving owner's objectives. Finance management is an important aspect of project/construction management functions. Project management has a great impact on both owner as well as contractor. Timely and relevant financial information provides a basis for better decisions, thus speeding the progress of the project.

3.3.12.1   Develop Financial Management Plan

Financial plan is;

• Estimating the total funds required to complete the project
• Identifying the source of funds (financial means to cover the project requirement)

- Preparing project cash flow (usage of funds)
- Making sure that the needed finance is available at right time
- Identifying the alternate process of securing the funds to mitigate risk or contingencies

Most construction projects begin with the recognition of new facility or refurbishment/repair of existing facility. In either case, the owner has to arrange the funds to develop/refurbish the facility. Once the need is identified, and feasibility study is conducted, project cost is estimated based on rough order of magnitude. This estimate helps owner for initial commitment, in principle, for project funding and manage the financial resources to build the facility. As the project phases advances to next stage, the level of accuracy is refined. Depending on the type of project delivery system and contract methods, more accurate and a definitive estimate is evolved. At every stage, the funding requirements for construction are reviewed with the earlier stage and adjusted, if needed. The contracted amount is considered as project cost baseline for both Owner and Contractor. However, both owner and contractor have to consider reserves while estimating the financial resources. These reserves, contingency and management, are included in a cost estimate to mitigate cost risk by allowing for future situations that are difficult to predict.

Figure 3.57 illustrates process to develop financial management plan, and Table 3.31 lists the contents of owner's financial plan.

3.3.12.2 Control Finance

In construction projects, fund requirements for the project and the availability of funds are reviewed, compared, managed, and controlled at each stage of the project life cycle. Generally, the initial cost estimates are based on analogous (rough order of magnitude). These estimates are further refined as project development stages progresses, and greater understanding of the project scope, schedule, quality, and resources are evolved. The contract price agreed between Owner and Contractor is the cost the Owner has to pay to the Contractor for the services rendered to construct/build the project/facility as specified in the contract and to complete as per agreed-upon schedule and quality.

The bid price (or negotiated price) is the contracted amount which is paid by the owner to the contractor as per the terms of contract. The contractor bases its expenses and cash flow on the basis of this figure. The bidding price includes:

- Direct cost

  - Material
  - Labor
  - Equipment

- Indirect cost

  - Overhead cost

- Risks and profit

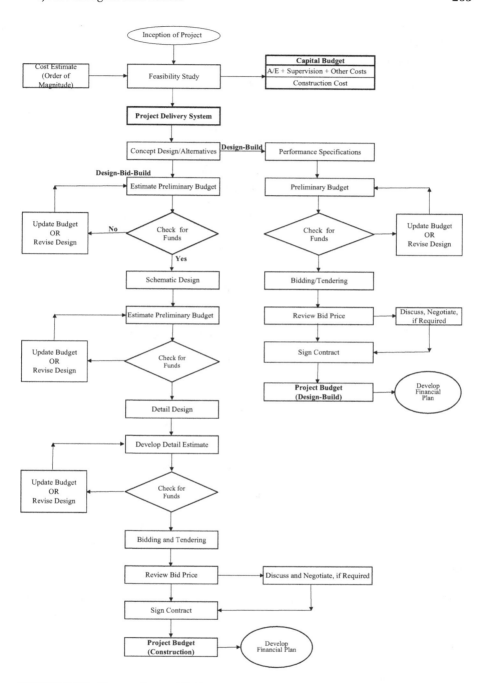

FIGURE 3.57   Process for establishing financial plan.

Source: Abdul Razzak Rumane. (2016). *Handbook of Construction Management.* Reprinted with permission of Taylor & Francis Group.

TABLE 3.31
Contents of Contractor's Financial Plan

Section	Topic
1	Description of project
2	Finance management organization
3	Estimated cost of project
4	Project schedule
5	Type of contracting method
6	Source of funds
7	Payment (receivable) schedule breakdown
	7.1 Advance payment
	7.2 Monthly progress payment
8	Accounts payable schedule breakdown
	8.1 Insurance, bonds
	8.2 Staff salaries
	8.3 Labor payment
	8.4 Material purchases
	8.5 Equipment purchases
	8.6 Equipment rent
	8.7 Sub-contractor payment
	8.8 Consumables
	8.9 Taxes
9	Retention
10	Cash flow
11	Project risk and response strategy
12	Contingency plan to secure funds from alternate source
13	Internal control
	13.1 Accounting system
	13.2 Accounts auditing
	13.3 Track changes in budget
14	Records management
	14.1 Project expenses
	14.2 Finance closure report

The contractor has to plan, manage, and control all the construction activities within the contracted amount.

The owner, however, adds contingency and management reserve to the base cost estimate to mitigate the risks. The major components of a project budget for the owner are;

- Base cost estimate (definitive cost at the end of detail design or bid amount)
- Contingency: It is the amount set aside to allow for responding to identified risks. Sometimes, it is known as known unknown.
- Management reserve: It is the amount set aside to allow for future situations that are unpredictable or could not have been foreseen. This includes changes to the scope of work or unidentified risks.

Monitoring and control of project cost are essential for both parties. Normally S-Curve is used as a reference to track the progress of the project over time. It also helps monitoring cash flow and determines the slippage in the project schedule and cost baseline. Contractor uses cost-loaded schedule for determining the contract payments during the project against the approved progress of works. Contractor can plan the project expenses against the expected payments and has to control the finances toward:

- Labor salaries
- Staff salaries
- Material purchases
- Equipment purchases
- Taxes and other regulatory expenses
- Retention

The owner has to arrange funds to pay progress payments on a regular basis. Funds required to paying the changes to the scope of work or approved claims can be managed from the reserves. The contractor's monthly expenses are different to the actual progress payment expected from the owner. During the initial stage of project execution, the contractor's expenses are much higher than the progress payment. To meet these expenses, the contractor has to arrange funding from other sources. In certain projects, the contractor receives advance payment from the owner. However, it is necessary that contractor plans and controls the finances for smooth and uninterrupted execution of the project. Figure 3.58 is an example expense (accumulated cost) and payment (accumulated net sale) schedule for a project. Plotting of cost versus sale helps the contractor to control the project finances.

The schedule is updated regularly to reflect ongoing changes in the progress of work and forecasted payments. Normally, it takes considerable time between the expenditure on resources and payment to be received from the owner for the work

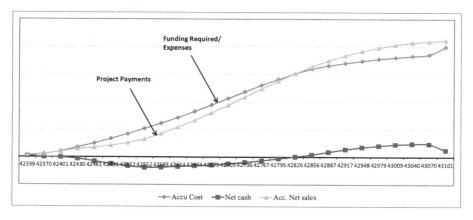

FIGURE 3.58   Contractor's cash flow.

contractor has done and getting approved for payment by the owner. The contractor has to secure funds to cover the cash flow difference.

During execution construction project, the Owner has to consider for payments toward unexpected changes such as:

- More work done (progress of work) than planned
- Claims
- Failure to adjust extra payments
- Default from funding agency

Similarly, the Contractor has to consider the following:

- Slow progress of work resulting in less payments than expected/planned
- Productivity is less than estimated
- Early delivery of material
- Delay in payment by the Owner
- Resources to do extra (additional) work required by the Owner

3.3.12.3   Administer and Record Finance

Project accounting system is required to manage the project needs and expenses and provide the financial information to all the interested stakeholders. Table 3.32 lists different logs maintained by finance department.

TABLE 3.32
Logs Maintained by Finance Department

Section	Log
1	Incoming and outgoing correspondence
2	Progress (interim) payment
3	Subcontractor payment
4	Material purchase
5	Equipment purchase
6	Procurement (general)
7	Procurement (consumables)
8	Letter of credit
9	Freight/transportation charges
10	Custom clearance
11	Equipment rent
12	Vehicle rent
13	Insurance, bonds, guarantees
14	Regulatory, license fee
15	Staff salaries
16	Labor salaries
17	Office rent
18	Camp rent
19	Cash in hand

3.3.13   CLAIM MANAGEMENT

Claim management is the process to mitigate the effects of the claims that occur during the construction process and resolve quickly and effectively. If the claims are not managed effectively, it can lead to disputes ending in litigation. Even under the most ideal circumstances, contract documents cannot provide full information; therefore, claims do occur in the construction projects. In construction projects, claim is defined as seeking adjustment or consideration by one party against other party with respect to:

• Extension of time
• Scope
• Method
• Payment

There are mainly three types of claims. These are:

1. Contractual claims: Claims that fall within the specific clause of the contract.
2. Extra-contractual claims: Claims that result from the breach of contract.
3. Ex-gratia claims: Claims that the contractor believes are his rights on moral ground.

Table 3.33 lists major causes of claims in construction projects.
The claims in construction projects can be attributed to:

1. Owner
2. Contractor

Figure 3.59 illustrates claim management process.

3.3.13.1   Identify Claims

Claims are common part of almost every construction project. With utmost care taken to coordinate all the activities and related contract documents, claims in construction projects may arise under any form of construction contracts. Table 3.34 illustrates effects of claims on construction projects.

Most construction contracts impose time deadlines for submitting claims. Normally, the claims to be made within 30 days after the claim arose and also to be recorded in the monthly progress report which the contractor submit to the owner. Claim identification involves proper interpretation of contract requirements and gathering complete information to substantiate the claim. The identified claim needs to be submitted with all the supportive documents for justification.

3.3.13.2   Prevent Claims

Table 3.35 illustrates preventive actions to mitigate effects of claims on construction projects.

TABLE 3.33
Major Causes of Construction Claims

Serial Number	Causes
I	**Owner Responsible**
I-1	Delay in issuance of notice to proceed
I-2	Delay in making the site available on time
I-3	Different site conditions
I-4	Project objectives are not well defined
I-5	Inadequate specifications
	a) Design errors
	b) Omissions
I-6	Scope of work not well defined
I-7	Conflict between contract documents
I-8	Change/modification of Design
I-9	Change in schedule
I-10	Addition of work
I-11	Omission of work
I-12	Delay in approval of subcontractor
I-13	Delay in approval of materials
I-14	Delay in approval of shop drawings
I-15	Delay in response to contractor's queries
I-16	Delay in payment to contractor
I-17	Lack of coordination among different contractor directly under the control of owner
I-18	Interference and change by the owner
I-19	Delay in owner-supplied material
I-20	Acceleration
II	**Contractor Responsible**
II-1	Delay to meet milestone dates
II-2	Noncompliance with specifications
II-3	Changes in specified process/methodology
II-4	Substitution of material
II-5	Noncompliance to regulatory requirements
II-6	Charges payable to outside party due to the cancellation of certain items/products
II-7	Material not meeting the specifications
II-8	Workmanship not to the mark
II-9	Suspension of work
II-8	Termination of work
III	**Miscellaneous**
III-1	New regulations
III-2	Weather conditions
III-3	Unforeseen circumstances

Source: Abdul Razzak Rumane. (2013). *Quality Tools for Managing Construction Projects.* Reprinted with permission of Taylor & Francis Group.

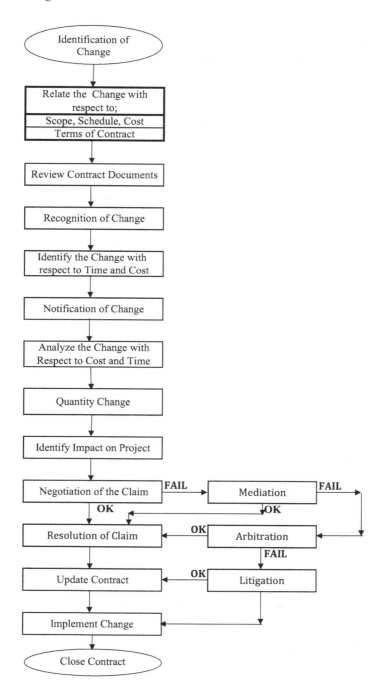

FIGURE 3.59 Claim management process.

Source: Abdul Razzak Rumane. (2016). *Handbook of Construction Management*. Reprinted with permission of Taylor & Francis Group.

TABLE 3.34

Effects of Claims on Construction Projects

Serial Number	Causes	Effects on Project
I	**Owner Responsible**	
I-1	Delay in issuance of notice to proceed	• Delay in completion of project
I-2	Delay in making the site available on time	• Delay in completion of project
I-3	Different site conditions	• Delay in the completion of project • Change in project cost
I-4	Project objectives are not well defined	• Delay in the completion of project • Change in project cost
I-5	Inadequate specifications c) Design errors d) Omissions	• Delay in completion of project • Change in project cost
I-6	Scope of work not well defined	• Delay in completion of project • Changes in project cost
I-7	Conflict between contract documents	• Changes in schedule
I-8	Change/modification of Design	• Change in project cost
I-9	Change in schedule	• Change in project cost
I-10	Addition of work	• Change in project cost • Change in the completion of project
I-11	Omission of work	• Change in project cost
I-12	Delay in the approval of subcontractor	• Delay in completion of project
I-13	Delay in approval of materials	• Delay in completion of project
I-14	Delay in approval of shop drawings	• Delay in completion of project
I-15	Delay in response to contractor's queries	• Delay in completion of project
I-16	Delay in payment to contractor	• Delay in completion of project
I-17	Lack of coordination among different contractors directly under the control of owner	• Changes in schedule
I-18	Interference and change by the owner	• Delay in completion of project • Change in project cost
I-19	Delay in owner-supplied material	• Delay in completion of project
I-20	Acceleration	• Change in project cost
II	**Contractor Responsible**	
II-1	Delay to meet milestone dates	• Delay in completion of project
II-2	Noncompliance with specifications	• Changes in project cost
II-3	Changes in specified process/ methodology	• Changes in project cost • Changes in project schedule
II-4	Substitution of material	• Changes in project cost

Serial Number		Causes	Effects on Project
	II-5	Noncompliance to regulatory requirements	• Changes in project cost
	II-6	Charges payable to outside party due to cancellation of certain items/products	• Changes in project cost
	II-7	Material not meeting the specifications	• Changes in project cost
	II-8	Workmanship not up to the mark	• Changes in project cost
	II-9	Suspension of work	• Changes in project cost
	II-8	Termination of work	• Changes in project cost
III	**Miscellaneous**		
	III-1	New Regulations	• Delay in project completion • Possible changes in cost
	III-2	Weather conditions	• Delay in project completion
	III-3	Unforeseen circumstances	• Delay in project completion

Source: Abdul Razzak Rumane. (2013). *Quality Tools for Managing Construction Projects.* Reprinted with permission of Taylor & Francis Group.

TABLE 3.35
Preventive Actions to Mitigate Effects of Claims on Construction Projects

Serial Number		Causes	Preventive Actions
I	**Owner Responsible**		
	I-1	Delay in issuance of notice to proceed	• Complete all the required documentation, permits before signing of contract
	I-2	Delay in making the site available on time	• Obtain legal documents and title deeds in time
	I-3	Different site conditions	• Conduct pre-design study/survey through specialist consultant/contractor • Collect historical data prior to start of design • Proper site investigations and if any existing utility services under the site • Ensure that the contractor is aware of geographic/geological conditions of the area.
	I-4	Project objectives are not well defined	• Business case to be properly defined • Designer to collect all required data and clarification from the owner
	I-5	Inadequate specifications e) Design errors f) Omissions	• Review of design drawings by the designer • Review of specifications by the designer • Designer to ensure design drawings are reviewed and coordinated

(Continued)

**TABLE 3.35
(Continued)**

Serial Number	Causes	Preventive Actions
		• Designer to ensure clear and complete Bill of Quantities
		• Specifications and documents are properly prepared as per standard format
		• Quality management system during design phase
		• Allow reasonable time to designer to prepare complete and clear drawings and specifications
I-6	Scope of work not well defined	• Construction documents are accurate, and all the items/activities are properly defined
		• Review all the documents
		• Prepare clear and unambiguous documents
I-7	Conflict between contract documents	• Review documents and design drawing and coordinate with all the trades
I-8	Change/modification of design	• Resolve issue without major effects on schedule
I-9	Improper schedule	• Designer to check schedule for all activities and precedence
		• Check for constraints and consistency
I-10	Change in schedule	• Keep to contracted schedule for the completion of project
I-11	Addition of work	• Review scope of work to ensure all the requirements are included
		• Establish proper mechanism to process changes
I-12	Omission of work	• Negotiate and resolve without any impact on the project
I-13	Delay in approval of subcontractor	• Take action within specified time to respond to the transmittals
		• Ensure that review period for transmittals is appropriate
I-14	Delay in approval of materials	• Take action within specified time to respond to the transmittals
		• Ensure that review period for transmittals is appropriate
I-15	Delay in approval of shop drawings	• Take action within specified time to respond to the transmittals
		• Ensure that review period for transmittals is appropriate
I-16	Delay in response to contractor's queries	• Take action within specified time to respond to the transmittals
		• Ensure that review period for transmittals is appropriate
I-17	Delay in payment to contractor	• Arrange funds to pay the progress payments as per specified time in the contract documents
I-18	Lack of coordination among different contractor directly under the control of owner	• Division of works and packages are properly coordinated and clearly identified for each contractor
		• Develop cooperative and problem-solving attitudes for successful completion of project

Serial Number		Causes	Preventive Actions
	I-19	Interference and change by the owner	• Avoid interference and define clearly the roles and responsibilities of each part. • Follow change order procedure specified in the contract documents
	I-20	Delay in owner-supplied material	• Owner to ensure timely delivery of material suitable for installation
	I-21	Acceleration	• Study contractual consequences before the issuance of instruction to accelerate or fast track the activities
II	**Contractor Responsible**		
	II-1	Delay to meet milestone dates	• Check the milestone activities and perform as included in the schedule • Identify constraints to complete milestone activities • Establish strategy to deal with tight schedule
	II-2	Noncompliance with specifications	• Ensure compliance to the specifications • Establish proper quality management system
	II-3	Bill of Quantities not matching with design drawings	• Contractor to check quantities while bidding
	II-4	Changes in specified process/methodology	• Follow specified method • Request for change if recommended by the manufacturer • Submit and get approval on method of statement for installation of works
	II-5	Substitution of material	• Follow relevant sections of contract specifications
	II-6	Noncompliance to regulatory requirements	• Study and keep track of regulatory requirement for the execution of works
	II-7	Charges payable to outside party due to the cancellation of certain items/products	• Settle the matter amicably
	II-8	Material not meeting the specifications	• Submit request for substitution and follow the contract requirements for substitution
	II-9	Workmanship not to the mark	• Use skilled manpower
	II-10	Suspension of work	• Follow contract conditions
	II-11	Termination of work	• Follow contract conditions
III	**Miscellaneous**		
	III-1	New regulations	• Follow change order procedures for such conditions
	III-2	Weather conditions	• Follow change order procedures for such conditions
	III-3	Unforeseen circumstances	• Follow change order procedures for such conditions

Source: Abdul Razzak Rumane. (2013). *Quality Tools for Managing Construction Projects.* Reprinted with permission of Taylor & Francis Group.

TABLE 3.36

Documents Required for the Analysis of Construction Claims

Serial Number	Name of Document
1	Contract documents
	1.1 Tender documents (related to claim, if required)
	1.2 Conditions of contract (general conditions, specific conditions)
	1.3 Particular specifications
2	Drawings
	1.1 Contract drawings
	1.2 Approved shop drawings
3	Construction schedule
	1.1 Contract schedule
	1.2 Approved construction schedule
4	Reports
	4.1 Daily report
	4.2 Monthly report
	4.3 Test reports
	4.4 Noncompliance report
	4.5 Material delivered at site
5	Minutes of meetings
	5.1 Progress meetings
	5.2 Coordination meetings
	5.3 Safety meetings
	5.4 Quality meetings
6	Submittal logs
	6.1 Material, sample
	6.2 Shop drawings
	6.3 Request for information
	6.4 Request for substitution (alternative)
7	Job site instruction
8	Variation order
9	Site progress records
	9.1 Photographs
	9.2 Videos
10	Payment request
11	Checklists
12	Correspondence with regulatory authorities

3.3.13.3 Resolve Claims

All the claims are to be resolved and contract to be closed. The method of resolution depends on the type of claim, size of the claim, severity of the claim, effects, and consequences of the claim on the project. The submitted claims are to be resolved in a justifiable manner. Following are the methods used to resolve the claim:

- Negotiation
- Mediation
- Arbitration
- Litigation

In all these cases, a comprehensive analysis is necessary to come to an amicable solution. Table 3.36 lists documents required for the analysis of construction claims.

4 Risk Management in Quality of Project Processes

4.1 INTRODUCTION

Risk is the likelihood (probability) of occurrence of an undesirable event that will have an impact (positive or negative) on objectives. All projects involve risks. There is always at least some level of uncertainty in a project's outcome. Most projects are complex and involve many participants.

Construction projects are mainly capital investment projects. They are customized and non-repetitive in nature. No two projects are alike. Construction projects being unique and non-repetitive in nature need specific attention to maintain the quality. Construction projects are custom oriented and custom designed, having specific requirements set by the customer/owner to be completed within finite duration and assigned budget.

Quality in construction projects is not only the quality of product and equipment used in the construction facility, but also the total management approach to complete the facility within specified budget and time.

Construction projects comprise a cross section of many different participants. These participants are both influenced by and depend upon each other in addition to "other players" involved in the construction projects.

Traditional construction projects have the involvement of mainly,

1) Owner
2) Designer
3) Contractor

Because of the participation involvement of many players to achieve the quality in construction projects, the construction projects have become more complex and technical, and the relationships and the contractual grouping of those who are involved are also more complex and contractually varied. Therefore, extensive efforts are required to reduce rework and costs associated with time, materials, and engineering.

The quality management of manufactured products is performed by the manufacturer's own team and has control over all the activities of the product life cycle, whereas construction projects have diversity of interaction and relationship between owners, architects/engineers, and contractors, as well as many others.

DOI: 10.1201/9781003245612-4

Quality in construction projects is not only the quality of products and equipment used in the construction. It is also the total management approach to complete the facility/project as per the scope of works to customer/owner satisfaction to be completed within specified schedule and within the budget to meet owner's defined purpose/requirements.

The products used in construction projects are expensive, complex, immovable, and long-lived. Construction projects are unique and non-repetitive in nature, which need specified attention to maintain the quality. Each project has to be designed and built to serve a specific need. Quality in construction projects typically involves ensuring compliance with minimum standards of material and workmanship in order to ensure the performance of the facility according to the design. Quality in a construction project is a cooperative form of doing the business that relies on the talents and capabilities of both labor and management to continually improve quality. The important factor in construction projects is to complete the facility as per the scope of works to customer/owner satisfaction within the specified schedule and as per agreed-upon budget to complete the work to meet the owner's defined purpose.

It is difficult to generalize project life cycle to system life cycle. However, considering that there are innumerable processes that make up the construction process, the technologies and processes, as applied to systems engineering, can also be applied to construction projects.

A systems engineering approach to construction projects helps to understand the entire process of project management and to manage and control its activities at different levels of various phases to ensure timely completion of the project with an economical use of resources to make the construction project most qualitative, competitive, and economical.

PMBOK® Guide-Fifth Edition identifies and describes five Project Management Process Groups required for successful completion of any project. These groups are made up of 13 knowledge areas for construction projects. Each of these having management areas consists of processes, tools, and techniques that are applied during the management of project to ensure the success of the project. Application of project management processes enhances the chance of success of the project.

The construction projects have varying risks. These risks are to be managed throughout the execution of project to prevent unwanted consequences and effects on the project

Risk management applications by identifying, analyzing, and treating of risk in each of these processes will help the organization to achieve the project objectives.

Following sections discuss the identification of potential risks in the major activities of process groups, probable effects on the project, ownership of the risks, and risk treatment, control measures to avoid and mitigate the occurrence of risk(s) in the project.

4.2   RISK MANAGEMENT IN QUALITY OF INITIATING PROCESS GROUP

Risk management in quality of initiating process group consists of mainly following major activities:

1. Identify major activities and processes of initiating process
2. Identify probable risk(s) in the identified activities
3. Communicate and consult with stakeholders
4. Define scope of risk
5. Registration of risk(s)
6. Assign/identify ownership of risk
7. Analyze the risk
8. Evaluate the risk
9. Document the risk
10. Communicate the risk
11. Formulating and selecting risk treatment options
12. Prepare and implement risk treatment options
13. Prepare and implement risk treatment plans
14. Approval of risk treatment plans
15. Monitor risk treatment activities
16. Review risk treatment activities
17. Monitor quality and effectiveness of process design
18. Monitor risk management process
19. Periodic review of risk management process
20. Incorporate and update organization's performance management
21. Report risk management activities
22. Document risk management activities
23. Continual process improvement

Figure 4.1 illustrates a flowchart for risk management in quality of initiating process group.

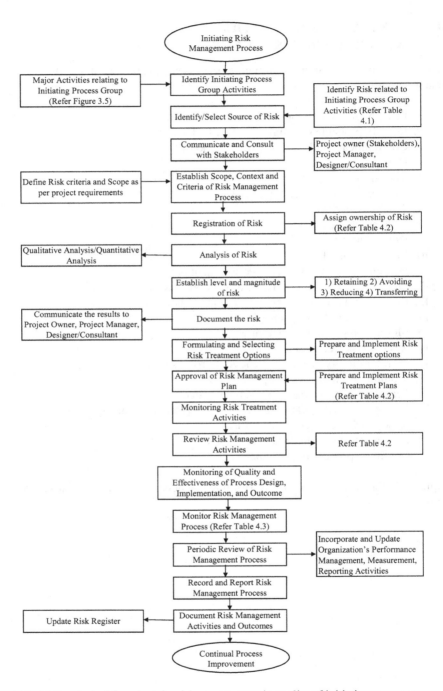

FIGURE 4.1 Typical flowchart for risk management in quality of initiating process group.

4.2.1 Risk in Activities Related to Quality of Initiating Process Group

Table 3.1 discussed in Chapter 3, Section 3.2.1, illustrates major activities relating to initiating process group.

Table 4.1 lists potential risks in the activities related to initiating process group.

TABLE 4.1
Potential Risks in Activities Related to Initiating Process Group

Serial Number	Activity	Major Elements	Potential Risk
1	1.1 Develop project charter		
		1. Project inception	Following points are not considered at project inception stage:
			1. Establishing a justification or business case for the project
			2. Establishing project goals and objectives, scope, and boundary conditions
			3. Estimating overall schedule and cost for the project
			4. Conducting cost/benefit analysis
			5. Completion within budget, that is, to meet the investment plan for the facility
			6. Management strategy/strategic plan as mandated by the management
			7. Regulatory requirements and compliance
			8. Market demand
			9. Customer needs
		2. Problem Statement/ need identification	Following points are not considered for the identification of need:
			1. Establishing properly defined goals and objectives as the most fundamental elements of project planning
			2. The need or opportunity that has triggered the project
			3. Establishment of main goals of the project
			4. How the project solves the current problems or improves the current situation or business growth perspective
			5. What specific benefits the project will deliver?
		3. Need analysis	Need analysis is not performed considering following major points:
			1. Organization's strategy/strategic plan
			2. Regulatory requirements
			3. Market demand
			4. Effect on environment
			5. Need is realistic and genuine
			6. Need to resolve a specific problem
			7. Is need a part of mandatory investment?
			8. Is need a part of Corporate Social Responsibility (CSR)?

(Continued)

**TABLE 4.1
(Continued)**

Serial Number	Activity	Major Elements	Potential Risk
		4. Need statement	Need statement has not considered following points:
			1. Project purpose and need
			2. Why project needed now?
			3. Supporting data for the determination of project
			4. Impact of the need
			5. Environmental impact
			6. Benefits of the project to the performing organization
			7. Hurdles
		5. Need feasibility	The feasibility study is not performed taking into consideration following major points:
			1. Technical viability
			2. Financial viability
			3. Business case
			4. Degree of risk involved
			5. Environmental impact
			6. Social impact
			7. Project details
		6. Project goals and objectives	1. Project goals and objectives are not SMART (Specific, Measurable, Achievable, Realistic, Time bound)
		7. Project deliverables	1. Project deliverables are not properly defined
		8. Design deliverables	1. Design deliverables for design phases are not properly defined and listed
	1.2 Develop preliminary scope statement		
		1. Project Terms of Reference (TOR)	1. TOR does not detail the services to be performed by the designer, service provider (consultant)
			2. TOR does not give the project team (designer) a clear understanding for the development of the project to satisfy owner's/stakeholder's requirements and performance characteristics of the project
		2. Contract documents	1. Contract documents are not developed to meet owner's and stakeholder's needs to the required level of criteria, quality, schedule, and budget
			2. Contract documents are not based on the type of project delivery system
			3. Contract documents are not based on the type of contract/pricing method
			4. Certain specifications are missing and are ambiguous
			5. Design errors/mistakes

Serial Number	Activity	Major Elements	Potential Risk
2	2.1 Identify stakeholders		
		1. Project delivery system	While selecting the project delivery system, the owner has not taken into consideration following points: 1. Size and complexity of the project 2. Type of project 3. Owner's level of project expertise 4. Establishing the scope and responsibilities for how the project is delivered to the owner
		2. Project life cycle	1. Project life cycle phases are not divided considering the innumerable processes that make up the construction process, the technologies, processes and complexity of the project to manage and control the project.
		3. Project team members	1. Proper selection method (Competency Matrix) is not followed to select project team members.

4.2.2 RISK EFFECT AND RISK TREATMENT IN ACTIVITIES RELATED TO QUALITY OF INITIATING PROCESS GROUP

Table 4.2 lists potential risks, probable effect on project, and risk treatment in activities related to initiating process group.

TABLE 4.2
Potential Risks, Probable Effect on Project, and Risk Treatment in Activities Related to Initiating Process Group

Serial Number	Activity	Major Elements	Potential Risk	Probable Effect on Project	Ownership of Risk	Risk Treatment/ Control Measures
1	1.1	Develop project charter				
		1. Project inception	Following points are not considered at project inception stage:			
			1. Establishing a justification or business case for the project	The owner may not get best facility, project	Project owner	Define business case with proper justification for the project

(Continued)

TABLE 4.2
(Continued)

Serial Number	Activity	Major Elements	Potential Risk	Probable Effect on Project	Ownership of Risk	Risk Treatment/ Control Measures
		2. Establishing project goals and objectives, scope, and boundaries	1. Changes in construction project. 2. Project investment plan may change.	Project owner	1. Set performance measures to achieve defined goals and objectives. 2. Establish properly defined goals and objectives, scope, and project boundaries	
		3. Estimating overall schedule and cost for the project	Delay in the completion of project as per schedule and within budget	Project Owner/ Project Manager/ Consultant	Work out and prepare estimated schedule and cost properly and accurately	
		4. Conducting cost/benefit analysis	Cost-effectiveness over the entire life cycle of project	Project Owner Project Manager/ Consultant	Conduct cost–benefit study and analysis	
		5. Completion within budget, that is, to meet the investment plan for the facility	Delay in the completion of project	Project Owner/ Project Manager/ Consultant	Estimate the project delivery within the owner-approved budget	
		6. Management strategy/ strategic plan as mandated by the management	Noncompliance with corporate policy	Project Manager/ Designer	Develop project in line with organization's strategy/strategic plans	
		7. Regulatory requirements and compliance	Delay in getting approval from statutory/ regulatory authority	Project Manager/ Consultant	Establish the need that meets government/ regulatory requirements	

Serial Number	Activity	Major Elements	Potential Risk	Probable Effect on Project	Ownership of Risk	Risk Treatment/ Control Measures
			8. Market demand	The project may not have maximum profit that will meet the investment plan	Project Manager/ Consultant	Perform feasibility study to assess the viability of project
			9. Customer needs			
	2.	Problem statement/ need identification	Following points are not considered:			
			1. Establishing properly defined goals and objectives as the most fundamental elements of project planning	Project team will not be able to develop the project/facility that will satisfy the owner's/end user's requirements and fulfill owner's needs	Project owner	Establish properly defined goals and objectives
			2. The need or opportunity that has triggered the project	The facility may not meet owner's need	Project owner	Establish the need that meets the requirements that has triggered the project
			3. Establishment of main goals of the project	May not be able to achieve fundamental elements of project	Project owner	Establish and define main goals of the project
			4. How the project solves the current problems or improves the current situation or business growth perspective?	The project may not meet the stated project objectives	Project owner/ Consultant	Establish and identify the need of the project, facility that solves the current problems and improves the current situation
			5. What specific benefits the project will deliver?	The project may not meet the stated need and customer satisfaction	Project owner	Identify and establish the project need that meets and satisfy owner's/end user's requirement

(Continued)

TABLE 4.2
(Continued)

Serial Number	Activity	Major Elements	Potential Risk	Probable Effect on Project	Ownership of Risk	Risk Treatment/ Control Measures
		3. Need analysis	Need analysis is not performed considering following major points:			
			1. Organization's strategy/ strategic plan	Nonconformance with corporate policy	Project owner/ Consultant	Analyze the need for organization's strategy/ strategic policy
			2. Regulatory requirements	Project may not be as per government regulation	Project owner/ Consultant	Analyze the need for regulatory requirements
			3. Market demand	Project may not result for best value of investment	Project owner/ Consultant	Analyze properly for market demand
			4. Effect on environment	Project may not comply with environmental aspects into business operations and standards	Project owner/ Consultant	Analyze for health, safety, and environmental requirements as per ISO 14000
			5. Need is realistic and genuine	Project may not have maximum investment benefits	Project owner	Identify and establish the need that is realistic and genuine
			6. Need resolve a specific problem	Project may not meet owner's requirements and satisfaction	Project owner	Establish and identify the need that resolves specific problem
			7. Is need a part of mandatory investment?	Project may contradict company policy	Project owner	Analyze that need is part of corporate mandatory policy
			8. Is need a part of Corporate Social Responsibility (CSR)?	Project may not serve CSR requirements	Project owner	Establish and identify the need that is part of CSR

Serial Number	Activity	Major Elements	Potential Risk	Probable Effect on Project	Ownership of Risk	Risk Treatment/ Control Measures
	4.	Need Statement	Need statement has not considered following points:			
			1. Project purpose and need	The owner's need statement does not define the requirements and objectives	Project owner/ Consultant	Identify and recognize the project purpose properly
			2. Why project needed now?	The owner's need statement does not provide a specifically focused requirement	Project owner/ Consultant	Perform feasibility study
			3. Supporting data for the determination of project	Supporting data is not collected and considered for need statement	Project owner/ Consultant	Collect all the supporting data to develop need statement
			4. Impact of the need	Impact of need not considered	Project owner	Consider the impact of need while developing need statement
			5. Environmental impact	Environmental impact not considered	Project owner	Consider environmental impact of need while developing need statement and comply with environmental protect agency requirements
			6. Benefits of the project to the performing organization	The project may not have best value for financial investment	Project owner	Establish and identify the project that will meet the owner's needs and benefits to the organization's objectives
			7. Hurdles	Political or financial hurdle not considered	Project owner	Consider all the hurdles that may result in negativity to the project needs

(Continued)

TABLE 4.2
(Continued)

Serial Number	Activity	Major Elements	Potential Risk	Probable Effect on Project	Ownership of Risk	Risk Treatment/ Control Measures
		5. Need feasibility	The feasibility study is not performed taking into consideration following major points:			
		1. Technical viability 2. Financial viability 3. Business case 4. Degree of risk involved	The feasibility study has not taken into account relevant technical, financial, economical, moral, social, environmental, and technical constraints that give sufficient information to enable the client to proceed or abort the project.	Project owner/ Consultant	1. Conduct feasibility study to analyze the ability to complete a project successfully, taking into account various factors such as i. Technical ii. Economical iii. Financial iv. Environmental v. Social vi. Time scale etc. and looking into the positive and negative effects of a project before investing the company resources, viz. time and money. 2. Include project details in the feasibility study report	
		5. Environmental impact 6. Social impact				

Serial Number	Activity	Major Elements	Potential Risk	Probable Effect on Project	Ownership of Risk	Risk Treatment/ Control Measures
			7. Project details	The feasibility study does not include project details	Project owner/ Consultant	
	6.	Project goals and objectives	Project goals and objectives are not SMART (Specific, Measurable, Achievable, Realistic, Time bound)	Affect the fundamental elements of project planning	Project owner	Provide clear goals and objectives to the project team with appropriate boundaries to make decisions about the project/facility that will satisfy the owner's/ end user's requirements fulfilling owner's need
	7.	Project deliverables	Project deliverables are not properly defined	May not achieve satisfactory and successful completion of the project	Project owner/ Consultant	1. Defines responsibility/ obligations each of the participants is expected to perform, such as scheduling, cost control, quality management, safety management, risk management, during various phases of construction project life cycle (concept design, schematic (preliminary) design, detailed design, construction, and testing, commissioning and handover).

(Continued)

TABLE 4.2
(Continued)

Serial Number	Activity	Major Elements	Potential Risk	Probable Effect on Project	Ownership of Risk	Risk Treatment/ Control Measures
						2. Establishes scope and responsibility for how the project is delivered to the owner.
	8.	Design deliverables	Design deliverables for design phases are not properly defined and listed	Designer may not develop the required deliverables at different phases of the project	Project owner/ Consultant	Define and list the design deliverables properly in the Terms of Reference (TOR), project charter
	1.2 Develop preliminary scope statement					
	1.	Project Terms of Reference (TOR)	1. TOR does not detail the services to be performed by the designer, service provider (consultant)	Designer may not develop all the required documents and design deliverables	Project owner/ Consultant	Develop TOR taking into consideration owner needs and requirements for satisfactory completion of project. The input from owner about the project goals and objectives to be defined clearly.
			2. TOR does not give the project team (designer) a clear understanding for the development of the project to satisfy owner's/ stakeholder's requirements and performance characteristics of the project	Project design may not fulfill the owner's requirements to develop the project that meets and fully satisfies the owner's need	Project owner/ Consultant	1. Develop TOR taking into consideration project quality requirements, applicable codes and standards, regulatory requirements, environmental requirements for successful completion of the project.

Serial Number	Activity	Major Elements	Potential Risk	Probable Effect on Project	Ownership of Risk	Risk Treatment/ Control Measures
						2. Develop TOR taking into consideration owner's/ stakeholder's needs and requirements for satisfactory completion of project.
	2. Contract documents	1. Contract documents are not developed to meet owner's and stakeholder's needs to the required level of criteria, quality, schedule, and budget	The contractor may not develop the facility that fully satisfies the owner's requirements and meets required project quality	Project manager (design)	1. Develop contract documents considering the input and requirements from all the relevant stakeholders to achieve project goals and objectives 2. Review all the documents, drawings for accuracy, and coordination with all the trades.
		2. Contract documents are not based on the type of project delivery system	1. Bidding price may differ with estimated project cost 2. Scope of project work is not properly defined 3. Request for variations 4. Affect the project quality	Project manager (design)	1. Define scope of work to meet owner's requirements 2. Develop contract document that matches the project delivery system and specifies properly the responsibility/ obligations of each team member

(Continued)

TABLE 4.2
(Continued)

Serial Number	Activity	Major Elements	Potential Risk	Probable Effect on Project	Ownership of Risk	Risk Treatment/ Control Measures
						3. Establish procedures, actions, and sequence of events to be carried out by each team member for successful completion of project
		3. Contract documents are not based on the type of contract/ pricing method	1. Different price estimation 2. Different bidding price	Project manager (design)	Define type of project contract/ pricing such as: 1. Fixed price lumpsum 2. Unit price 3. Cost reimbursement 4. Guaranteed maximum price, etc.
		4. Certain specifications are missing and are ambiguous	1. Frequent Request for Information (RFI) by the contractor 2. Variation orders 3. Delay in project schedule	Project manager (design)	1. Review the documents to identify missing and ambiguous specifications and include all the missing specifications for successful completion of the project to meet owner's intended need

Serial Number	Activity	Major Elements	Potential Risk	Probable Effect on Project	Ownership of Risk	Risk Treatment/ Control Measures
						2. Prepare contract documents in simple and clearly written language that is unambiguous and convenient to understand by all the concerned parties.
			5. Design errors/ mistakes	1. Affect the project quality 2. Variation order	Project manager (design)	Review design drawings for mistake proofing to eliminate the errors such as: 1. Information 2. Mismanagement 3. Omission 4. Selection
2	2.1	Identify stakeholders				
		1. Project delivery system	While selecting the project delivery system, the owner has not taken into consideration following points:			
			1. Size and complexity of the project	May not achieve successful completion of project delivery	Project owner	Consider the following points while selecting an appropriate project delivery system: 1. Size and complexity of the project 2. Type of project 3. Location of project

(Continued)

TABLE 4.2
(Continued)

Serial Number	Activity	Major Elements	Potential Risk	Probable Effect on Project	Ownership of Risk	Risk Treatment/ Control Measures
		2. Type of project	Project delivery system may match the type of project such as: • process type of projects • non-process type of projects	Project owner	4. Owner's level of construction expertise (human resources available with owner and Owner's knowledge of construction management practices) 5. Owner's interest to exert influence/ control over the design 6. Owner's interest to exert influence/ control over the management of planning 7. Owner's interest to exert influence/ control over the management of construction 8. Owner's interest to exert influence/ control over the management of project and the end user(s)

Serial Number	Activity	Major Elements	Potential Risk	Probable Effect on Project	Ownership of Risk	Risk Treatment/ Control Measures
			3. Owner's level of project expertise	Owner may not be able to achieve project delivery to meet the intended need of the project.	Project owner	
			4. Establishing the scope and responsibilities for how the project is delivered to the owner	Owner may not be able to achieve successful project delivery	Project owner	
	2. Project life cycle		1. Project life cycle phases are not divided considering the innumerable processes that make up the construction process, the technologies, processes, and complexity of the project to manage and control the project.	1. Project may not be conveniently monitored and managed 2. Affect the cost of project design and development, production/ construction, system operation and support, system retirement, and material disposal 3. Affect project acquisition time	Project owner	1. Divide project life cycle phases considering the complexity and size of the project to conveniently manage the project 2. Divide each phase of the project to improve the control and planning of the project at every stage before a new phase starts.

(Continued)

TABLE 4.2
(Continued)

Serial Number	Activity	Major Elements	Potential Risk	Probable Effect on Project	Ownership of Risk	Risk Treatment/ Control Measures
		3. Project team members	1. Proper selection method (Competency Matrix) is not followed to select project team members.	1. The team members may not be able to perform their duties for achieving a successful completion of project and desired deliverables	Project owner	Follow proper selection procedure for team members to select qualified members as per the project requirements

4.2.3 RISK MONITORING IN ACTIVITIES RELATED TO QUALITY OF INITIATING PROCESS GROUP

Table 4.3 is an example process of risk monitoring of initiating process group activity.

TABLE 4.3
Risk Monitoring of Initiating Process Group Activity

Serial Number	Description	Action
1	Risk ID	Initiating Process Group -1 (1.2)
2	Description of Risk	1. Project Terms of Reference
		1.1 TOR does not detail the services to be performed by the designer, service provider (consultant)
		1.2 TOR does not give the project team (designer) a clear understanding for the development of the project
3	Response	1.1 Develop TOR taking into consideration owner needs and requirements for a satisfactory completion of project. The input from owner about the project goals and objectives to be defined clearly
		1.2 Develop TOR taking into consideration project quality requirements, applicable codes and standards, regulatory requirements, and environmental requirements for successful project

Serial Number	Description	Action			
4	Strategy of response	Avoidance	Transfer	Mitigation X	Acceptance
5	Monitoring and control	Risk owner Project Owner/ Consultant	Review date D/M/Y	Critical issue Development of TOR specifying design deliverables and other requirements to meet owner's requirements	
6	Estimated impact on the project	Delay in design development by 3 months			
7	Actual impact on the project	3 months' delay			
8	Revised response	Reschedule design development time to adjust delay to prepare revised TOR			
9	Record update date	D/M/Y			
10	Communication to stakeholders	Information sent to all stakeholders on D/M/Y			

4.3 RISK MANAGEMENT IN QUALITY OF PLANNING PROCESS GROUP

Figure 4.2 illustrates a flowchart for risk management in quality of initiating process group.

4.3.1 RISK IN ACTIVITIES RELATED TO QUALITY OF PLANNING PROCESS GROUP

Table 3.2 discussed in Chapter 3, Section 3.2.2 illustrates major activities relating to planning process group.

Table 4.4 lists potential risks in the activities related to planning process group.

4.3.2 RISK EFFECT AND RISK TREATMENT IN ACTIVITIES RELATED TO QUALITY OF PLANNING PROCESS GROUP

Table 4.5 lists potential risks, probable effect on project, and risk treatment in activities related to planning process group

Figure 4.3 illustrates project staffing process, and Figure 4.4 illustrates candidate selection process.

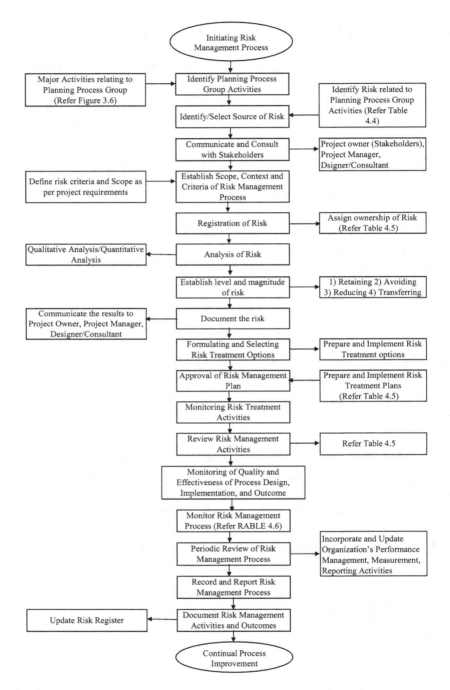

FIGURE 4.2 Typical flow chart for risk management in quality of planning process group.

TABLE 4.4
Potential Risks in Activities Related to Planning Process Group

Serial Number	Activities		Major Elements	Potential Risk
1	1.1	Project baseline plan		
			1. Preliminary plans	1. Project activities for decomposition of work packages are not properly identified
				2. Sequencing of activities and logic dependencies not considered
				3. Project activity duration in which activity is to be performed is not properly considered
				4. Resources required to perform the activity are not properly estimated
				5. Schedule is not analyzed to ensure that the timing of each activity is aligned with the resources
				6. Key milestones missing
2	2.1	Responsibilities Matrix		
			1. Owner, Designer, Contractor, Other Stakeholders	1. The responsibility and needs of stakeholders are not addressed while planning the responsibility matrix
				2. Roles and responsibilities of stakeholders are not planned to consider what information or report is to be communicated to the stakeholders?
	2.2	Stakeholders requirement (work progress)		
			1. Design progress	1. The involvement of stakeholders in approval and review of project activities/ work progress is not considered
			2. Construction progress	
			3. Testing, commissioning, and handover	2. The activities that need reporting project status/ performance are not considered
	2.3	Change reporting		
			1. Updated schedule	Stakeholder engagement to report schedule update, variation, changes in scope, cost variation is not considered
			2. Variation report	
			3. Cost variation	
	2.4	Project updates		
				Responsibility matrix has not considered reporting of project updates
	2.5	Status reports		
			1. Status logs	Responsibility matrix has not considered reporting of status log, performance reports, and issue log
			2. Performance reports	
			3. Issue log	

(Continued)

TABLE 4.4
(Continued)

Serial Number	Activities		Major Elements	Potential Risk
	2.6	Meetings		
			1. Kick-off meeting	Responsibility matrix has not considered
			2. Progress meetings	attendance/reporting of meetings held at different stages of the project
			3. Coordination meetings	
			4. Other meetings	
	2.7	Payments		
			1. Payment status	Responsibility matrix has not considered reporting of payment status
3	3.1	Establish scope baseline plan		The scope does not include the activities to formulate and define the client's needs by establishing project goals and objectives and performance characteristics of the project
	3.2	Collect requirements		
			1. Need statement	The owner's need is not well defined, indicating the minimum requirements of quality and performance, required completion date, and an approved main budget
			2. Project goals and objectives	Lack of input from owner about the project goals and objectives
			3. Project Terms of Reference (TOR)	1. TOR is not developed detailing the scope of works by the designer 2. An accurate and comprehensive TOR is not prepared to give clear understanding for project development
			4. Owner's preferred requirements	Detailed list of owner's requirements related to different trades is not prepared by the consultant
	3.3	Project scope documents		
			1. Design development • Concept design • Schematic design • Detail design	Project documents does not explain activities to formulate and define the client's need by establishing project goals and objectives
			2. Final design	The project scope documents do not explain the boundaries of the project, establish project responsibilities for each team member, and set up procedures for how completed works will be verified and approved

Serial Number	Activities	Major Elements	Potential Risk
		3. Bill of Quantity	Bill of Quantity may not match with that of design drawings and project deliverables
		4. Project specifications	1. Scope of work is not properly established 2. All the related specifications are not included in the project scope documents
		5. Construction documents	1. Documents do not match with project delivery system 2. Documents are not as per the type of contract/pricing methodology 3. Regulatory requirements are not taken into consideration
		6. Project deliverables	Design deliverables are not taken into consideration while preparing the construction documents
3.4	Organizational Breakdown Structure		
		1. Project delivery system	Project delivery system is not selected, considering size and complexity of the project and has not defined each participant's responsibility/obligations during various phases of the project
		2. Organizing	1. The relationship between project team members and stakeholders is not considered 2. Identification and classifications of activities in an organized and structured manner
		3. Staffing	1. Staffing is not done by manning the organizational structure through proper and effective human resources 2. Selection and development of proper people to fill the roles designed into the organization structure 3. Acquisition of human resources is not done properly
		4. Project design	Project design organization considering the responsibilities of various participant is not developed properly
3.5	Work Breakdown Structures		
		1. Project life cycle	Project life cycle phases are not divided based on systems engineering approach and further should be subdivided into the WBS principle to reach a level of complexity where each element/activity can be treated as a single unit that can be conveniently managed.
		2. Work packages	Major deliverables of work elements are not subdivided into a level into small and manageable components (activities)

(Continued)

TABLE 4.4
(Continued)

Serial Number	Activities	Major Elements	Potential Risk
4	4.1 Bill of Quantity		
		1. Quantities' take off	1. Activities to be performed in the project are not properly tabulated and listed and contain the following information related to activities: • Activity name • Activity identification number • Brief description of the activity 2. BOQ activities may not match project-specific activities identified by the drawings and specifications 3. Measurement system may be missing
		2. Sequencing of activities	Relationship and dependency among the activities are not identified and taken into consideration
		3. Estimate activity resources	Different types of resources such as: • Human resources (Manpower) • Equipment • Material required in a certain quantity to perform and complete the activity are not accounted to estimate the activity duration.
		4. Estimate duration of activity	While estimating the duration of activity, following points are not considered: • Subcontractor's work • Availability of particular resources • Availability of funds to perform the activity • Availability of space to perform certain types of activity simultaneously
	4.2 Identify Project Assumption		
		1. Dependencies	Project logics dependencies and relationships direct or indirect are not considered while sequencing of the activities to be performed
		2. Risks and constraints	Schedule is not reviewed and analyzed to make sure the timing of each activity is aligned with resources and to ensure schedule accuracy with the relevant assumptions, constraints, and milestones.
		3. Milestone	Key milestones and high-level project activities are not considered

Serial Number	Activities	Major Elements	Potential Risk
	4.3 Develop baseline schedule		1. Baseline schedule does not meet the project objectives and the scope of work and is accepted by the stakeholders as a benchmark for tracking and measuring the project performance and progress 2. All the activities are not properly identified and listed to meet the scope of work 3. Assumptions and constraints are not sound and true 4. Regulatory requirements are not considered
	4.4 Develop schedule		
		1. Pre-design stage	Inappropriate schedule to complete pre-design stage work
		2. Design development • Concept design • Schematic design • Detail design	1. Impractical schedule to complete design at each phase 2. Appropriate levels of schedule details for planning, scheduling, monitoring, controlling, and reporting on the overall project are not considered
		3. Contract documents	Impractical schedule to complete construction documents phase
		4. Bidding/ tendering and contract award	Impractical schedule to complete bidding/ tendering and contract award phase
		5. Construction phase	Impractical schedule to complete construction phase
		6. Testing, commissioning, and handover	Impractical schedule to complete testing, commissioning, and handover phase
	4.5 Construction schedule		
		1. Contractor's construction schedule	1. Major activities for preparation of schedule are not established 2. Estimated time duration for each activity is not established 3. Proper sequencing of the task is not considered 4. Requirements of resources needed for satisfactory completion of project are not considered 5. Milestone date and constraints are not considered 6. Cost loading is not considered
5	5.1 Estimate cost		
		1. Conceptual estimate	1. Conceptual cost may exceed the cost estimated during the feasibility study. 2. Cost estimates are not refined taking into considerations for the conceptual alternatives as this is required by the owner to determine the capital cost of construction.

(Continued)

**TABLE 4.4
(Continued)**

Serial Number	Activities	Major Elements	Potential Risk
		2. Preliminary estimate	1. The estimated cost is not based on elemental parametric methodology.
			2. The preliminary estimate is not prepared by estimating the cost of activities and resources.
		3. Detail estimate	1. Detailed costing is not done based on work packages, Bill of Quantities (BOQ)
			2. The cost estimate may exceed the owner's capability of financing the project.
		4. Definitive estimate	The cost estimated is not based on detail costing methodology taking into consideration project activities as detailed in BOQ
5.2	Estimate budget		
		1. Prepare budget	The budget may not be of definitive nature taking into consideration project scope, detail drawings, specifications, BOQ, and project schedule
5.3	Determine project cost baseline		
		1. S-Curve	1. S-Curve may not have taken into consideration all the points to exactly predict the amount that will be spent over the established project schedule
			2. S-Curve may not have taken into consideration all the points to measure exactly the project performance and predict the expenses over project duration.
		2. Cost loading	The cost loading is not based on the estimated cost in a time-phased manner to perform all the known activities to establish an authorized baseline
		3. Resource loading	All the required resources (material, manpower, and equipment) are not brought together considering the correct quantity at correct time for each activity for the required period
5.4	Estimate cost		
		1. Estimate project resources cost	Project resources (material, manpower, and equipment) cost is not estimated properly
		2. Estimate project material cost	Project material cost is not estimated properly
		3. Estimate project equipment cost	Project equipment cost is not estimated properly
		4. Bill of Quantities	Estimated cost is not based taking into consideration all the activities mentioned in the BOQ
		5. BOQ price analysis	Price analysis of certain activities from the BOQ is not included

Serial Number	Activities	Major Elements	Potential Risk
	5.5 Contracted project value		
		1. Progress payments	Contractor's progress payment is not based on the checklist-approved works
	5.6 Change order procedure		
		1. Change order	1. Process to resolve contractor-initiated change order is not followed
			2. Process to resolve owner-initiated change order is not followed
		2. Cost variation	1. Proposal for cost variation is not approved by the owner
			2. Cost variation may result in cost overrun of the acceptable limits
6	6.1 Project quality management plan		
		1. Quality codes and standards to be compiled	Relevant quality codes and standards are not considered
		2. Design criteria	Design criteria are not properly defined and established
		3. Design procedure	Design procedure is not established
		4. Quality Matrix (design stage)	Responsibilities matrix for quality at design stages is not established
		5. Well-defined specification	1. Well-defined specifications for all the materials, products, components, and equipment to be used to construct the facility are not established
			2. Inadequate and ambiguous specifications
		6. Detailed construction drawings	1. Conflict with different trades
			2. Conflict with detailed drawings and specifications
		7. Quality Matrix (Construction phase)	Responsibilities matrix for site quality control is not prepared and followed
		8. Construction process	
		9. Detailed work procedures	1. All the works at site are not performed as per approved shop drawings
			2. The execution/installation works is not carried out by the following mentioned procedures:
			• Preparatory phase
			• Start-up phase
			• Execution phase
			3. Method statement for execution/installation of work is not followed
		10. Quality Matrix (Inspection and testing during execution)	Responsibilities matrix for quality control during execution/installation is not prepared and followed properly

(Continued)

TABLE 4.4
(Continued)

Serial Number	Activities	Major Elements	Potential Risk
		11. Defect prevention/ rework	Appropriate preventive actions are not taken to avoid repetition of non-conformance work
		12. Quality Matrix (Testing and handing over—start up)	Responsibilities matrix for quality control during testing and handover is not prepared and followed
		13. Regulatory requirements	Regulatory requirements are not identified, and approval procedure is not established
		14. Quality assurance/ quality control procedures	Major activities to be performed for quality assurance and quality control procedure are not identified
		15. Reporting quality assurance/quality control problems	Quality control organization chart is not prepared
		16. Stakeholders' quality requirements	Stakeholders' quality requirements are not identified, and responsibilities matrix is prepared
7	7.1	Project human resources	
		1. Construction/ project manager	1. Selection procedure is not followed 2. Professional qualification, knowledge, and skills are not considered while selecting the manager
		2. Designer's team	The designer's team is selected not taking into consideration following points: • Demonstrated competence in similar nature and type of project • Professional qualifications of team members • Experience in the type of services required
		3. Supervision team	1. The selected supervision team members are not familiar with their responsibilities to supervise, monitor and control, and implement the procedure specified in the contract documents and ensure successful completion of the project 2. Selected team members not performing as expected
	7.2	Construction resources	
		1. Contractor's core team	Contractor may not be able to engage core team members as per the required qualification and experience for different positions
		2. Construction material	1. Unable to get specified material 2. Failure/delay of material delivery
		3. Construction equipment	1. Low productivity of equipment/plants

Serial Number	Activities	Major Elements	Potential Risk
		4. Construction labor	1. Low productivity of labor
			2. Insufficient skilled workforce
		5. Subcontractor (s)	Selected subcontractor is not competent to execute the subcontracted works
8	8.1 Communication plan		
		1. Communication Matrix	The communication matrix is not providing enough guidelines indicating what information to communicate, which team member initiates, who will receive and take appropriate action, when to communicate, and the method of communication
	8.2 Communication methods		
		1. Design progress	Communication method does not identify and describe internal and external stakeholders involved to receive and take required action
		2. Work progress	
		3. Project issues	
		4. Project variations	
		5. Authorities	
	8.3 Submittal procedures		
		1. Submittal procedure	1. Proper correspondence and reporting method are not established.
		2. Progress payments	2. The stakeholders who have interests and expectations and who influence the project are not identified for receiving the submittals
		3. Progress reports	
		4. Minutes of meetings	
		5. Other meetings	
	8.4 Documents		
		1. Design documents	The detailed procedure for submitting the documents as specified in the contract documents is not established
		2. Contract documents	
		3. Construction documents	
		4. As-built documents	
		5. Authority-approved documents/ drawings	
	8.5 Logs		
		1. Issue log	Contractor/supervisor is not maintaining the required logs
		2. Correspondence with stakeholders	
		3. Correspondence with team members	
		4. Regulatory authorities	

(Continued)

TABLE 4.4
(Continued)

Serial Number	Activities	Major Elements	Potential Risk
9	9.1 Risk Identification		
		1. During inception	Lack of input from the owner about the project goals and objectives
		2. During design	1. Design errors/omissions
			2. Design deliverables not matching the owner's needs and needs for successful completion of project
		3. During bidding	Bid value exceeding the estimated definitive cost
		4. During construction	1. Delay in mobilization
			2. Inappropriate construction method
			3. Scope/design changes
			4. Inappropriate schedule
			5. Availability of resources
			6. Errors and omissions in contract documents
		5. During testing and commissioning	1. Major items for testing and commissioning are not identified
			2. Inspection and test plan is not properly established
		6. During handing over	Delay in acceptance/takeover of the project
	9.2 Managing Risk		
		1. Risk register	Risk register is not maintained and updated regularly
		2. Risk analysis	Qualitative and quantitative analysis is not performed using appropriate tools and techniques
		3. Risk response	Proper risk response is not identified to reduce/mitigate threats to the identified risk
10	10.1 Project Delivery System		
		1. Selection of CM	Following points are not considered while selecting CM:
			1. Thorough knowledge and understanding of construction process
			2. Knowledge of various tools and techniques used in construction project
			3. Skills to oversee and manage complex construction projects
		2. Selection of designer	Following points are not considered while selecting designer:
			1. Experience in similar types and size of projects
			2. Competent team members having design experience
			3. Design production capacity

Serial Number	Activities	Major Elements	Potential Risk
	10.2 Bidding and Tendering		
		1. Pre-qualification of contractors	1. The contractor selection procedure is not properly followed
			2. Bidder selection procedure such as low bid, quality-based is not defined
		2. Issue tender documents	Tendering process is not followed
		3. Acceptance of tender	Review and evaluation of bid procedure are not followed properly
11	11.1 Environmental Compatibility		
	11.2 Safety Management Plan		
		1. Safety consideration in design	Designer has not taken safety considerations
		2. HSE plan for construction site safety	Site safety and HSE requirements are not listed in the contract documents
		3. Emergency evacuation PLAN	Emergency evacuation plan at project site is not mentioned in the contract documents
	11.3 Waste Management Plan		Waste management plan is not included in the contract documents
12	12.1 Financial Planning		
		1. Payments to Designer (Consultant), Construction/Project Manager, Contractor	Owner is not maintaining records for payments to designer, management staff, and contractor's progress payment
		2. Material procurement	Material procurement payments are not paid as per agreed-upon conditions
		3. Equipment procurement	Equipment procurement payments are not paid as per agreed-upon conditions
		4. Project staff salaries	Staff salaries are not paid regularly
		5. Bonds, insurance, guarantees	Bonds, insurance, guarantees are not paid in time
		6. Cash flow	Cash flow chart is not maintained
13	13.1 Claim Identification		
		1. Design errors	Change in scope of work
		2. Additional works	Variation/change order
		3. Delays in payment	Claim by the contractor
	13.2 Claim Quantification		
		1. Change order procedures	Change
			Variation order procedure is not followed properly
		• Cost	
		• Time	

TABLE 4.5

Potential Risks, Probable Effect on Project, and Risk Treatment in Activities Related to Planning Process Group

Serial Number	Activities	Major Elements	Potential Risk	Probable Effect on Project	Ownership of Risk	Risk Treatment/ Control Measures
1	1.1 Project Baseline Plan	1. Preliminary plans	1. Project activities for decomposition of work packages are not properly identified	1. Proper identification of logical relationship among the activities	Project Manager/ Planning and Controlling Manager (PCM)	Prepare the preliminary schedule considering the following:
			2. Sequencing of activities and logic dependencies not considered	2. Affect the estimation of amount of time required to complete all project activities with the available resources		1. Identification of work packages
						2. Sequential order of the planned activities
			3. Project activity duration in which activity is to be performed is not properly considered	3. Errors in resource estimation will affect the successful completion of project		3. Time (duration) required to complete each activity
			4. Resources required to perform the activity are not properly estimated	4. Affect schedule accuracy		4. Resources estimation for each activity as per resource calendar
			5. Schedule is not analyzed to ensure that the timing of each activity is aligned with the resources			5. Review and analyze schedule
			6. Key milestones missing			6. Identify and include key milestones in the schedule (refer Figure 3.30 in Chapter 3)
2	2.1 Responsibilities Matrix	1. Owner, Designer, Contractor, other Stakeholders	1. The responsibility and needs of stakeholders is not addressed while planning the responsibility matrix	1. The stakeholders will not be able to address the needs/expectations and issues of the stakeholders	Project owner (Consultant) Project Manager (Design)	1. Establish responsibilities matrix to address the needs of stakeholders effectively by involving the relevant stakeholders
			2. Roles and responsibility of stakeholders are not planned to consider what information or report to be communicated to the stakeholders	2. The stakeholder will not get information as to what challenges or threats will they have toward the project		2. Identify and plan the responsibilities of stakeholders for communicating and working together to address the needs/ expectations and issues of the stakeholders and include in the matrix

2.2	Stakeholders Requirement (Work Progress)			
	1. Design progress	1. The involvement of stakeholders in approval and review of project activities/ work progress is not considered 2. The activities that need reporting project status/ performance are not considered	1. The stakeholders will not be able to predict how the project progress will be affected 2. The stakeholders will not get information or reports about the status/ performance of the project	Project Manager (Design)	1. Develop stakeholder engagement plan considering the roles and responsibilities of the stakeholder needs, expectations, and influence on the project and predicting how the project will be affected 2. Consider all the activities that need to be addressed
	2. Construction progress				
	3. Testing, commissioning, and handover				
2.3	Change Reporting				
	1. Updated schedule	Stakeholder engagement to report schedule update, variation, changes in scope, and cost is not considered	All the related information and reports will not be properly distributed to the concerned stakeholders	Project Manager (Design)	Prepare stakeholder engagement to ensure all the related information, project status/performance, changes in scope, schedule, and cost variation orders and reports are properly distributed to the concerned stakeholder
	2. Variation report				
	3. Cost variation				
2.4	Project Updates	Responsibility matrix has not considered reporting of project updates	The concerned stakeholders will not be aware of their responsibilities	Project Manager (Design)	Prepare responsibilities matrix that includes reporting of project progress update
2.5	Status Reports				
	1. Status logs	Responsibility matrix has not considered maintaining status log, performance reports, and issue log	Project status will not be known at any required time	Project Manager (Design)	Prepare responsibility matrix that includes stakeholder engagement and distribution of project-related issues, project progress/performance reports, and all the related logs are maintained

(Continued)

TABLE 4.5
(Continued)

Serial Number	Activities	Major Elements	Potential Risk	Probable Effect on Project	Ownership of Risk	Risk Treatment/ Control Measures
		2. Performance reports				
		3. Issue log				
	2.6 Meetings					
		1. Kick-off meeting	Responsibility matrix has not considered attendance/ reporting of meetings held at different stages of the project	Information or reports to be communicated will not be known	Project Owner	Prepare the responsibilities matrix to ensure it includes the distribution of minutes of meetings to the concerned stakeholders
		2. Progress meetings			Project Manager (Design)	
		3. Coordination meetings			Project Manager (Design)	Prepare the responsibilities matrix to ensure it includes the distribution of minutes of meetings to the concerned stakeholders
		4. Other meetings				
	2.7 Payments					
		1. Payment status	Responsibility matrix has not considered reporting of payment status	Stakeholders will not be aware of payment status and required forecasted payments	Project Manager (Design)	Prepare the responsibilities matrix to ensure the distribution/ reporting of payment status to the concerned stakeholders

3	3.1	Establish Scope Baseline Plan	The scope does not include the activities to formulate and define the client's needs, by establishing project goals and objectives and performance characteristics of project	The project team will not have guidelines to for making decisions about change request during the project and remain focused on task	Project Manager (Design)	1. Establish scope baseline that describes the performance capabilities of the project to meet the owner's needs 2. Develop scope baseline considering technical specifications, working drawings, BOQ, and general/ particular conditions
	3.2	Collect Requirements				
		1. Need statement	The owner's need is not well defined, indicating the minimum requirements of quality and performance, required completion date, and an approved main budget	Project will not meet owner's requirements	Project Owner/ Project Manager (Design)	Define the need clearly indicating the project scope and quality performance are properly documented
		2. Project goals and objectives	Lack of input from owner about the project goals and objectives	The project/facility will not satisfy owner's/end users' requirements and fulfil owner's needs	Project Owner/ Project Manager (Design)	Prepare project goals and objectives taking into consideration the final recommendations/ outcome of the feasibility study to provide project team with appropriate boundaries to take decisions about the project and ensure that the project will satisfy owner's requirements
		3. Project Terms of Reference (TOR)	1. TOR is not developed detailing the scope of works by the designer 2. An accurate and comprehensive TOR is not prepared to give a clear understanding for the project development	1. The design will not fully meet the owner's requirements 2. Designer will not have clear understanding for the development of project	Project Owner/ Project Manager (Design)	1. Develop TOR that gives the designer a clear understanding for the development of project. 2. Develop an accurate and comprehensive TOR detailing the scope of works (refer Figure 3.10 in Chapter 3 for development of TOR)

(Continued)

TABLE 4.5
(Continued)

Serial Number	Activities	Major Elements	Potential Risk	Probable Effect on Project	Ownership of Risk	Risk Treatment/ Control Measures
		4. Owner's preferred requirements	Detailed list of owner's requirements related to different trades is not prepared by the consultant	The project/facility will not fully meet the owner's requirements	Project Manager (Design)	Designer to collect owner's preferred requirements and other details, TOR requirements to develop the project design
	3.3 Project Scope Documents					
		1. Design development • concept design • Schematic design • Detail design	Project documents do not explain activities to formulate and define the client's need, by establishing project goals and objectives	The developed design is not accurate to the possible extent, free of errors, and minimum omissions, thus resulting in the project/facility not meeting owner's requirements	Project Manager (Design)	Develop project scope documents taking into considerations TOR, owner's goals and objectives, project delivery system, and other related documents (please refer Figure 3.11 in Chapter 3)
		2. Final design	The project scope documents do not explain the boundaries of the project, establish project responsibilities for each team member, and set up procedures for how completed works will be verified and approved	The size, shape, levels, performance characteristics, and technical requirements of all the individual components not meeting the owner's needs and requirements	Project Manager (Design)	Develop project scope documents explaining project boundaries, project performance requirements, constructability, functionality and usage of project, responsibilities of each project team to complete the project meeting the specifications and satisfy the owner's needs
		3. Bill of Quantity	Bill of Quantity may not match with that of design drawings and project deliverables	1. Errors in the estimate of project cost/budget 2. Errors in bidding price by the contractor	Project Manager (Design)	Prepare BOQ based on detail design and specifications to estimate detailed costing of the project

	Risk	Responsibility	Impact	Action
4. Project specifications	1. Scope of work is not properly established 2. All the related specifications are not included in the project scope documents	Project Manager (Design)	1. It may result in unnecessary disruption of work that will impact on schedule and cost 2. It will affect the quality of the project	1. Define scope of work properly to prepare project specifications for performance quality, acceptable to be used by the contractor as a measure of quality compliance during construction process 2. Include all the related specifications in the scope documents
5. Construction documents	1. Documents do not match with project delivery system 2. Documents not as per the type of contract/pricing methodology 3. Regulatory requirements are not taken into consideration	Project Manager (Design)	1. Team members will not be able to perform to achieve project as per owner's expectations and needs 2. Errors in cost estimation of the project 3. Approval of regulatory authorities	Prepare construction documents based on: 1. Project delivery system, 2. Type of contracting/pricing, 3. Regulatory requirements to meet owner's needs and successful completion of project
6. Project deliverables	Design deliverables are not taken into consideration while preparing the construction documents	Project Manager (Design)	Bidding price will differ the estimated project cost	Include all the related deliverables required for establishing project scope documents in the TOR

3.4 Organizational Breakdown Structure

	Risk	Responsibility	Impact	Action
1. Project delivery system	1. Project delivery system is not selected considering the size and complexity of the project and has not defined each participant's responsibility/obligations during various phases of the project	Project Owner/ Consultant	The owner will not be able to achieve satisfactory completion of project from inception to the occupancy/ handover	Select project delivery system taking into consideration the following: 1. Size and complexity of the project 2. Type of project (process, non-process) 3. Owner's level of expertise in execution of project 4. Owner's interest to exert influence/ control over the project at different stages of the project

(Continued)

TABLE 4.5
(Continued)

Serial Number	Activities Major Elements	Potential Risk	Probable Effect on Project	Ownership of Risk	Risk Treatment/ Control Measures
	2. Organizing	1. The relationship between project team members and stakeholders is not considered	1. The team members will not know the required activities to be performed to achieve a specific goal or a set of goal and objectives	Project Owner/ Project Manager	1. Develop organizational breakdown structure of the project team members for managing the designated/ assigned scope of work
		2. Identification and classifications of activities in an organized and structured manner	2. The team members will not know the assigned tasks to put plans into action		2. Define roles and responsibilities of all team members
	3. Staffing	1. Staffing process is not established by manning the organizational structure through proper and effective human resources	1. Incompetent human resources	Project Owner/ Project Manager (Design)	Follow project staffing process properly (refer Figure 4.3)
		2. Selection and development of proper people to fill the roles designed into the organization structure	2. Selected team members will not be able to perform assigned roles and responsibilities		
		3. Acquisition of human resources is not done properly	3. Selected team members may not be suitable for the specific job for which he/ she is selected		
	4. Project design	Project design organization considering the responsibilities of various participants is not developed properly	The project design deliverable will differ or have errors with the TOR requirements	Project Manager (Design)	Follow candidate selection procedure properly (refer Figure 4.4)

3.5 Work Breakdown Structures

1. Project life cycle	Project life cycle phases are not divided based on systems engineering approach and further be subdivided into the WBS principle to reach a level of complexity where each element/ activity can be treated as a single unit that can be conveniently managed	Easy understanding the entire process of project management and conveniently managing and controlling the activities at different levels of various phases	Project Owner/ Consultant	Divide project life cycle, into number of phases, developed on systems engineering approach, that will help conveniently manage and control the project at each phase/stage and sub\ divide each phase on work breakdown principle
2. Work packages	Major deliverables of work elements are not subdivided to a level into small and manageable components (activities)	1. Difficult to conveniently manage the components/ elements/activities 2. Each activity will not be of specific purpose	Project Owner/ Consultant/ Project Manager (Design)	Divide major deliverables to a level of details appropriate to meet the managing and controlling requirements of the specific activity to be performed in the project

4 4.1 Bill of Quantity

1. Quantities' take off	1. Activities to be performed in the project are not properly tabulated and listed and contain the following information related to activity: • Activity name • Activity identification number • Brief description of the activity 2. BOQ activities may not match project- specific activities identified by the drawings and specifications 3. Measurement system may be missing	1. Errors in estimated definitive cost 2. Errors in calculation of bidding price 3. Amendment to construction documents, tender documents	Project Manager/ Quantity Surveyor (Design)	Include all the related information and activities identified by the drawings and specifications in the BOQ tabulation
2. Sequencing of activities	Relationship and dependency among the activities are not identified and taken into consideration	Errors in sequence of work progress from start till completion of the project	Project Manager/ PCM (Design)	Consider project logic relationship and dependency among all the project activities while sequencing the activities to be performed

(Continued)

TABLE 4.5
(Continued)

Serial Number	Activities	Major Elements	Potential Risk	Probable Effect on Project	Ownership of Risk	Risk Treatment/ Control Measures
		3. Estimate activity resources	Different types of resources such as: • Human resources (manpower) • Equipment • Materials required in a certain quantity to perform and complete the activity are not accounted to estimate the activity duration.	1. Improper construction schedule 2. Information about utilization of resources to perform the assigned activity/task will not be available	Project Manager/ PCM (Design)	Estimate activity resources based on resource productivity, related experience, availability of the resources, availability of exact number of resources, and availability of fund to carry out a particular activity/job
		4. Estimate duration of activity	While estimating the duration of activity, following points are not considered: • Subcontractor's work • Availability of particular resources • Availability of funds to perform the activity • Availability of space to perform certain types of activity simultaneously	Schedule preparation will be affected due to nonavailability of approximate time, duration required to complete each activity	Project Manager/ PCM (Design)	Estimate activity duration between the start and finish of the activity/task to approximate the amount of time required to complete project activities in a sequential manner
4.2	Identify Project Assumption					
		1. Dependencies	Project logics dependencies and relationships direct or indirect are not considered while sequencing of the activities is to be performed	Network diagram will not represent activities with proper interrelationship and to know exact work progress from start till completion of the project	Project Manager/ PCM (Design)	Identify direct or indirect relationship and dependency among project activities to develop network diagram
		2. Risks and constraints	Schedule is not reviewed and analyzed to make sure the timing of each activity is aligned with resources and to ensure schedule accuracy with the relevant assumptions, constraints, and milestones	Schedule will not have accuracy	Project Manager/ PCM (Design)	Review and analyze the basic schedule to ensure that the timing of each activity is aligned with resources and to ensure schedule accuracy with the relevant assumptions, constraints, and milestones

	Cause	Risk	Responsibility	Action
3. Milestone	Key milestones and high-level project activities are not considered	Identification of specific point(s) having significant importance in the project schedule	Project Manager/ PCM (Design)	Identify key milestones and high-level activities having significant importance in the project schedule in order to monitor the progress/status of the project
4.3 Develop Baseline Schedule	1. Baseline schedule does not meet the project objectives and the scope of work and is accepted by the stakeholders as a benchmark for tracking and measuring the project performance and progress 2. All the activities are not properly identified and listed to meet the scope of work 3. Assumptions and constraints are not sound and true 4. Regulatory requirements are not considered	1. Actual performance (status) of the project will be difficult 2. Measuring project performance activities will be difficult 3. Schedule will be improper 4. Difficult to know regulatory requirements and its impact at various stages	Project Manager/ PCM (Design)	1. Develop baseline schedule to meet the project objectives and all the scope of work, accepted by the stakeholders as a benchmark for tracking and measuring the project performance and progress 2. Identify and include all the activities in the baseline schedule to assess actual performance of the project 3. Identify and consider the assumptions and constraints that are accurate and true 4. Determine and include the variance and forecast to complete the project 5. Consider regulatory requirements
4.4 Develop Schedule 1. Pre-design stage	Inappropriate schedule to complete pre-design stage work	Design will not be completed as per agreed-upon time	Project Manager/ PCM (Design)	Develop schedule to ensure that all the related activities/tasks can be performed smoothly during this stage
2. Design development • Concept design • Schematic design • Detail design	1. Impractical schedule to complete design at each phase 2. Appropriate levels of schedule details for planning, scheduling, monitoring, controlling, and reporting on the overall project are not considered	1. Design completion as per agreed-upon time and TOR requirements 2. Design progress at various stages/phases will not be known	Project Manager/ PCM (Design)	1. Develop schedule to perform the works during each stage/phase in an organized and structure manner to accomplish all the key activities/tasks and to achieve overall project performance activities 2. Develop schedule suitable for monitoring and controlling of activities during all the stages/phases of project

(Continued)

TABLE 4.5
(Continued)

Serial Number	Activities	Major Elements	Potential Risk	Probable Effect on Project	Ownership of Risk	Risk Treatment/ Control Measures
3.		Contract documents	Impractical schedule to complete construction documents phase	Delay in completion of construction documents	Project Manager/ Contract Administrator (Design)	Prepare work schedule to ensure it meets the time required for preparation and completion of contract documents
4.		Bidding/ tendering and contract award	Impractical schedule to complete bidding/tendering and contract award phase	Delay in completion of bidding and tendering phase requirements	Project Manager/ PCM (Design)	Prepare work schedule to ensure it meets the time required for the completion of bidding and tendering phase activities
5.		Construction phase	Impractical schedule to complete construction phase	1. Delay in completion of construction phase activities 2. Delay in start of the testing and commissioning of the project	Project Manager/ PCM (Design)	Establish, prepare construction schedule to ensure that the periods for completion for all the sections of the works and activities will be accomplished to achieve project objectives and works of the project are carried out in an organized and structured manner
6.		Testing, commissioning, testing, commissioning, and handover and handover phase	Impractical schedule to complete	Delay in handover/ acceptance of the project	Project Manager/ PCM (Design)	Establish testing, commissioning, and handover schedule to ensure all the activities are performed as per agreed-upon plan

4.5 Construction Schedule

Item	Cause	Risk	Responsible	Action
1. Contractor's construction schedule	1. Major activities for preparation of schedule are not established 2. Estimated time duration for each activity is not established 3. Proper sequencing of the task is not considered 4. Requirements of resources needed for satisfactory completion of project are not considered 5. Milestone date and constraints are not considered 6. Cost loading is not considered	1. Monitoring of project progress and status will be difficult 2. Tracking and measuring of project performance and progress	Project Manager/ PCM (Design)	Establish contractor's construction schedule considering: 1. All the major activities 2. Time duration of each activity 3. Sequencing and relationship between project activities and their dependency and precedence 4. Resources and time duration for each resource 5. Milestone and constraints 6. Cost loading of each activity, resource (refer Figure 3.30)

5 Estimate Cost

5.1

Item	Cause	Risk	Responsible	Action
1. Conceptual estimate	1. Conceptual cost may exceed the cost estimated during the feasibility study 2. Cost estimate is not refined taking into considerations the conceptual alternatives as this is required by the owner to determine the capital cost of construction	1. Inaccurate definitive cost estimation 2. Funding requirements for the project	Project Manager/ PCM/QS (Design)	1. Use analogous/ parametric tools/ methodology to determine conceptual cost estimate during conceptual design phase 2. Update cost estimate for conceptual alternatives 3. Estimate cost considering all the requirements listed in TOR
2. Preliminary estimate	1. The estimated cost is not based on elemental parametric methodology. 2. The preliminary estimate is not prepared by estimating the cost of activities and resources	Preparation of accurate definitive cost estimation	Project Manager/ PCM/QS (Design)	1. Develop preliminary cost estimate using elemental parametric tools/ methodology to determine preliminary cost estimate during preliminary design phase 2. Consider cost of activities and resources for estimating preliminary cost estimate

(Continued)

TABLE 4.5
(Continued)

Serial Number	Activities	Major Elements	Potential Risk	Probable Effect on Project	Ownership of Risk	Risk Treatment/ Control Measures
		3. Detail estimate	1. Detailed costing is not done based on work packages, Bill of Quantities (BOQ) 2. The cost estimate may exceed the owner's capability of financing the project	Preparation of accurate definitive cost estimation	Project Manager/ PCM/QS (Design)	1. Determine detail cost estimate using elemental parametric/detail costing tools/ methodology and BOQ items during detail design phase 2. Match the cost estimate with owner's budget
		4. Definitive estimate	The cost estimated is not based on detail costing methodology taking into consideration project activities as detailed in BOQ	Inaccurate owner budget toward the project	Project Manager/ PCM/QS (Design)	Prepare definitive cost estimate using detailed costing methodology and BOQ activities
	5.2 Estimate Budget					
		1. Prepare budget	The budget may not be of definitive nature taking into consideration project scope, detail drawings, specifications, BOQ, and project schedule	Owner will not know the exact funds for the project	Project Manager/ PCM/QS (Design)	1. Prepare project budget aggregating in a time-phased manner to establish an authorized baseline to finance performance, funding all the known activities to achieve the intended project objectives 2. Prepare budget taking into considerations BOQ and all the related parameters 3. Prepare budget taking into consideration resource calendar 4. Set contingency

5.3 Determine Project Cost Baseline

Task	Risk	Impact	Responsibility	Action
1. S-Curve	1. S-Curve may not have taken into consideration all the points to exactly predict the amount that will be spent over the established project schedule 2. S-Curve may not have taken into consideration all the points to measure exactly the project performance and predict the expenses over project duration	Project financing during construction phase	Project Manager/ PCM/QS (Design)	1. Develop cost baseline based on the construction budget to be used as basis against which the overall cost performance of the project is measured, monitored, and controlled 2. Consider all the points, parameters to develop S-Curve 3. Determine basis of cost baseline
2. Cost loading	The cost loading is not based in a time-phased manner on the estimated cost to perform all the known activities to establish an authorized baseline	Prediction of the amount that will be spent over the established project schedule	Project Manager/ PCM/QS (Design)	Load each project activity in the approved schedule with the budget cost of the activity in the BOQ in a time-phased manner for the prediction of appropriate level of funding for smooth execution of construction activities
3. Resource loading	All the required resources (material, manpower, and equipment) are not brought together considering the correct quantity at correct time for each activity for the required period	Delay in execution of project due to lack of resources in time	Project Manager/ PCM/QS (Design)	Prepare the plan considering all the resources fully coordinated and brought together to complete the project on time and within budget

5.4 Estimate Cost

Task	Risk	Impact	Responsibility	Action
1. Estimate project resources cost	Project resources' (material, manpower, and equipment) cost is not estimated properly	Project financing during project execution will be different than estimated	Project Manager/ PCM/QS (Design)	Estimate project resources cost properly taking into consideration all the related factors
2. Estimate project material cost	Project material cost is not estimated properly	Project financing during project execution will be different than estimated	Project Manager/ PCM/QS (Design)	Estimate project material cost properly taking into consideration all the related factors

(Continued)

TABLE 4.5
(Continued)

Serial Number	Activities	Major Elements	Potential Risk	Probable Effect on Project	Ownership of Risk	Risk Treatment/ Control Measures
		3. Estimate project equipment cost	Project equipment cost is not estimated properly	Project financing during project execution will be different than estimated	Project Manager/ PCM/QS (Design)	Estimate project equipment cost properly taking into consideration all the related factors
		4. Bill of Quantities	Estimated cost is not based taking into consideration all the activities mentioned in the BOQ	Project financing during project execution will be different than estimated	Project Manager/ PCM/QS (Design)	Estimate definitive cost taking into consideration BOQ
		5. BOQ price analysis	Price Analysis of certain activities from the BOQ is not included	Project financing during project execution will be different than estimated	Project Manager/ PCM/QS (Design)	Identify important activities listed in BOQ for price analysis (refer Figure 3.38)
5.5	Contracted Project Value					
		1. Progress payments	Contractor's progress payment procedure is not specified in the contract documents	1. Delay in approval of progress payment 2. Progress payment to the contractor will defer	Project Manager/ PCM/QS (Design)	Establish progress payment procedure (refer Figure 4.5)
5.6	Change Order Procedure					
		1. Change order	1. Process to resolve contractor-initiated change order is not followed 2. Process to resolve owner-initiated change order is not followed	Delay in the approval of change order	Project Owner/ Project Manager (Design)	Specify the process to resolve contractor-initiated scope change in the contract documents (refer Figure 3.22 in Chapter 3)
		2. Cost variation	1. Proposal for cost variation is not approved by the owner 2. Cost variation may result in cost overrun of the acceptable limits	1. Delay in resolving the variation issue 2. Additional finance for successful completion of project	Project Owner/R.E./ QS	Specify the process for cost variation in the contract documents (refer Figure 3.24 in Chapter 3 and Figure 4.6)

6 6.1 Project Quality Management Plan

Item	Risk event	Impact	Responsibility	Action
1. Quality codes and standards to be compiled	Relevant quality codes and standards are not considered	Project quality may not meet as per expected codes and standards	Project Manager/Quality Manager (Design)	Identify all the relevant codes and standards to be complied and apply the same while developing the project quality requirements
2. Design criteria	Design criteria is not properly defined and established	Lack of accuracy in the design	Project Manager/Quality Manager (Design)	Identify and establish design criteria to develop design to meet owner's requirements
3. Design procedure	Design procedure is not established	Delay in completion of design	Project Manager (Design)	Establish design procedure considering organizational policy, base design, design review, and coordination method
4. Quality matrix (design stage)	Responsibilities matrix for quality at design stages is not established	Stakeholders not aware of their roles and responsibilities	Project Manager (Design)	Establish responsibilities matrix for quality-management-related personnel
5. Well-defined specification	1. Well-defined specifications for all the materials, products, components, and equipment to be used to construct the facility are not established 2. Inadequate and ambiguous specifications	Project may not meet the owner's requirements	Project Manager (Design)	Prepare the specifications taking into consideration well-defined specification for the materials, products, components, and equipment to be used to construct the project to meet the TOR, owner's requirements
6. Detailed construction drawings	1. Conflict with different trades 2. Conflict with detailed drawings and specifications	Delay in execution of project	Project Manager (Design)	Review the design to resolve interdisciplinary conflict among 1. different trades and 2. drawings
7. Quality Matrix (Construction Phase)	Responsibilities matrix for site quality control is not prepared and followed	1. Relevant stakeholders not aware of their roles and responsibilities 2. Project quality will be affected	Project Manager (Design)	Establish responsibilities matrix specifying responsibilities for site quality control

(Continued)

TABLE 4.5
(Continued)

Serial Number	Activities	Major Elements	Potential Risk	Probable Effect on Project	Ownership of Risk	Risk Treatment/ Control Measures
8.		Construction process	Sequence of construction activities not being followed	Delay in the completion of project	Project Manager (Design)	Establish construction procedure to ensure that works are performed as per approved shop drawings, sequence of work activities, method statement, and approved quality control plan (refer Figure 4.7)
9.		Detailed work procedures	1. All the works at site are not performed as per approved shop drawings 2. The execution/installation works are not carried out by following procedures: • Preparatory phase • Start-up phase • Execution phase 3. Method statement for execution/ installation of work is not followed	Delay in the execution of project activities	Project Manager (Design)	1. Specify to perform/execute the works at site as per approved shop drawings and approved material to achieve project performance 2. Specify applying three phases of control such as preliminary, start-up, and execution of each activity as specified under project quality management requirements
10.		Quality matrix (inspection, testing during execution)	Responsibilities matrix for quality control during execution, installation is not prepared and followed	Team members will not know their roles and responsibilities	Project Manager (Design)	Establish and specify responsibilities matrix for inspection and testing of executed/installed works
11.		Defect prevention/ rework	Appropriate preventive actions are not taken to avoid repetition of nonconformance of work	1. Repetitive rework 2. Delay in project progress	Project Manager (Design)	Specify procedure for taking appropriate preventive action to avoid repetition of nonconformance of work
12.		Quality matrix (testing and handing over—start up)	Responsibilities matrix for quality control during testing and handover is not prepared and followed	Team members will not know their roles and responsibilities	Project Manager (Design)	Establish and specify responsibilities matrix for quality control during testing, commissioning, and handover phase

13. Regulatory requirements	Regulatory requirements are not identified, and approval procedure is not established	Delay in project completion	Project Manager (Design)	Specify the regulatory requirements approval procedure while establishing quality criteria
14. Quality assurance/ quality control procedures	Major activities to be performed for quality assurance and quality control procedure are not identified	Quality of the project	Project Manager (Design)	Develop quality management plan by identifying and including all the activities in the quality assurance/ quality control procedure to validate project performance requirements
15. Reporting quality assurance/ quality control problems	Quality control organization chart is not prepared	Responsibilities of project team members	Project Manager (Design)	Includes reporting of quality-related activities in the quality matrix
16. Stakeholders quality requirements	Stakeholders' quality requirements are not identified, and responsibilities matrix is prepared	Team members will not know their roles and responsibilities	Project Manager (Design)	Identify stakeholders' quality control requirements and include in the quality plan, responsibilities matrix
7 Project Human Resources				
7.1				
1. Construction/ Project Manager	Nonavailability of project staff as per the professional qualification, knowledge, and skills specified in the contract documents	1. Incompetent staff 2. Selected staff will not be able to perform assigned work within the given time frame	Project Manager (Design)	1. Specify qualification and skills requirements to engage construction/ project manager 2. Search and select the project staff having technical experience, skills in the similar type of project, and project-specific requirements to perform achieving successful completion of project (refer Figure 4.4)

(Continued)

TABLE 4.5
(Continued)

Serial Number	Activities	Major Elements	Potential Risk	Probable Effect on Project	Ownership of Risk	Risk Treatment/ Control Measures
		2. Designer's team	Nonavailability of designer's team having: • Demonstrated competence in similar nature and type of project • Professional qualifications of team members • Experience in the type of services required	1. Incompetent design team 2. Errors in design of project 3. Project quality not meeting as per owner's requirements	Project Manager (Design)	Search and select the designer team having technical experience in the similar type of project and technical skills and capabilities in relevant areas to perform achieving project design that meets owner's requirements, specifications (refer Figure 4.4)
		3. Supervision team	1. Difficult to select and approve supervision team members are not familiar with their responsibilities to supervise, monitor, and control, and implement the procedure specified in the contract documents and ensure successful completion of the project 2. Selected team members not performing as expected	1. Incompetent supervision team 2. Quality of execution work	Project Manager (Supervision)	1. Search and select the supervision team members having technical experience in the similar type of project and technical skills and capabilities to supervise similar types and size of projects 2. Provide training to selected members (refer Figure 4.4)
7.2	Construction Resources					
		1. Contractor's core team	Nonavailability of core team members as per the required qualification and experience for different positions	Proper execution of project works	Project Manager (Contractor)	1. Search for qualified team members to perform and achieve project objective successfully 2. Select contractor's core staff as per the qualifications mentioned in the contract documents
		2. Construction material	1. Unable to get specified material 2. Failure/delay of material delivery	Delay in getting approval of alternate material	Project Manager (Design)	1. Make extensive search to get specified material 2. Submit approved equal material as per the substitute clause in case nonavailability of specified material

3. Construction equipment	1. Low productivity of equipment/plants	Delay in project execution	Project Manager (Design)	1. Make extensive search to get specified equipment 2. Submit approved equal material as per the substitute clause in case nonavailability of specified material
4. Construction labor	1. Low productivity of labor 2. Insufficient skilled workforce	Delay in project execution	Project Manager (Construction)	1. Search competent labor 2. Select labors having skills, qualifications to perform the assigned work
5. Subcontractor(s)	Selected subcontractor is not competent to execute the subcontracted works	1. Delay in execution of works 2. Affect workmanship	Project Manager (Construction)	1. Declare names of the subcontractor(s) in the tender documents having the expertise in the related field, able to perform the designated works 2. Monitor and supervise regularly the workmanship and work progress of subcontractor
			Project Owner/ Project Manager (Design)	Project Owner to approve qualified subcontractor(s)
8 8.1 Communication Plan				
1. Communication matrix	The communication matrix is not providing enough guidelines indicating what information to communicate, which team member initiates, who will receive and take appropriate action, when to communicate, and the method of communication	Smooth flow of communication among all the relevant stakeholders will be affected	Project Manager (Design)	Establish communication matrix to address the needs of project stakeholders and their roles and responsibilities on the project and communicate the required information or reports

(Continued)

TABLE 4.5
(Continued)

Serial Number	Activities	Major Elements	Potential Risk	Probable Effect on Project	Ownership of Risk	Risk Treatment/ Control Measures
	8.2	Communication Methods				
		1. Design progress	Communication method does not identify and describe internal and external stakeholders involved to receive and take required action	Difficulties to effectively manage the interest, expectations, and influence of stakeholders to ensure the completion of a successful project	Project Manager (Design)	Establish communication matrix for smooth flow of communication among related stakeholders during design phases/stages
		2. Work Progress			Project Manager (Design)	Establish communication matrix for smooth flow of communication during project execution
		3. Project Issues				
		4. Project Variations				
		5. Authorities				
	8.3	Submittal Procedures	1. Proper correspondence and reporting method is not established.	1. Difficulties in understanding of product specifications, contract drawings, and installation methods	Project Manager (Design)	Establish matrix for administration and communication considering the following:
		1. Submittal procedure	2. The stakeholders who have interests and expectations, and who influence the project are not identified for receiving the submittals	2. Unable to ensure proper functioning of the contract compliance requirement		1. Submittal procedure taking into consideration the requirements listed under contract documents
		2. Progress payments				2. Stakeholders' interest
		3. Progress Reports				
		4. Minutes of Meetings				
		5. Other Meetings				

8.4 Documents

1. Design documents	The detailed procedure for submitting the documents as specified in the contract documents is not established	Reviewing, verifying, and approval of material, contract drawings, shop drawings, specifications, samples will be affected	Project Manager (Design)	Specify the procedure to distribute contract documents among the relevant stakeholders for information or action as per contract requirements
2. Contract documents			Project Manager (Design)	
3. Construction documents			Project Manager (Design)	
4. As-built documents	The detailed procedure for submitting the documents as specified in the contract documents is not established	Reviewing, verifying, and approval of material, contract drawings, shop drawings, specifications, samples will be affected	Project Manager (Design)	Specify the procedure for submitting as-built drawing, regulatory approved documents
5. Authority approved documents/drawings				

8.5 Logs

1. Issue log	The detailed procedure to maintain logs is not specified	Stakeholders will not get related information about project progress	Project Manager (Design)	Establish submission procedure for logs and project control documents
2. Correspondence with stakeholders				
3. Correspondence with team members				Establish correspondence method among project team members
4. Regulatory authorities		Delay in execution of project	Project Manager (Design)	Establish procedure for authority-approval-related correspondence

(Continued)

TABLE 4.5
(Continued)

Serial Number	Activities	Major Elements	Potential Risk	Probable Effect on Project	Ownership of Risk	Risk Treatment/ Control Measures
9	9.1	Risk Identification				
		1. During inception	Lack of input from the owner about the project goals and objectives	Errors in design and specifications	Project Manager (Design)	Collect owner's input and requirements about project goals and objectives
		2. During design	1. Design errors/omissions	Design not meeting owner's requirements	Project Manager (Design)	1. Consider relevant data/information, TOR requirements, regulatory requirements, codes, and standards while developing the design
			2. Design deliverables not matching the owner's needs and needs for successful completion of the project			2. Develop design deliverables to match owner's requirements
		3. During bidding	Bid value exceeding the estimated definitive cost	Delay in managing extra funds	Project Manager (Design)	Estimating definitive cost maintaining accuracy
		4. During construction	1. Delay in mobilization	Delay in project execution, completion	Construction Manager (Contractor)	Follow approved construction schedule
			2. Inappropriate construction method			
			3. Scope/design changes			
			4. Inappropriate schedule			
			5. Availability of resources			
			6. Errors and omissions in contract documents			
		5. During testing and commissioning	1. Major items for testing and commissioning are not identified	Errors in testing and commissioning	Project Manager (Design)	1. Identify and specify testing and commissioning items required in the contract documents
			2. Inspection and test plan is not properly established			2. Establish testing and commissioning plan

6. During handing over	Delay in acceptance/takeover of the project	Delay in move-in plan	Project Manager (Design)	Establish schedule for handing over of the project

9.2 Managing Risk

1. Risk register	Risk register is not maintained and updated regularly	All the risks will not be known	Project Manager (Design)	Specify in contract documents to maintain and update risk register on regular basis
2. Risk analysis	Qualitative and quantitative analysis is not performed using appropriate tools and techniques	Analyzing errors	Project Manager (Design)	Perform risk analysis for the identified list to know the impact on the project
3. Risk response	Proper risk response is not identified to reduce/mitigate threats to the identified risk	Project performance and quality of the project	Project Manager (Design)	Determine the action to be taken in order to address the identified and assessed risk(s)

10 Project Delivery System

10.1

1. Selection of CM	Following points are not considered while selecting CM: 1. Thorough knowledge and understanding of construction process required for the project and to perform the assignment within the given frame of time 2. Knowledge of various tools and techniques used in construction project 3. Skills to oversee and manage complex construction projects	Selected candidate may not have right knowledge, skills, and experience that are required for the project and to perform the assignment within the given frame of time	Project Owner (Consultant)	Establish and follow the candidate selection procedure (refer Figure 4.4)
2. Selection of designer	Following points are not considered while selecting the designer: 1. Experience in similar types and size of projects 2. Competent team members having design experience 3. Design production capacity	Design deliverables may not match as per TOR	Project Owner (Consultant)	Select designer having experience in similar type, size of projects and having technical, engineering, designing, and management experience as established through pre-qualification questionnaire for selecting the designer (A/E)

(Continued)

TABLE 4.5
(Continued)

Serial Number	Activities	Major Elements	Potential Risk	Probable Effect on Project	Ownership of Risk	Risk Treatment/ Control Measures
	10.2	Bidding and Tendering				
		1. Pre-qualification of contractors	1. The contractor selection procedure is not properly followed	Unqualified bidder is selected	Project Owner (Tendering Team)	Perform pre-qualification of bidders by short listing of bidders with pre-qualification questionnaires and the response
			2. Bidder selection procedure such as low bid, quality-based is not defined			
		2. Issue tender documents	Tendering process is not followed	May not get qualified contractor	Project Owner (Tendering Team)	Follow tendering process considering the following points: 1. Bid notification 2. Distribution of tender documents 3. Pre-bid meeting 4. Issuing addendum, if any 5. Bid submission
		3. Acceptance of tender	Review and evaluation of bid procedure are not followed properly	Incompetent contractor is selected	Project Owner (Tendering Team)	Review and evaluate submitted bids and select the contractor based on the procurement strategy
11	11.1	Environmental Compatibility				
	11.2	Safety management plan				
		1. Safety consideration in design	Designer has not taken safety considerations into account	Unsafe design	Project Manager (Design)	Consider health, safety, and environmental compatibility while developing the design
		2. HSE Plan for construction site safety	Site safety and HSE requirements are not listed in the contract documents	More accidents at site	Project Manager (Design)	Establish HSE plan and implement the same to avoid accidents and safety violations

3. Emergency evacuation plan	Emergency evacuation plan at project site is not mentioned in the contract documents	More fatalities at site	Project Manager (Design)	Establish emergency evacuation plan
11.3 Waste management plan	Waste management plan is not included in the contract documents	Safety hazards	Project Manager (Design)	Establish waste management plan
12				
12.1 Financial Planning				
1. Payments to Designer (Consultant), Construction/Project Manager, Contractor	Owner is not maintaining records for payments to designer, management staff and contractor's progress payment	Payment schedule will be affected	Project Owner	Establish financial plan estimating the total funds required and project cash flow (usage of funds) at every phase/stage of project
2. Material procurement	Material procurement payments are not paid as per agreed-upon conditions	Default in payment promises	Construction Manager (Contractor)	Estimate the fund required for payments toward procurement of material
3. Equipment procurement	Equipment procurement payments are not paid as per agreed-upon conditions	Default in payment promises	Construction Manager (Contractor)	Estimate the fund required for payments toward procurement of equipment
4. Project staff salaries	Staff salaries are not paid regularly	Default in payment promises	R.E. Construction Manager (Contractor)	Estimate funds required for staff salary Estimate the fund required for payments to be made toward staff salaries
5. Bonds, insurance, guarantees	Bonds, insurance, guarantees are not paid in time	Delay in the processing of contractual requirements	Project Owner (Tendering Committee)	Select the contractor complying with tendering procedure
6. Cash flow	Cash flow chart is not maintained	Default in payment promises	All the concerned parties	Maintain cash flow for a smooth functioning of the contract

(Continued)

TABLE 4.5
(Continued)

Serial Number	Activities	Major Elements	Potential Risk	Probable Effect on Project	Ownership of Risk	Risk Treatment/ Control Measures
13	13.1 Claim Identification					
		1. Design errors	Change in scope of work	Delay in project execution	Project Manager (Design)	Resolve the claim as per the terms in the contract documents
		2. Additional works	Variation/change order	Delay in project execution	Project Owner Contractor	Resolve the claim as per the terms in the contract documents
		3. Delays in Payment	Claim by the contractor	Overrun of project cost	Project Owner	Make approved payment as per the contract conditions
	13.2 Claim quantification					
		1. Change order procedures • Cost • Time	Change variation order procedure is not followed properly	Conflict with the contract documents	Project Manager (Design)	Resolve the conflict amicably to settle the variation order

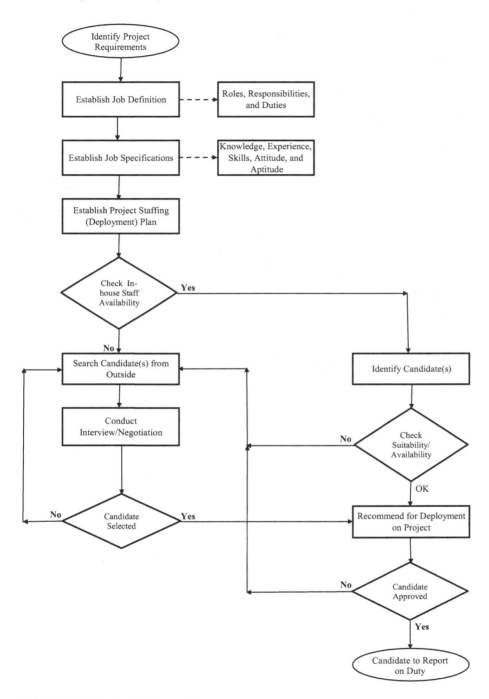

FIGURE 4.3   Project staffing process.

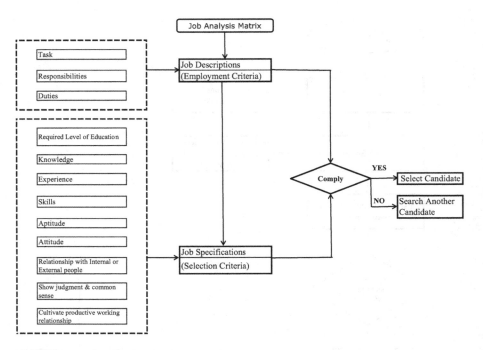

FIGURE 4.4 Candidate selection procedure.

Figure 4.5 illustrates progress payment approval process.

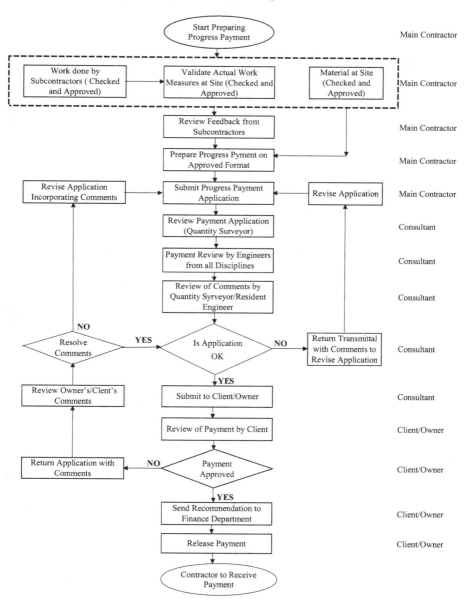

FIGURE 4.5 Progress payment approval process.

Figure 4.6 illustrates process to resolve scope change (owner initiated)

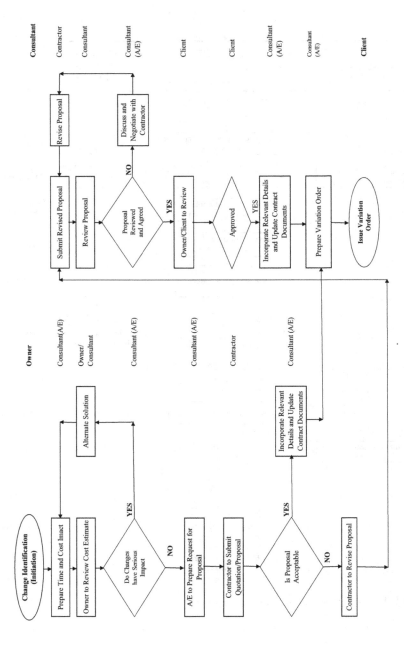

FIGURE 4.6   Process to resolve scope change (owner initiated).

Source: Abdul Razzak Rumane. (2016). *Handbook of Construction Management.* Reprinted with permission of Taylor & Francis Group.

Figure 4.7 illustrates sequence of execution of works

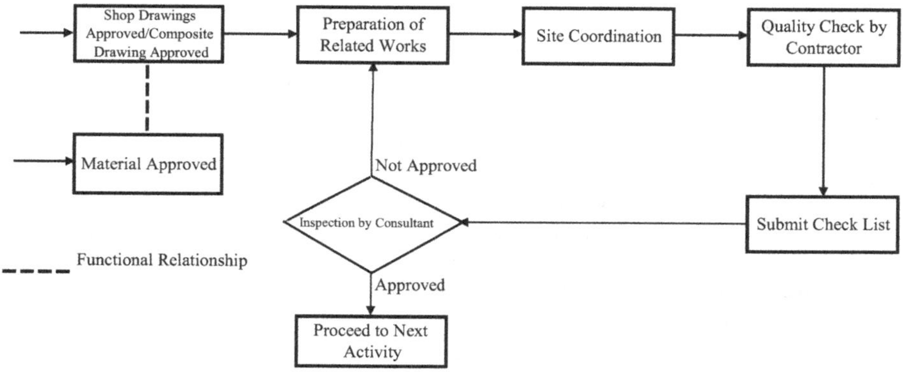

FIGURE 4.7 Sequence of execution of works.

Source: Abdul Razzak Rumane. (2017). *Quality Management in Construction Projects*, second edition. Reprinted with permission from Taylor & Francis Group Company.

4.3.3 Risk Monitoring in Activities Related to Quality of Planning Process Group

Table 4.6 is an example process of risk monitoring of planning process group activity.

TABLE 4.6
Risk Monitoring of Planning Process Group Activity

Serial Number	Description	Action			
1	Risk ID	Planning process group—3 (3.3)			
2	Description of risk	1. Project scope document (final design) 1.1 The project scope documents does not explain the boundaries of the project, establishes project responsibilities for each team member, and sets up procedures for how completed works will be verified and approved			
3	Response	1.1 Develop project scope documents explaining project boundaries, project performance requirements, constructability, functionality and usage of project, responsibilities of each project team to complete the project meeting the specifications, and satisfy the owner's needs			
4	Strategy of response	Avoidance	Transfer	Mitigation X	Acceptance

(Continued)

TABLE 4.6
(Continued)

Serial Number	Description	Action		
5	Monitoring and control	Risk owner	Review date	Critical issue
		Project Manager (Design) Consultant	D/M/Y	Accuracy of project design to meet owner's requirements and needs
6	Estimated impact on the project	Interruption of work execution for clarification/information about the responsibilities of team members		
7	Actual impact On The project	3 months delay		
8	Revised response	Review project scope documents clearly specifying the roles and responsibilities for successful completion of project to meet owner's needs		
9	Record update date	D/M/Y		
10	Communication to stakeholders	Information sent to all stakeholders on D/M/Y		

4.4   RISK MANAGEMENT IN QUALITY OF EXECUTING PROCESS GROUP

Figure 4.8 illustrates a flowchart for risk management in quality of executing process group.

4.4.1   RISK IN ACTIVITIES RELATED TO QUALITY OF EXECUTING PROCESS GROUP

Table 3.3 discussed in Chapter 3, Section 3.2.2 illustrates major activities relating to executing process group.

Table 4.7 lists potential risks in the activities related to executing process group.

4.4.2   RISK EFFECT AND RISK TREATMENT IN ACTIVITIES RELATED TO QUALITY OF EXECUTING PROCESS GROUP

Table 4.8 lists potential risks, probable effect on project, and risk treatment in activities related to executing process group.

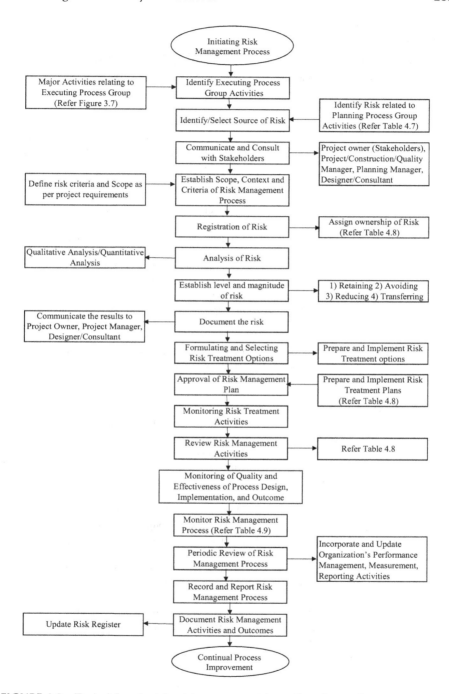

FIGURE 4.8 Typical flowchart for risk management in quality of executing process group.

TABLE 4.7

Potential Risks in Activities Related to Executing Process Group

Serial Number	Activities	Major Elements	Potential Risk
1	1.1 Design Development		
		1. Concept design	1. Lack of owner input (owner-preferred requirements)
			2. Relevant data/information on existing conditions on project site and surrounding areas not collected or is incomplete
			3. Environmental studies not performed and considered for designing
		2. Schematic design	1. Regulatory authorities' requirements are not taken into consideration
			2. Related product data and information collected are incomplete or incorrect
			3. Inadequate and ambiguous specifications
			4. Schematic design deliverables are not as per TOR requirements
		3. Detail design	1. Incomplete design drawings and related information
			2. Design drawings are not duly coordinated for conflict between different trades
			3. Design not meeting owner requirements and the scope of work as per TOR
			4. Difficulties in dealing with specifications and standards concerning existing conditions and client requirements
	1.2 Construction		
		1. Notice to proceed	1. Delay in conducting kick-off meeting
			2. Delay in transfer of site after issuing notice to proceed
		2. Mobilization	1. Delay in mobilization
			2. Different site conditions
		3. Submittals	1. Delay in submittal and approval of contractor's construction schedule
			2. Delay in submittals of management plan
			3. Submittals are not properly monitored
		4. Execution	1. Contractor has lack of knowledge and experience in execution of similar type of construction projects
			2. Resources/procurement are not managed properly
		5. Corrective actions	Delay in taking corrective action
		6. Project deliverables	1. Project execution scope is not established
			2. Construction phase deliverables not established
	1.3 Implement Changes		
		1. Approved changes	Relevant documents not updated
		2. Preventive actions	Preventive actions not taken to prevent the occurrence of nonconformances in future

Serial Number	Activities	Major Elements	Potential Risk
		3. Defect repairs	Delay in defect repair
		4. Rework	Delay in carrying out rework of the rejected activity
		5. Update scope	Scope not updated by incorporating the relevant change(s)
		6. Update plans	Plans not updated by incorporating the relevant change(s)
		7. Update contract documents	Contract documents not updated by incorporating the relevant change(s)
2	2.1	Project Status/Performance Report	
		1. Updated plans	1. Progress performance is not updated on regular basis
			2. Stakeholders are not receiving updated plan
	2.2	Payments	
		2. Progress payments	1. Progress payment does not match the approved work executed at site
			2. Stakeholders are not receiving copy of progress payment
	2.3	Change requests	
		1. Site work instruction	Relevant stakeholders are receiving site works instruction
		2. Change orders	Relevant stakeholders are not receiving change orders
		3. Schedule	Relevant stakeholders not receiving change request for schedule
		4. Materials	Relevant stakeholders not receiving request for approval of substitute material
	2.4	Conflict Resolution	
		1. Delay in conflict resolution	
	2.5	Issue Log	
		1. Issue log is not maintained properly	
		2. Issues are not reported to relevant stakeholders	
3		Scope Management	
		1. Incomplete scope of work	
		2. Incomplete specifications	
		2. Errors and omissions in contract documents	
4		Schedule Management	
		1. All the activities are not identified and listed to meet the scope of work	
		2. Identification and isolation of most important and critical information/activities are not done	
		3. Key milestones are not listed	
		4. Inappropriate construction schedule	
		4. Delay in change order negotiations	
		5. Failure/delay of material/equipment delivery	
		6. Forecasting and predicting the failure of progress of activities to be performed are not done	

(Continued)

TABLE 4.7
(Continued)

Serial Number	Activities	Major Elements	Potential Risk
5	Cost Management		
		1. Variation in construction material cost	
		2. Variation in resources cost	
		3. Inflation	
		4. Delay in payment	
		5. Cost and time control to monitor work progress and comparing actual work done against planned work are not done on regular basis	
		6. Factors that influence the changes in cost baseline are not identified	
		7. Cash flow forecast is not identified and established	
6	6.1 Quality Assurance		
		1. Design compliance to TOR	Owner's needs and requirements are not included in the scope of works (TOR)
		2. Design coordination with all disciplines	Interdisciplinary coordination not done while developing the design
		3. Material approval	1. Material approval and procurement procedure is not followed
			2. Approved material and equipment are not installed
		4. Shop drawing approval	1. Shop drawings are not approved
			2. Works are not executed as per approved shop drawings
		5. Method approval	1. Installed/executed works not conforming to specifications
			2. Installation method is not approved and rejected
		6. Method statement	Method statement is not submitted for approval
		7. Mock-up	1. Mock-up is not prepared and submitted for approval
			2. Mock-up is not approved
		8. Quality audit	Quality audit is not performed as specified in the quality procedure/manual
		9. Functional and technical compatibility	1. The project is not meeting the functional requirements of the owner
			2. The executed works are not technically compatible with specification requirements and owner's goals and objectives
7	7.1 Project Staff		
		1. Project/ Construction Manager Staff	1. Lack of knowledge to resolve conflict by searching an alternate solution
			2. Lack of knowledge of quality management techniques
			3. Lack of strong and responsive leadership skills
			4. Lack of problem-solving skills

Serial Number	Activities	Major Elements	Potential Risk
		2. Supervision Staff	1. Ineffective and lack of supervision (monitoring and controlling) experience by supervision staff (Consultant) 2. Not having adequate knowledge to review contract drawings and resolve technical discrepancies in the contract documents
	7.2	Project Manpower	
		1. Core staff	1. Lack of competency in the specific field for which they are assigned 2. Selected core staff does not possess the qualification and related experience mentioned in the contract documents
		2. Site staff	Failure of site staff to perform as expected
		3. Workforce	1. Low productive workforce 2. Availability of skilled workforce
	7.3	Team Management	
		1. Team behavior	1. Ground rules for cohesion among team members are not established 2. Teams performance is not tracked 3. Team members may not have shared vision
		2. Conflict resolution	1. The causes of disagreement not identified 2. The team leader (project manager) has no alternate solution to resolve the conflict 3. Appropriate method to resolve conflict is not applied 4. Frequent meetings and status review sessions are not conducted
		3. Demobilization Project Workforce	Demobilization/release of project team members is not properly planned and organized
	7.4	Construction Resources	
		1. Material	1. Material management process is not followed 2. Procurement log is not maintained 3. Failure/delay of material delivery
		2. Equipment	1. Nonavailability of equipment as per schedule
		3. Low productivity labor	1. Delay in completion of project
		4. Subcontractor(s)	1. Incompetent subcontractor 2. Proper selection method not followed to select the subcontractor 3. Extensive subcontracting
8	8.1	Submittals	
		1. Shop drawings	1. Shop drawings are not approved 2. Shop drawing preparation/ review and approval procedure is not established/ followed

(Continued)

TABLE 4.7
(Continued)

Serial Number	Activities	Major Elements	Potential Risk
		2. Material	1. Delay in approval of material submittals
			2. Quality of material
		3. Change orders	1. Change order submittal procedure is not followed
			2. Delay in change order negotiations
		4. Payments	1. Delay in payment
	8.2	Documentation	
		1. Status log	Project status logs are not maintained/updated properly
		2. Issue log	Issue logs are not maintained/ updated properly
		3. Minutes of meetings	Minutes of meetings are not distributed to the relevant stakeholders
		4. Contract documents	Procedure specified in the contract is not followed to manage the documents
		5. Specifications	Specifications are not updated to incorporate approved change order/specifications
		6. Payments	1. Payment request is not submitted as per approved communication matrix
			2. Corresponding documents are not attached with progress payment request
			3. Delay in issuance of payment certificate
	8.3	Correspondence	
		1. Stakeholders	Communication matrix for correspondences with stakeholders is not established
		2. Regulatory authorities	Procedure for correspondence with regulatory authorities not established
		3. Correspondence among team members	1. Site administrative and communication matrix is not established
			2. Stakeholders' responsibilities matrix not established
9	9.1	Manage Risk	
		1. Risk register	Risk register is not updated every time a new risk is identified, or relevant actions are taken
		2. Risk response	1. Delay in taking action on risk
10	10.1	Contract Documents	
		1. Notice to proceed	Delay in beginning of project work after the issuance of notice to proceed
	10.2	Selection of subcontractor(s)	
			1. Delay in submitting the names of subcontractors for approval
			2. Subcontractor submission and approval log is not maintained
	10.3	Selection of Materials, Systems and Equipment	
			1. Material selection as per specifications and approval procedure is not followed
			2. Long-lead items are not identified
			3. Details of procurement data relating to all the approved products, equipment, and systems are not maintained

Serial Number	Activities	Major Elements	Potential Risk
	10.4	Execution of Works	
		1. Construction work is not executed as per approved shop drawings	
		2. Method statement is not followed	
		3. Approved material is not installed	
		4. Routine inspection/supervision during the construction process is not carried out	
11	11.1	HSE Management Plan	
		1. Site safety	Necessary measures are not followed to ensure safety of all those working at construction site
		2. Preventive and mitigation measures	1. The authority of the company is not giving importance to preventive and mitigation measures to maintain safe and healthy working conditions that will safeguard employees
			2. Safety trainings are not conducted
			3. Safety meetings are not conducted
		3. Temporary firefighting	1. Temporary firefighting system is not installed at site
			2. Temporary firefighting system is not approved by the relevant authority
			3. Temporary firefighting system is not working
		4. Environmental protection	ISO 14000 environmental management system is not in place
		5. Waste management	Waste management plan is not established
		6. Safety hazards	1. Hazards management and precautionary measures are not operative
			2. Precautionary measures to avoid leakage of hazardous material are not established
12		Financial Management	
		1. Financial management plan is not established	
		2. Cash flow is not as per estimated S-Curve	
		3. Monitoring and control of project cost are not done properly	
13		Claim Management	
		1. Identified claims are not submitted with all the supportive documents for justification	
		2. Preventive actions are not taken to mitigate claims	
		3. Claim management process is not followed to resolve the claims	
		4. Delay in resolving the claim	

TABLE 4.8

Potential Risks, Probable Effect on Project, and Risk Treatment in Activities Related to Executing Process Group

Serial Number	Activities	Major Elements	Potential Risk	Probable Effect on Project	Ownership of Risk	Risk Treatment/ Control Measures
1	1.1 Design Development					
		1. Concept design	1. Lack of owner input (owner-preferred requirements) 2. Relevant data/information on existing conditions on project site and surrounding areas not collected or is incomplete 3. Environmental studies not performed and considered for designing 4. Technical and functional capability	1. Affect the design development to meet owner's requirements 2. Impact on the planning and design of project 3. Environmental effects on project design 4. Constructability 5. Cost-effectiveness on the entire life cycle of the project	Project Manager (Design)	Develop concept design considering following major items: 1. Owner's requirements 2. Scope of work/ requirements mentioned in the TOR 3. Data/information about site conditions, topographical survey, geotechnical investigations 4. Environmental studies 5. Energy conservation 6. Sustainability 7. Quality codes and standards
		2. Schematic design	1. Regulatory authorities' requirements are not taken into consideration 2. Related product data and information collected are incomplete or incorrect 3. Inadequate and ambiguous specifications 4. Schematic design deliverables are not as per TOR requirements	1. Design not meeting regulatory requirements 2. Design deliverables not meeting owner's requirements 3. Conflict in design development 4. Schematic design not meeting TOR	Project Manager (Design)	Develop schematic design considering following major items; 1. Regulatory authorities' requirements 2. Concept design deliverables and review comments and related data 3. TOR requirements and design deliverables 4. LEED requirements 5. Environmental issues 6. Technical and functional compatibility 7. Value engineering study

	Risk	Impact	Responsibility	Recommendations
3. Detail design	1. Incomplete design drawings and related information 2. Design drawings are not duly coordinated for conflict between different trades 3. Design not meeting owner requirements and the scope of work as per TOR 4. Difficulties in dealing with specifications and standards concerning existing conditions and client requirements	1. Design not meeting owner's requirements and needs 2. Omission and errors during execution of the activities 3. Not meeting quality of the project 4. Project quality not meeting the required specifications and standards	Project Manager (Design)	Develop detail design considering following major items: 1. Schematic design deliverables and review comments 2. Interdisciplinary coordination 3. Specifications and contract documents 4. Owner's and TOR requirements 5. Site investigations and existing conditions 6. Regulatory authorities' requirements 7. HSE requirements 8. Constructability 9. Existing site conditions

1.2 Construction

	Risk	Impact	Responsibility	Recommendations
1. Notice to Proceed	1. Delay in conducting kick-off meeting 2. Delay in transfer of site after issuing notice to proceed	1. Delay in start of construction phase 2. Delay in start of construction	Project Owner (Consultant)	1. Conduct kick-off meeting once the notice to proceed is issued to begin the project work 2. Obtain legal documents and title deeds in time to transfer the site
2. Mobilization	1. Delay in mobilization 2. Different site conditions	1. Delay in construction activities 2. Extra work	Construction Manager (Contractor)	1. Start mobilization activities immediately after necessary permits are obtained from the relevant authorities and as per construction schedule 2. Investigate site conditions prior to starting the relevant activity

(Continued)

TABLE 4.8
(Continued)

Serial Number	Activities	Major Elements	Potential Risk	Probable Effect on Project	Ownership of Risk	Risk Treatment/ Control Measures
		3. Submittals	1. Delay in submittal and approval of contractor's construction schedule 2. Delay in submittals of management plan 3. Submittals are not properly monitored	Delay in start of relevant activity	Construction Manager (Contractor)	1. Prepare project construction plans based on the contracted time schedule of the project and submit for approval 2. Prepare and submit management plans as per contract requirements 3. Submit core staff for approval 4. Submit subcontractor(s) for approval 4. Maintain submittal logs
		4. Execution	1. Contractor has lack of knowledge and experience in execution on similar type of construction project 2. Resources/procurement is not managed properly	1. Poor workmanship 2. Delay in execution of works	Construction Manager (Contractor)	1. Monitor the execution of works regularly 2. Execute works to comply with the specifications and satisfy the quality as per owner's requirements 3. Engage skilled manpower having experience in similar projects 4. Manage the resources as per approved schedule
		5. Corrective Actions	Delay in taking corrective action	Delay in work execution as per schedule	Construction Manager (Contractor)	Take corrective action for the non-approved (rejected) works and establish appropriate preventive actions plan to avoid repetition of nonconformance work

6. Project Deliverables	1. Project execution scope is not established; 2. Construction phase deliverables not established	Smooth execution of project	Construction Manager (Contractor)	1. Establish project execution requirements and develop project execution scope for successful completion of project; 2. Establish construction phase deliverables
1.3 Implement Changes				
1. Approved changes	Relevant documents not updated	Project progress	Resident Engineer	Update relevant documents incorporating the approved changes
2. Preventive Actions	Preventive actions not taken to prevent the occurrence of nonconformances in future	1. Repetitive errors in the execution; 2. Delay in project	Construction Manager (Contractor)	Take preventive action to prevent the occurrence of nonconformance work and reduce rework
3. Defect Repairs	Delay in defect repair	Delay in project	Construction Manager (Contractor)	Take immediate action to repair the rejected (defective) work
4. Rework	Delay in carrying out rework of the rejected activity	Delay in project	Construction Manager (Contractor)	Establish process to avoid the rejection of work and taking immediately preventive action
5. Update Scope	Scope not updated by incorporating the relevant change (s)	Quality of project	Construction Manager (Contractor)	Regularly update the scope incorporating approved changes
6. Update Plans	Plans not updated by incorporating the relevant change (s)	Project progress status	R.E./PCM	Regularly monitor and update the schedule incorporating approved changes
			Construction Manager (Contractor)	Regularly update the schedule incorporating approved changes
7. Update Contract Documents	Contract documents not updated by incorporating the relevant change (s)	Project progress status	R.E.	Regularly update contract documents incorporating approved changes

(Continued)

TABLE 4.8
(Continued)

Serial Number	Activities	Major Elements	Potential Risk	Probable Effect on Project	Ownership of Risk	Risk Treatment/ Control Measures
2	2.1 Project Status/ Performance Report					
		1. Updated Plans	1. Progress performance is not updated on regular basis	Project progress status	Resident Engineer	Monitor and control schedule to update following information: 1. Project status 2. Project progress 3. Forecasting
			2. Stakeholders are not receiving updated plan		Construction Manager (Contractor)	Monitor and control schedule to update following information: 1. Project status 2. Project progress 3. Forecasting
					Construction Manager (Contractor)	Regularly update contract documents incorporating approved changes
	2.2 Payments					
		2. Progress Payments	1. Progress payment does not match the approved work executed at site	Delay in payment of progress payment	Construction Manager (Contractor)	1. Prepare request for progress payment taking into consideration related approved executed works (refer Figure 4.5) 2. Contractor to have contingency plan
			2. Stakeholders are not receiving copy of progress payment			2. Distribute request for progress payment to the concerned stakeholders as per the administration matrix
					R.E.	Approve progress payment that matches the contract conditions

Process	Cause	Impact	Responsible	Action
			Project Owner	Release progress payment recommended by R.E.
2.3 Change Requests				
1. Site work instruction	Relevant stakeholders are not receiving Site Works Instruction (SWI)	Stakeholders not aware of impact on project	Project Owner	Distribute SWI to relevant stakeholders as per the matrix and contract document requirements (refer Figure 3.25 in Chapter 3)
2. Change orders	Relevant stakeholders are not receiving change orders	Stakeholders not aware of impact on project	R.E.	Distribute change orders to relevant stakeholders as per the matrix and contract document requirements
3. Schedule	Relevant stakeholders not receiving change request for schedule	Stakeholders not aware of impact on project progress	R.E.	Distribute schedule to relevant stakeholders as per the matrix and contract document requirements
4. Materials	Relevant stakeholders not receiving request for the approval of substitute material	Stakeholders not aware of impact on project quality	R.E.	Distribute material approval submittals to relevant stakeholders as per the matrix and contract document requirements
2.4 Conflict Resolution		Delay in relevant project activity	Construction Manager (Contractor)	Manage the changes in the project and resolve in accordance with the condition of contract
2.5 Issue Log	1. Issue log is not maintained properly 2. Issues are not reported to relevant stakeholders	1. Project status 2. Stakeholders not aware of project issues	R.E./ Document Controller	1. Maintain issue logs regularly 2. Distribute information about the project issues to the concerned stakeholders to provide necessary information about the project

(Continued)

TABLE 4.8 (Continued)

Serial Number	Activities	Major Elements	Potential Risk	Probable Effect on Project	Ownership of Risk	Risk Treatment/ Control Measures
3	Scope Management					
		1. Incomplete scope of work 2. Incomplete specifications 2. Errors and omissions in contract documents		Project quality	Project Manager (Design)	1. Establish project scope to include activities to meet client's needs, project goals and objectives, boundaries of the project, responsibilities of contractor and set up procedures for how completed project will be verified and approved 2. Develop complete specifications 3. Review contract documents to minimize errors and omissions
4	Schedule Management					
		1. All the activities are not identified and listed to meet the scope of work 2. Identification and isolation of most important and critical information/activities are not done 3. Key milestones are not listed 3. Inappropriate construction schedule 4. Delay in change order negotiations 5. Failure/delay of material/equipment delivery 6. Forecasting and predicting the failure of progress of activities to be performed are not done		Project progress and status	Construction Manager (Contractor)/ PCM	Develop construction schedule in an organized and structured manner showing periods for all sections and activities to be completed as specified in the contract documents taking into consideration relationships and dependency and resources calendar (refer Figure 4.9)

	Description	Risk	Responsibility	Control Measures
5 Cost Management	1. Variation in construction material cost 2. Variation in resources' cost 3. Inflation 4. Delay in payment 5. Cost and time control to monitor work progress and compare actual work done against planned work is not done on regular basis 6. Factors that influence the changes in cost baseline are not identified 7. Cash flow forecast is not identified and established	Cost management	Construction Manager (Contractor)/ PCM	1. Monitor the status of project to update the project cost and managing changes to baseline. 2. Manage project delivery within approved budget by: • Comparing and monitoring project progress • Identification and control of approved variations • Identification of cash flow forecast • Comparing current budget with the forecast for remining works
6 6.1 Quality Assurance				
1. Design	compliance included in the scope of works (TOR)	Owner's needs and requirements are not compliance to TOR Project not meeting owner's requirements	Construction Manager (Contractor)/ Quality Manager	Prepare quality control plan based on project-specific requirements to meet the performance standards specified in the contract documents for achieving construction quality (refer Figure 3.43 in Chapter 3)
2. Design coordination	Interdisciplinary coordination not done while developing the design with all disciplines	Difficulties during construction	Project Manager (Contractor)	Prepare composite and coordination drawings
3. Material approval	1. Material approval and procurement procedure is not followed 2. Approved material and equipment are not installed	Material not meeting specification requirements	Construction Manager (Contractor)	Follow material approval procedure (refer Figure 4.10)
4. Shop Drawing approval	1. Shop drawings are not approved 2. Works are not executed as per approved shop drawings	Delay in execution of works	Construction Manager (Contractor)	Follow shop drawing approval procedure (Figure 4.11)

(Continued)

TABLE 4.8
(Continued)

Serial Number	Activities	Major Elements	Potential Risk	Probable Effect on Project	Ownership of Risk	Risk Treatment/ Control Measures
5.		Method approval	1. Installed/executed works not conforming to specifications 2. Installation method is not approved and rejected	Rejection of works	Construction Manager (Contractor)	Submit method for installation and get approval to ensure work compliance with specification requirements
6.		Method statement	Method statement is not submitted for approval	Nonconformance of works	Construction Manager (Contractor)	Prepare method statement and get the approval from R.E.
7.		Mock-up	1. Mock-up is not prepared and submitted for approval 2. Mock-up is not approved	Nonconformance of works	Construction Manager (Contractor)	Prepare mock-up for the specified activity and get approval
8.		Quality Audit	Quality audit is not performed as specified in the quality procedure/ manual	Nonconformance of works	Construction Manager (Contractor)/ Quality Manager	Perform quality audit as specified in the quality procedures and ensure that system is followed by the quality personnel
9.		Functional and technical compatibility	1. The project is not meeting the functional requirements of the owner 2. The executed works are not technically compatible with specification requirements and owner's goals and objectives	Nonconformance of project quality	Project Manager (Design)	1. Develop project design meeting the functional requirements having technical compatibility and suitable for usage. 2. Contractor to execute the works to meet owner's requirements

7

7.1 Project Staff

		Risk	Consequence	Responsible	Response
1.	Project/ construction manager staff	1. Lack of knowledge to resolve conflict by searching an alternate solution 2. Lack of knowledge of quality management techniques 3. Lack of strong and responsive leadership skills 4. Lack of problem-solving skills	1. Project performance not as expected 2. Delay in resolving the conflict and issues	Construction Manager (Contractor)	1. Select competent staff having knowledge, expertise to perform achieving successful project 2. Follow candidate selection procedure (refer Figure 4.4) and project team acquisition process (refer Figure 3.44)
2.	Supervision Staff	1. Ineffective and lack of supervision (monitoring and controlling) experience by supervision staff (Consultant) 2. Not having adequate knowledge to review contract drawings and resolve technical discrepancies in the contract documents	1. Quality of executed works 2. Executed works not conforming to specified requirements	Project Manager (Supervision)/ R.E.	1. Select competent staff 2. Follow candidate selection procedure (refer Figure 4.4) and project team acquisition process (refer Figure 3.44)

7.2 Project Manpower

		Risk	Consequence	Responsible	Response
1.	Core staff	1. Lack of competency in the specific field for which they are assigned 2. Selected core staff does not possess the qualification and related experience mentioned in the contract documents	1. Nonconformance of executed works 2. Project performance	Construction Manager (Contractor)	1. Search competent staff 2. Follow candidate selection procedure (refer Figure 4.4) and project team acquisition process (refer Figure 3.44)
2.	Site staff	Failure of site staff to perform as expected	Supervision of project works and workmanship	Construction Manager (Contractor)	1. Search competent staff 2. Follow candidate selection procedure (refer Figure 4.4) and project team acquisition process (refer Figure 3.44)
3.	Workforce	1. Low productive workforce 2. Availability of skilled workforce	Delay in completion of relevant activities	Construction Manager (Contractor)	Follow candidate selection procedure (refer Figure 4.4) and project team acquisition process (refer Figure 3.44)

(Continued)

TABLE 4.8
(Continued)

Serial Number	Activities	Major Elements	Potential Risk	Probable Effect on Project	Ownership of Risk	Risk Treatment/ Control Measures
7.3	Team Management					
		1. Team behavior	1. Ground rules for cohesion among team members is not established 2. Teams performance is not tracked 3. Team members may not have shared vision	1. Conflict, confrontation, among team members 2. Project performance 3. The team is not a cohesive unit	Resident Engineer	1. Establish ground rules by maintaining cohesion among the team members 2. Keep track of team member performance, resolve issues, and manage changes to meet project performance optimization 3. Create shared vision among team members
					Construction Manager (Contractor)	1. Establish ground rules by maintaining cohesion among the team members 2. Keep track of team member performance, resolve issues, and manage changes to meet project performance optimization 3. Create shared vision among team members
		2. Conflict Resolution	1. The causes of disagreement not identified 2. The team leader (project manager) has no alternate solution to resolve the conflict 3. Appropriate method to resolve conflict is not applied 4. Frequent meetings and status review sessions are not conducted	1. Delay in resolving the issue 2. Delay in convincing the parties to accept the proposed solution 3. Delay in problem resolution 4. Disunity among team members	R.E.	Follow conflict management process (refer Figure 3.45 in Chapter 3)

Item	Risk/Cause	Consequence	Responsibility	Action/Mitigation
3. Demobilization Project Workforce	Demobilization/release of project team members are is properly planned and organized	Delay in demobilization	Construction Manager (Contractor); R.E.; Construction Manager (Contractor)	Follow conflict management process (refer Figure 3.45 in Chapter 3); Demobilize project team members to close the project; Demobilize project team members to close the project
7.4 Construction Resources				
1. Material	1. Material management process is not followed 2. Procurement log is not maintained 3. Failure/delay of material delivery	1. Delay in getting approved material 2. Procurement status 3. Delay in project execution	Construction Manager (Contractor)	Follow material management process (Figure 3.46 in Chapter 3)
2. Equipment	1. Nonavailability of equipment as per schedule	Delay in execution activity	Construction Manager (Contractor)	Search and locate approved equal equipment as per contract documents
3. Low productivity labor	1. Delay in the completion of project	1. Delay in the completion of project 2. Activity execution cost exceeding the estimated cost	Construction Manager (Contractor)	1. Select competent labor 2. Provide training
4. Subcontractor(s)	1. Incompetent subcontractor 2. Proper selection method not followed to select the subcontractor 3. Extensive subcontracting	1. Project workmanship 2. Delay in project execution schedule 3. Work coordination	Construction Manager (Contractor)	1. Select subcontractor having expertise in performing similar types of projects 2. Monitor the workmanship and work progress regularly
8 Submittals				
8.1				
1. Shop drawings	1. Shop drawings are not approved 2. Shop drawing preparation/review and approval procedure is not established/followed	Delay in execution of project works	Construction Manager (Contractor)	Follow shop drawing approval procedure (refer Figure 4.11)
2. Material	1. Delay in approval of material submittals 2. Quality of material	Delay in execution of project works	Construction Manager (Contractor)	Follow material approval procedure (refer Figure 4.10)

(Continued)

TABLE 4.8 (Continued)

Serial Number	Activities	Major Elements	Potential Risk	Probable Effect on Project	Ownership of Risk	Risk Treatment/ Control Measures
		3. Change orders	1. Change order submittal procedure is not followed	Delay in project work	Construction Manager (Contractor)	Follow change order procedure (refer Figure 3.22 in Chapter 3 and Figure 4.6)
			2. Delay in change order negotiations			
		4. Payments	1. Delay in payment	Project progress	Project Owner	Establish payment procedure
	8.2 Documentation					
		1. Status Log	Project status logs are not maintained/ updated properly	Knowledge of project progress	R.E.	Maintain status log for all incoming and outgoing documents
		2. Issue Log	Issue logs are not maintained/ updated properly	Knowledge of project progress	R.E.	Maintain issue log and update with current status
		3. Minutes of Meetings	Minutes of meetings are not distributed to the relevant stakeholders	Project status to stakeholders	R.E.	Establish procedure for circulation of minutes of meeting
		4. Contract Documents	Procedure specified in the contract is not followed to manage the documents	Delay in resolving the conflict	R.E.	Establish and follow the procedure to manage contract documents as specified in the contract
					Construction Manager (Contractor)	Establish and follow the procedure to manage contract documents as specified in the contract
		5. Specifications	Specifications are not updated to incorporate approved change order/ specifications	Quality of project	R.E.	Establish procedure to update approved changes in the specifications

Item	Cause	Effect	Responsibility	Action
6. Payments	1. Payment request is not submitted as per approved communication matrix 2. Corresponding documents are not attached with progress payment request 3. Delay in issuance of payment certificate	Delay in approval of progress payment	Construction Manager (Contractor)	Submit progress payment request attached with all the relevant documents as per contract requirements to avoid delay in approval of progress payment
8.3 Correspondence				
1. Stakeholders	Communication matrix for correspondences with stakeholders is not established	Affect smooth flow of information about project	R.E.	Establish communication matrix for proper communication among all the concerned stakeholders
2. Regulatory Authorities	Procedure for correspondence with regulatory authorities not established	Delay in liaison with regulatory authorities	Construction Manager (Contractor)	Establish method of communication between regulatory authorities
3. Correspondence among Team Members	1. Site administrative and communication matrix is not established 2. Stakeholder's responsibilities matrix not established	Stakeholders will not be aware of their involvement for necessary action/information	R.E. / Construction Manager (Contractor)	Establish matrix for site administration and communication / Establish matrix for site administration and communication
9 Manage Risk				
9.1 1. Risk Register	Risk register is not updated every time a new risk is identified, or relevant actions are taken	Details of all the identified risks will not be known	R.E. / Construction Manager (Contractor)	Identify and update risk register on a regular basis / Identify and update risk register on a regular basis
2. Risk Response	Actions not determined	Proper addressing for developing options and action on the identified, assessed risks	R.E. / Construction Manager (Contractor)	1. Identify the response for the identified risk 2. Select the risk response for identified risk / 1. Identify the response for the identified risk 2. Select the risk response for identified risk

(Continued)

TABLE 4.8
(Continued)

Serial Number	Activities	Major Elements	Potential Risk	Probable Effect on Project	Ownership of Risk	Risk Treatment/ Control Measures
10	10.1 Contract Documents					
		1. Notice to Proceed	Delay in beginning of project work after issuance of notice to proceed	Delay in execution of activities as per schedule	Project Owner	1. Complete all related documents and obtain permits prior to issuance of notice to proceed
						2. Mutually agree with the date/date of enterprise and issue the notice to proceed
					Construction Manager (Contractor)	Start site works as per the agreed-upon date and as per contract documents
	10.2 Selection of Subcontractor(s)					
		1. Delay in submitting the names of subcontractors for approval		1. Delay in start of subcontracted works	Construction Manager (Contractor)	1. Submit subcontractors for approval in a timely and orderly manner, as planned in the approved schedule
		2. Subcontractor submission and approval log is not maintained		2. Nonavailability of approval status		2. Maintain subcontractor approval log
	10.3 Selection of Materials, Systems, and Equipment					
		1. Material selection as per specifications and approval procedure is not followed		1. Delay in the approval of material	Construction Manager (Contractor)	1. Follow the material submittal procedure as per contract documents (refer Figure 4.10)
		2. Long-lead items are not identified		2. Delay in execution of relevant activity(ies)		2. Identify long-lead items
		3. Details of procurement data relating to all the approved products, equipment, and systems are not maintained		Nonavailability of procurement status		3. Prepare product specification for procurement enquiry

10.4 Execution of Works

Item	Cause	Risk	Responsibility	Mitigation
	Construction work is not executed as per approved shop drawings 2. Method statement is not followed 3. Approved material is not installed 4. Routine inspection/supervision during the construction process is not carried out	Rejection of works by supervisor resulting in delay in completion of project	Construction Manager (Contractor)	Execute the construction works as per: 1. Approved shop drawings in accordance with contract documents and specifications 2. Following approved method statement 3. Installing approved material 4. Performing inspection throughout the execution of project

11 11.1 HSE Management Plan

Item	Cause	Risk	Responsibility	Mitigation
1. Site safety	Necessary measures are not followed to ensure safety of all those working at construction site	More accidents at site	R.E./Safety Officer Construction Manager (Contractor)	Establish site safety plan Follow site safety procedures by taking necessary measures to ensure safety of all those working at the construction site
2. Preventive and Mitigation Measures	1. The authority of the company is not giving importance to preventive and mitigation measure to maintain safe and healthful working conditions that will safeguard employees 2. Safety trainings are not conducted 3. Safety meetings are not conducted	1. Unsafe conditions at site 2. Workforce unaware of importance of safety 3. Follow-up of safety measures	Construction Manager (Contractor)	Enforce safety measures and accident prevention program.
3. Temporary Firefighting	1. Temporary firefighting system is not installed at site 2. Temporary firefighting system is not approved by the relevant authority 3. Temporary firefighting system is not working	Unsafe site conditions	Construction Manager (Contractor)	Install temporary firefighting system and ensure that it is operating all the time

(Continued)

TABLE 4.8
(Continued)

Serial Number	Activities	Major Elements	Potential Risk	Probable Effect on Project	Ownership of Risk	Risk Treatment/ Control Measures
		4. Environmental Protection	ISO 14000 environmental management system is not in place	Unsafe site condition	Construction Manager (Contractor)	Establish environmental management system
		5. Waste Management	Waste management plan is not established	Unsafe site condition	Construction Manager (Contractor)	Establish waste management system
		6. Safety Hazards	1. Hazards management and precautionary measures are not operative 2. Precautionary measures to avoid leakage of hazardous material are not established	Dangerous situation at site	Construction Manager (Contractor)	Take necessary preventive measures for the storage of hazardous material
12	Financial Management		1. Financial management plan is not established 2. Cash flow is not as per estimated S-Curve 3. Monitoring and control of project cost is not done properly	Managing regular payments	Project Owner / R.E.	Establish financial management plan considering all the payments to be paid to the project teams (Designer, contractor, etc.) Maintain payment records and cash flow statement
13	Claim Management		1. Identified claims are not submitted with all the supportive documents for justification 2. Claim management process is not followed to resolve the claims 3. Delay in resolving the claim 4. Preventive actions are not taken to mitigate claims	1. Delay in approval of the claim 2. Rejection of claims 3. Delay in claim resolution	Construction Manager (Contractor)	1. Submit claims with all the supportive documents 2. Amicably resolve the claim as per the condition of contract 3. Take preventive action to mitigate effects of claims on construction project

Figure 4.9 illustrates logic flowchart for the development of contractor's construction schedule.

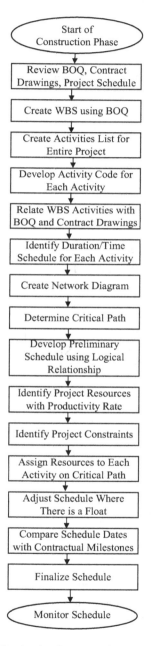

FIGURE 4.9   Logic flowchart for the development of contractor's construction schedule.

Source: Abdul Razzak Rumane. (2013). Quality Tools for Managing Construction Projects. Reprinted with permission of Taylor & Francis Group.

Figure 4.10 illustrates material/product/system approval procedure.

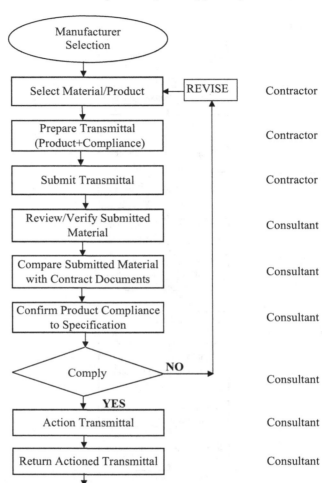

FIGURE 4.10   Material/product/system approval procedure.

Source: Abdul Razzak Rumane. (2013). Quality Tools for Managing Construction Projects. Reprinted
with permission of Taylor & Francis Group.

Figure 4.11 illustrates shop drawing preparation and approval procedure.

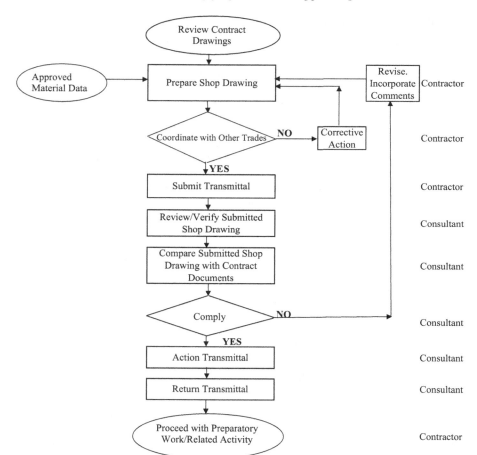

FIGURE 4.11   Shop drawing preparation and approval procedure.

Source: Abdul Razzak Rumane. (2013). Quality Tools for Managing Construction Projects. Reprinted with permission of Taylor & Francis Group.

4.4.3   Risk Monitoring in Activities Related to Quality of Executing Process Group

Table 4.9 is an example process of risk monitoring of Executing Process Group activity.

TABLE 4.9

Risk Monitoring of Executing Process Group Activity

Serial Number	Description	Action			
1	Risk ID	Executing Process Group—10 (10.4)			

2	Description of Risk	1. Execution of works
		    1.1. Construction work is not executed as per approved shop drawings
		    1.2. Method statement is not followed
		    1.3. Approved material is not installed
		    1.4. Routine inspection/supervision during the construction process is not carried out			

3	Response	Execute the construction works as per:
		    1.1. Approved shop drawings in accordance with contract documents and specifications
		    1.2. Following approved method statement
		    1.3. Installing approved material
		    1.4. Performing inspection throughout the execution of project			

4	Strategy of response	Avoidance	Transfer	Mitigation X	Acceptance
5	Monitoring and control	Risk owner	Review date	Critical issue	
		Construction Manager (Contractor)	D/M/Y	1. Workmanship of executed works 2. Rejection of executed works	
6	Estimated impact on the project	1. Delay in execution of project 2. Extra cost for rework			
7	Actual impact on the project	3 months delay			
8	Revised response	Execute the work with approved shop drawings, method and installing approved material			
9	Record update date	D/M/Y			
10	Communication to stakeholders	Information sent to all stakeholders on D/M/Y			

4.5 RISK MANAGEMENT IN QUALITY OF MONITORING AND CONTROL PROCESS GROUP

Figure 4.12 illustrates a flowchart for risk management in quality of monitoring and control process group.

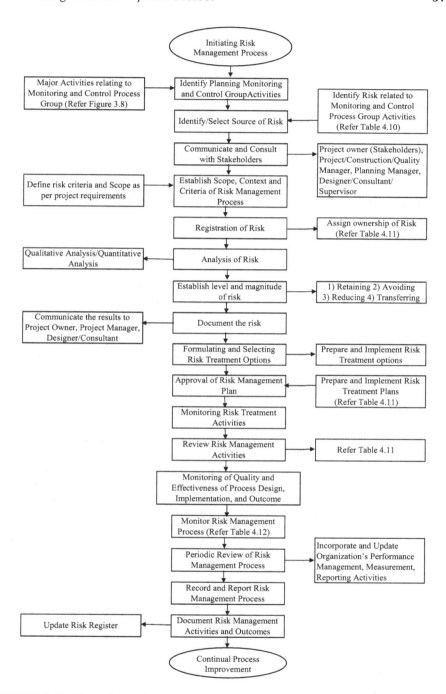

FIGURE 4.12 Typical flowchart for risk management in quality of monitoring and control process group.

4.5.1   RISK IN ACTIVITIES RELATED TO QUALITY OF MONITORING AND CONTROL PROCESS GROUP

Table 3.4 discussed in Chapter 3, Section 3.2.2, illustrates major activities relating to Monitoring and Control Process group.

Table 4.10 lists potential risks in the activities related to Monitoring and Control Process group.

TABLE 4.10
Potential Risks in Activities Related to Monitoring and Control Process Group

Serial Number	Activities	Major Elements	Potential Risk
1	1.1 Project Performance		
		1. Design performance	1. Design not meeting stakeholders' requirements
			2. Design does not support owner's goals and objectives and meet owner's requirements
			3. Design not meeting regulatory authorities' requirements
			4. Design not meeting the specific codes and standards
			5. Design does not meet technical and functional capability
			6. Design does not suit constructability
			7. Environmental impact not considered
		2. Construction Performance	1. Poor workmanship
			2. Constructed facility may not meet the intended use/need of the owner/end user
			3. Constructed facility may not have supportability during maintenance/maintainability
			4. Constructed facility may not have aesthetic harmony between the structure and surrounding nature and built environment
		3. Project start-up	1. Electromechanical works/systems/equipment are not fully functional and operate as specified after energizing
			2. Individual work/system are not functional and useful for the owner/end user after start up and commissioning
		4. Forecasted schedule	Current situations of all activities/tasks, milestones, sequencing, resources, duration, constraints, and project update are not monitored to establish and control forecasted schedule

Serial Number	Activities	Major Elements	Potential Risk
		5. Forecasted cost	1. Factors that influence the changes to the cost baseline are not identified to detect cost variance
			2. Cost variance factors are not established to prepare forecasted cost
		6. Issues	Issues are not responded efficiently and effectively to resolve successfully to ensure that the project deliverables meet stakeholder expectations
	1.2 Change Management		
		1. Design changes	Design changes/modifications are not identified in time to avoid disruption of work and its impact on time and cost
		2. Design errors	Design errors may result in change order
		3. Change requests	Proper submission process is not followed to resolve change requests
		4. Scope change	Scope change may result in claim
		5. Variation orders	Variation order proposal is not reviewed properly to avoid major impact on the project
		6. Site Work Instruction (SWI)	Site work instruction is issued without proper documents along with SWI
		7. Alternate material	Alternate material may not perform as per the specified material
		8. Specified methods	1. Delay in the completion of project
			2. Delay in the modification of specified methods
	1.3 Change Analysis		
		1. Review and evaluate changes	1. Changes are not reviewed, evaluated as per established reliable change control system
			2. Changes are not reviewed, evaluated for negative impact on the project
		2. Approve, delay, reject changes	1. Delay in processing of change request
			2. Delay in issuance of change order
		3. Corrective actions	1. Delay in corrective actions or defect repair
		4. Preventive actions	1. Delay in implementation of preventive actions
	1.4 Compliance to Contract Documents		
		1. Difference/errors in contract documents	
		2. Delay to clarify differences/errors observed in contract documents	
		3. Nonavailability of specified resources (material, systems, equipment)	
2	2.1 Project Performance		
		1. Progress reports	Progress reports/status is not reported/distributed to the relevant stakeholders
		2. Updates	Project updates are not distributed to the relevant stakeholders
		3. Safety report	Safety report is not distributed, and appropriate action is taken
		4. Risk report	Risk report is not distributed to the risk owner in order for appropriate action to be taken

(Continued)

TABLE 4.10
(Continued)

Serial Number	Activities	Major Elements	Potential Risk
	2.2	Project Updates	
		1. Contract documents	Approved changes are not updated regularly
	2.3	Payments	
		1. Payment certificate	Delay in approval of progress payment and issuance of payment certificate
	2.4	Change Requests	
		1. Site Work Instruction	Delay in issuing Site Works Instruction
		2. Change orders	Delay in change order proposal review and issuance of change order
	2.5	Issue Log	
		1. Anticipated problems	Anticipated problems are not identified and listed in the Issue Log
	2.6	Minutes of Meetings	
		1. Progress meetings	1. Minutes of progress meeting are not prepared as per the specified format
			2. Minutes of progress meeting are not distributed to all the attendees and relevant stakeholders
		2. Other meetings	Minutes of different types of meetings held during the project are not distributed to all the attendees and relevant stakeholders
3	3.1	Validate Scope	
		1. Conformance to TOR	The design has not taken care of all the requirements listed under TOR
		2. Review of design documents	The design is not reviewed for accuracy of drawings, interdisciplinary coordination, and documents prior to proceeding with next stage
		3. Conformance to contract documents	1. All the requirements listed/ described under TOR are not utilized for the development of contract documents to suit the project delivery system
			2. Differences/errors in contract documents
		4. Approval of changes	Approved changes are not recorded and updated
		5. Authorities' approval of deliverables	Authorities' approval is not obtained
		6. Stakeholders' approval of deliverables	Project deliverables may not meet stakeholders' expectations
		7. Quality audit	Quality audit not performed in a timely manner
	3.2	Scope Change Control	
		1. Variation orders	A reliable change control system is not established to manage changes to the project scope

Serial Number	Activities	Major Elements	Potential Risk
		2. Change orders	Change orders are not processed properly, considering who initiated change to resolve scope change
	3.3	Performance Measures	
			Performance measures are not defined to achieve the goals and objectives/owner's requirements of the project
4	4.1	Schedule Monitoring	
		1. Project status	Project control mechanism is not used to determine the current status of the schedule, the precise effect of these deviations on the plan that causes the schedule changes, and to replan and reschedule the status of project
	4.2	Schedule Control	
		1. Progress curve	The progress curve/network diagram is not replanned and rescheduled considering the revised duration of unfinished activities
	4.3	Schedule Changes	
		1. Approved changes	The schedule is not updated reflecting the approved changes, current situation of all activities/ tasks, milestones, sequencing, resources, duration, and constraints
	4.4	Progress Monitoring	
		1. Planned versus actual	As-built schedule (actual) is not prepared comparing the baseline (planned) schedule providing the following information: • Completion of each activity • Activities scheduled to start but not yet started • Remaining duration to complete each activity • Milestones not yet reached
	4.5	Submittals Monitoring	
		1. Subcontractors	Subcontractors' submittal and approval log is not maintained
		2. Material	Procurement log-E2 is not maintained
		3. Shop Drawings	Shop drawings status log is not maintained
5	5.1	Cost Control	
		1. Work performance	Cost control is not identifying the activities that influence the work performance and changes to the cost baseline

(Continued)

TABLE 4.10
(Continued)

Serial Number	Activities	Major Elements	Potential Risk
		2. S-Curve	1. Cost baseline changes are reflected in the S-Curve 2. Unexpected changes are not updated in the S-Curve
		3. Forecasted cost	1. Project progress in not monitored and compared with cost baseline to identify cash flow forecasts
	5.2	Change Orders	

Nonidentification of the factors that influence the changes in the cost baseline

	5.3	Progress Payment	

1. Delay in the payment of progress payment
2. Submitted progress payment is not as per contractual entitlement
3. Progress payment request is not as per the approved executed works

	5.4	Variation Orders	

1. Variation orders are not properly evaluated and processed
2. Delay in the approval of variation orders

6	6.1	Control Quality	
		1. Quality metrics	Responsibilities matrix for quality control is not established
		2. Quality checklist	Proper sequencing and method statement for installation of works are not followed prior to the submission of checklist
		3. Material inspection	1. Off-site inspection of material is not carried out 2. Material inspection checklist for material received at site is not submitted
		4. Work inspection	Work inspection is not continuously carried out throughout the construction period either by the contractor, construction supervision team, or appointed inspection agency
		5. Rework	Delay in completion of rework
		6. Testing	1. Testing and commissioning plan is not developed
		7. Regulatory compliance	The constructed facility does not fully comply with regulatory requirements
7	7.1	Conflict Resolution	

1. Conflict management process is not followed
2. The training in communication and interpersonal skills is not given to cross-functional team members

	7.2	Performance Analysis	

Project team members' performance track is not maintained to ensure project performance optimization

	7.3	Material Management	

Material management process for project is not followed

Serial Number	Activities	Major Elements	Potential Risk
8	8.1	Meetings	
		1. Progress meetings	Progress meetings are not conducted on regular basis as per specified frequency in the contract documents
		2. Coordination meetings	Coordination meetings are not conducted
		3. Safety meetings	Safety meetings are not conducted
		4. Quality meetings	Quality meetings are not conducted
	8.2	Submittal Control	
		1. Drawings	Log for submittal of shop drawings is not maintained and updated
		2. Material	Log for submittal of material is not maintained and updates
	8.3	Documents' Control	
		1. Correspondence	1. Communication matrix is not established
			2. Proper documentation is not used for communication with external stakeholders
9	9.1	Monitor and Control Risk	
		1. Scope change risk	1. The source of risk is not identified
			2. Identified risk is not registered
			3. Timely implementation of risk response to the identified risk is not done
		2. Schedule change risk	1. The source of risk is not identified
			2. Identified risk is not registered
			3. Timely implementation of risk response to the identified risk is not done
		3. Cost change risk	1. The source of risk is not identified
			2. Identified risk is not registered
			3. Timely implementation of risk response to the identified risk is not done
		4. Mitigate risk	Early actions are not taken to reduce/mitigate the probability of risk
		5. Risk audit	Risk audit is not performed to risk management system
10	10.1	Inspection	
		Inspection of works is not carried out regularly to ensure full compliance with the contract documents	
	10.2	Checklists	
		Daily checklist status report is not maintained	
	10.3	Handling of Claims, Disputes	
		1. Unable to resolve claims and disputes	
		2. Delay in resolving the claims, disputes	
11	11.1	Prevention Measures	
		1. Accidents' avoidance/ mitigation	Safety and accident prevention program is not followed

(Continued)

**TABLE 4.10
(Continued)**

Serial Number	Activities	Major Elements	Potential Risk
		2. Firefighting System	Firefighting system is not working
		3. Loss prevention measures	Fundamental concepts and methodology of the program to identify all loss exposures, evaluate the risk of each exposure, plan how to handle each risk, and manage according to plan are not established and followed
	11.2 Application of Codes and Standards		
	Applicable safety codes and standards are not followed		
12	12.1 Financial Control		
		1. Payments to project team members	Delay in payments to project team members
		2. Payments to Contractor(s)/ Sub Contractor(s)	1. Cost-loaded schedule for determining the contract payments during the project against the approved progress of works by contractor and subcontractor is not followed
			2. The funds to pay progress payments on regular basis are not arranged
		3. Material purchases	Payments toward material purchases are not paid as per the terms in purchase order
		4. Variation order payment	Approved payment toward variation order is not in time
		5. Insurance and bonds	Delay in payment of insurance and bonds
	12.2 Cash Flow		
	S-Curve is not used to track the progress of the project over time and is not updated		
13	13.1 Claim Prevention		
		1. Proper design review	Design review process is not followed
		2. Unambiguous contract Documents' language	Contract documents are not reviewed for unambiguousness and clearly written to enable all team members to understand
		3. Practical schedule	Impractical planning and schedule
		4. Qualified contractor(s)	Contractor has lack of knowledge and experience on similar type of projects
		5. Competent project team members	1. Incompetent project team members
			2. Failure of team members to perform as expected
		6. RFI review procedure	Delay in processing and response to RFI
		7. Negotiations	Negotiations failed to resolve the issue
		8. Appropriate project delivery system	1. Project delivery system is selected not considering size, type, and complexity of project
			2. Project delivery system is selected not considering the level of owner's expertise

4.5.2 Risk Effect and Risk Treatment in Activities Related to Quality of Monitoring and Control Process Group

Table 4.11 lists potential risks, probable effect on project, and risk treatment in activities related to Monitoring and Control Process Group.

TABLE 4.11
Potential Risks, Probable Effect, and Risk Treatment in Activities Related to Monitoring and Control Process Group

Serial Number	Activities	Major Elements	Potential Risk	Probable Effect on Project	Ownership of Risk	Risk Treatment/ Control Measures
1	1.1 Project Performance	1. Design performance	1. Design not meeting stake holders' requirements 2. Design does not support owner's goals and objectives and meet owner's requirements 3. Design not meeting regulatory authorities' requirements 4. Design not meeting the specific codes and standards 5. Design does not meet technical and functional capability 6. Design does not suit constructability 7. Environmental impact not considered	1. Project quality not meeting stakeholder's requirements 2. Project not meeting owner's goals and objectives 3. Difficult to get regulatory approval for the project 4. Project not meeting the specified requirements 5. Project not suitable for specified usage 6. Project difficult to construct 7. Project affects environment	Construction Manager (Contractor)	1. Identify any obstacles encountered during execution of project and apply measures to mitigate these difficulties and to ensure that goals and objectives of project are being met. 2. Raise RFI for related matters such as design discrepancies, noncompliance to regulatory requirements, environmental issues, etc., to resolve the discrepancies in the project design

(Continued)

TABLE 4.11 (Continued)

Serial Number	Activities	Major Elements	Potential Risk	Probable Effect on Project	Ownership of Risk	Risk Treatment/ Control Measures
		2. Construction Performance	1. Poor workmanship 2. Constructed facility may not meet the intended use/ need of the owner/end user 3. Constructed facility may not have supportability during maintenance/ maintainability 4. Constructed facility may not have aesthetic harmony between the structure and surrounding nature and built environment	1. Quality of project 2. Project not suitable for intended usage 3. Difficulties during maintenance 4. Different aesthetic look around the project environment	Construction Manager (Contractor)	1. Use competent, skilled workforce 2. Monitor works regularly 3. Execute the project as per approved shop drawings with approved material as per the specifications to meet the intended use of the project 4. Raise Request for Information (RFI) and modify the works taking into consideration supportability/ maintainability 5. Raise Request for Information (RFI) and modify the works as per the instruction from the project owner
		3. Project start up	1. Electromechanical works/ systems/equipment are not fully functional and operate as specified after energizing 2. Individual work/system are not functional and useful for the owner/end user after start-up and commissioning	1. Delay in start-up of the project 2. Project not suitable for usage as per the owner's needs	Construction Manager (Contractor)	1. Arrange factory inspection/ test reports/pre-shipment inspection 2. Inspect and test during installation/ execution of works on regular basis

4. Forecasted schedule	Current situation of all activities/ tasks, milestones, sequencing, resources, duration, constraints, and project update are not monitored to establish and control forecasted schedule	Difficult to monitor project status and to recognize any obstacles encountered during execution of project	Construction Manager (Contractor)/ PCM	1. Monitor and control the schedule reflecting the current situation of all activities/ tasks, milestones, sequencing, resources, duration, constraints and project update.(refer Figure 3.37 in Chapter 3) 2. Identify millstone activities and perform as included in the schedule
5. Forecasted cost	Factors that influence the changes to the cost baseline are not identified to detect cost variance and are not established to prepare forecasted cost	Difficult to know the status of financial obligations	Construction Manager (Contractor)/ PCM	1. Develop cost baseline based on the contracted construction budget. 2. Prepare S-Curve by loading each activity in the approved schedule with the approved project budget cost in the BOQ (cost loading) 3. Monitor the status of the project to update the project cost and managing changes to the baseline
6. Issues	Issues are not responded to efficiently and effectively to get resolved successfully to ensure that the project deliverables meet stakeholder expectations	Project status not known to the stakeholders	Construction Manager (Contractor)/ PCM	Manage status logs and follow submittal procedure to distribute submittals to the relevant stakeholders

(Continued)

TABLE 4.11
(Continued)

Serial Number	Activities	Major Elements	Potential Risk	Probable Effect on Project	Ownership of Risk	Risk Treatment/ Control Measures
	1.2	Change Management				
		1. Design changes	Design changes/modifications are not identified in time to avoid disruption of work and its impact on time and cost	Delay in project execution	Construction Manager (Contractor)	Identify discrepancies/ errors and changes in the specified scope and use Request for Information (RFI) to clarify/resolve the issue (refer Figure 3.21 in Chapter 3)
		2. Design errors	Design errors may result in change order	Overrun of project estimated cost	Project Manager (Design)	Resolve scope change at the earliest (refer Figure 3.22 in Chapter 3)
		3. Change requests	Proper submission process is not followed to resolve change requests	Delay in approval of change order	Construction Manager (Contractor)	Follow the process to resolve the change request (refer Figure 3.23 and Figure 3.22 in Chapter 3 and Figure 4.6 as applicable)
		4. Scope change	Claim/variation request by the contractor	1. Additional time for the completion of project 2. Additional cost for completion of project	Project Manager (Design)	1. Review scope change 2. Follow the process to resolve the scope change (refer Figure 3.22 in Chapter 3 and Figure 4.6) and resolve the issue expeditiously
		5. Variation orders	Variation order proposal is not reviewed properly to avoid major impact on the project	1. Delay in approval of variation order 2. Additional cost toward project	R.E.	Review variation order proposal as per the conditions in the contract documents (refer Figure 3.24 in Chapter 3)

6. Site Work Instruction (SWI)	Site works instructions is issued without proper documents along with SWI	Delay in review of SWI	Project Owner/R.E.	Issue SWI with all the necessary documents followed by variation order (refer Figure 3.25 in Chapter 3)
7. Alternate material	Alternate material may not perform as per the specified material	Achieving specified project quality	Construction Manager (Contractor)	Submit approved equal material that meets the specification requirements for approval
8. Specified methods	1. Delay in the completion of project 2. Delay in modification of specified methods	1. Delay in handover/takeover of the project 2. Delay in execution of relevant activity(ies)	Construction Manager (Contractor)	1. Follow specified method/manufacturer-recommended method for installation/execution of works 2. Submit method statement for approval by R.E. (Construction Supervisor)
1.3 Change Analysis				
1. Review, evaluate changes	1. Changes are not reviewed, evaluated as per established reliable change control system 2. Changes are not reviewed, evaluated for negative impact on the project	1. Project scope, schedule, and cost 2. Achieving specified project quality, schedule, and cost	R.E.	Review and evaluate changes as per contract conditions
2. Approve, delay, reject changes	1. Delay in processing of change request 2. Delay in issuance of change order	Delay in execution of relevant activity(ies)	R.E.	Review, evaluate, and approve the changes if comply with contract documents
3. Corrective actions	1. Delay in corrective actions or defect repair	Delay in project completion	Construction Manager (Contractor)	Take corrective action immediately after the receipt of nonconformance report
4. Preventive actions	1. Delay in implementation of preventive actions	Delay in project completion	Construction Manager (Contractor)	Take preventive action to prevent, mitigate occurrence of nonconforming work

(Continued)

TABLE 4.11
(Continued)

Serial Number	Activities	Major Elements	Potential Risk	Probable Effect on Project	Ownership of Risk	Risk Treatment/ Control Measures
	1.5	Compliance to contract documents				
		1. Difference/errors in contract documents		1. Delay in execution of project	R.E.	Resolve any RFI raised by contractor without any delay to avoid unnecessary disruption of work
		2. Delay to clarify differences/errors observed in contract documents		2. Delay in execution of project		
		2. Nonavailability of specified resources (material, systems, equipment)		3. Delay in project execution due to additional time to approve substitute material	Construction Manager (Contractor)	Contractor to identify such discrepancies and submit RFI to clarify differences/errors observed
2	2.1	Project Performance				
		1. Progress reports	Progress reports/status are not reported /distributed to the relevant stakeholders	Stakeholders not aware of project information and progress status	R.E.	1. Monitor project performance regularly 2. Report the necessary information, project status, progress reports to the relevant stakeholders
		2. Updates	Project updates are not distributed to the relevant stakeholders	Stakeholders not aware of project progress status	R.E.	Distribute the update for information to the relevant stakeholders

3. Safety report	Safety report is not distributed, and appropriate action is taken	Stakeholders unable to take necessary action	R.E.	Distribute safety reports to concerned stakeholders for their information and further action
4. Risk report	Risk report is not distributed to the risk owner for appropriate action to be taken	Stakeholders not aware of risks on project and its impact	R.E.	Distribute identified risk to the relevant stakeholders
2.2 Project Updates				
1. Contract documents	Approved changes are not updated regularly	Difficult to monitor and control the project	R.E.	Update the contract documents with approved changes regularly
2.3 Payments				
1. Payment certificate	Delay in approval of progress payment and issuance of payment certificate	Contractor's cashflow is affected	R.E.	Review, evaluate, and take action on the submitted payment request within the specified time without any delay
2.4 Change Requests				
1. Site Work Instruction	Delay in issuing Site Works Instruction	Delay in execution of works	Project Owner/R.E.	Resolve RFI expeditiously to avoid its effect on construction schedule by giving instructions (SWI) to proceed with the changes (refer Figure 3.22 in Chapter 3 and Figure 4.6 as applicable)
2. Change orders	Delay in change order proposal review and issuance of change order	Delay in the execution of particular activity(ies)	Project Owner/R.E.	Assess the cost and time related to change order proposal and take appropriate action without any delay
2.5 Issue Log				
1. Anticipated problems	Anticipate problems are not identified and listed in the issue log	Delay in project progress	R.E.	Anticipate/forecast any adverse problem that will have impact on the problem and include in the issue log for further action

(Continued)

TABLE 4.11
(Continued)

Serial Number	Activities	Major Elements	Potential Risk	Probable Effect on Project	Ownership of Risk	Risk Treatment/ Control Measures
	2.6	Minutes of Meetings				
		1. Progress meetings	1. Minutes of progress meeting are not prepared as per the specified format 2. Minutes of progress meeting is not distributed to all the attendees and relevant stakeholders	1. Lack of complete information to the stakeholders 2. Stakeholders are not aware of project information status	R.E.	Prepare and distribute the minutes of meetings as per approved format and as per communication matrix
		2. Other meetings	Minutes of different types of meetings held during the project are not distributed to all the attendees and relevant stakeholders	Stakeholders are not aware of the project progress and status	R.E.	Distribute the minutes of meetings to related stakeholders as per the communication matrix
3	3.1	Validate Scope				
		1. Conformance to TOR	The design has not taken care of all the requirements listed under TOR	Project not meeting owner's requirements and needs	Project Manager (Design)	1. Develop project design taking care of all the requirements listed under TOR 2. Develop project design taking care of requirements of all the related stakeholders
		2. Review of design documents	The design is not reviewed for accuracy of drawings, interdisciplinary coordination, and documents prior to proceeding with next stage	1. Design errors 2. Additional time to resolve coordination problems 3. Delay in execution of project	Project Manager (Design)	Review design at each phase of the design development phases to ensure the design meeting the design deliverables and meet owner's requirements

3. Conformance to contract documents	1. All the requirements listed/described under TOR are not utilized for development of contract documents to suit the project delivery system 2. Differences/errors in contract documents	1. Conflict among procedures, actions, and sequence of events and participants responsibilities for smooth execution of project 2. Conflict in installation and integration of various activities	Project Manager (Design)	1. Develop contract documents based on the requirements described in the TOR to suit the project delivery system 2. Review the documents to avoid, mitigate errors/omissions and conflict among different trades
4. Approval of changes	Approved changes are not recorded and updated	Stakeholders not aware of project status	R.E.	Record the approved changes and update the scope and other related data
5. Authorities' approval of deliverables	Authorities' approval is not obtained	Delay in the approval of completed project	Construction Manager (Contractor)	Establish the procedure for submission and approval by Regulatory Authorities
6. Stakeholders' approval of deliverables	Project deliverables may not meet stakeholders' expectations	Project not meeting owner's/stakeholders' requirements	Project Manager (Design)	Develop stakeholders' responsibilities matrix and procedure to distribute reports that has stakeholders' involvement
7. Quality audit	Quality audit not performed in a timely manner	Loss of opportunities for improvements	Construction Manager (Contractor)/ Quality Manager	Perform quality audit as per specified quality policy
3.2 Scope Change Control				
1. Variation orders	A reliable change control system is not established to manage changes to the project scope	Unnecessary disruption of works resulting delay in project	R.E.	Establish change control procedure (refer Figure 3.20 in Chapter 3)

(Continued)

TABLE 4.11
(Continued)

Serial Number	Activities	Major Elements	Potential Risk	Probable Effect on Project	Ownership of Risk	Risk Treatment/ Control Measures
		2. Change orders	Change orders are not processed properly, considering who initiated change, to resolve scope change	Impact on time and cost of the project	R.E.	Follow change order process (refer Figure 3.22 in Chapter 3 and Figure 4.6 as applicable)
	3.3	Performance Measures				
		Performance measures are not defined to achieve the goals and objectives/ owner's requirements of the project		Project delivery not meeting owner's requirements	Project Manager (Design)	Develop project design to meet the owner's requirements and quality of project delivery, by defining performance measures to be achieved by the contractor
4	4.1	Schedule Monitoring				
		1. Project status	Project control mechanism is not used to determine the current status of the schedule, the precise effect of these deviations on the plan that causes the schedule changes, and to replan and reschedule the status of project	Project progress, status reports will not be accurate	R.E./PCM	Prepare schedule taking into consideration following points: 1. Properly identifying the activities and listing with logical sequence of the activities to be performed 2. Duration of each activity 3. Correctly estimating resources and duration 4. Assumptions and constraints 5. Identifying schedule reserves/ contingencies

4.2 Schedule Control			
1. Progress curve	The progress curve/network diagram is not replanned and rescheduled considering the revised duration of unfinished activities	Stakeholders will not be aware of current status, project progress to take appropriate actions	R.E./PCM	Report the necessary information in detail that can be interpreted to provide information to the concerned stakeholders
4.3 Schedule Changes			
1. Approved changes	The schedule is not updated reflecting the approved changes, current situation of all activities/tasks, milestones, sequencing, resources, duration, and constraints	Stakeholders will not be aware of current status of the schedule identifying the influence factors that cause schedule changes	R.E./PCM	Update the schedule with following information: 1. Percentage of the completion of each activity based on approved checklist 2. Approved changes 3. Current situation of all the activities
4.4 Progress Monitoring			
1. Planned versus actual	As-built schedule (actual) is not prepared comparing the baseline (planned) schedule providing the following information: • Completion of each activity • Activities scheduled to start but not yet started • Remaining duration to complete each activity • Milestones not yet reached	Measuring, verifying project progress and execution of the project activities and necessary information	R.E./PCM	Prepare as-built schedule with following information: 1. Completion of each activity 2. Activities scheduled to start but not yet started 3. Activities scheduled to complete but under progress 4. Remaining duration to complete each activity 5. Milestones not yet reached 6. Problems and issues

(Continued)

TABLE 4.11
(Continued)

Serial Number	Activities	Major Elements	Potential Risk	Probable Effect on Project	Ownership of Risk	Risk Treatment/ Control Measures
4.5	Submittals Monitoring					
		1. Subcontractors	Subcontractors' submittal and approval log is not maintained	Stakeholders unaware of subcontractors' approval status	R.E./Document Controller	1. Establish submittal procedure a per contract documents
						2. Establish communication matrix
						3. Manage submittals and approval logs (E-1)
						3. Monitor subcontractor approval log
					Construction Manager (Contractor)	1. Establish submittal procedure as per contract documents (log E-1)
						2. Manage subcontractors' approval logs
		2. Material	Procurement log-E2 is not maintained	Stakeholders not aware of procurement status	R.E./Document Controller	1. Manage submittal procedure as per communication matrix
						2. Maintain procurement log E-2
					Construction Manager (Contractor)	Maintain submission and approval of procurement log E-2
		3. Shop drawings	Shop drawings' status log is not maintained	Stakeholders not aware of shop drawing status	R.E./Document Controller	1. Manage submittal procedure as per communication matrix
						2. Maintain shop drawings' approval log
					Construction Manager (Contractor)	Maintain shop drawing submittal and approval log

5   5.1   Cost control

	Risk	Consequence		Actions
1. Work performance	Cost control is not identifying the activities that influence the work performance and changes to the cost baseline	Stakeholders will not know the amount that will be spent over the established project schedule	R. E./PCM	1. Establish project cost baseline in a time-phased manner that can be used as a baseline against which the overall cost performance of the project is measured, monitored, and controlled 2. Establish cost control taking into consideration work performance reports
2. S-Curve	Cost baseline changes are not reflected in the S-Curve 2. Unexpected changes are not updated in the S-Curve	Accurate information and prediction about project funding requirements	R.E/PCM	1. Develop S-Curve to correctly measure the project performance and predict the expenses over project duration 2. Update changes in the S-Curve
3. Forecasted cost	Project progress in not monitored and compared with cost baseline to identify cash flow forecasts	Unable to recognize variance from the approved plan to take corrective action	R.E/PCM	Develop S-Curve on earn value methodology to: 1. Compare and monitor project progress (measuring progress) 2. Identify and control approved variations 3. Identify cash flow forecast 4. Compare current budget forecast for remaining works 5. Provide budget and schedule variance 6. Set contingency

(Continued)

TABLE 4.11
(Continued)

Serial Number	Activities	Major Elements	Potential Risk	Probable Effect on Project	Ownership of Risk	Risk Treatment/ Control Measures
5.2	Change Orders	Nonidentification of the factors that influence the changes in the cost baseline		Project delivery within the approved budget	R.E./PCM	1. Identify the factors that influence the changes to the cost baseline 2. Determine if the changes are beneficial to the project
5.3	Progress Payment	1. Delay in payment of progress payment 2. Submitted progress payment is not as per contractual entitlement 3. Progress payment request is not as per the approved executed works		1. Contractor's payment plans will be affected 2. Delay in approval of progress payment 3. Rejection of progress payment request	1. Project Owner R.E. 2. Construction Manager (Contractor)	1. Release payment of approved progress payment as per contract documents 2. Review and take appropriate action on the progress payment submittal 3. Submit progress payment with all the necessary documents as specified in the contract
5.4	Variation Orders	1. Variation orders are not properly evaluated and processed 2. Delay in approval of variation orders		1. Unnecessary disruption of execution of works 2. Delay in execution of project activity(ies)	R.E./QS	Review and evaluate request for variation as per contract documents (refer Figure 3.23 and Figure 3.24 in Chapter 3)

6 6.1 Control Quality

	Risk	Responsibility	Action
1. Quality metrics	Responsibilities matrix for quality control is not established	Stakeholders not aware of responsibilities and obligations	R.E./Quality Manager Construction Manager (Contractor)	Establish quality matrix for all the quality-related personnel Establish quality matrix for contractor's responsibilities to manage construction quality
2. Quality checklist	Proper sequencing and method statement for installation of works is not followed prior to submission of checklist	Rejection of works by the supervisor results in delay of project	Construction Manager (Contractor)	Establish sequencing of execution of works (refer Figure 4.6)
3. Material inspection	1. Off-site inspection of material is not carried out 2. Material inspection checklist for material received at site is not submitted	Material received at site may not be as specified	Construction Manager (Contractor)	Establish material inspection procedure off-site as well after receipt of material at site as per contract specifications
4. Work inspection	Work inspection is not continuously carried out throughout the construction period either by the contractor, construction supervision team, or appointed inspection agency	Work not conforming to specified quality of works	Construction Manager (Contractor)	Perform the inspection of installed/executed works at regular basis throughout the construction of the project.
5. Rework	Delay in completion of rework	Delay in completion of project	Construction Manager (Contractor)	Repair nonconformed/rejected work without any delay
6. Testing	1. Testing and commissioning plan is not developed	Testing and commissioning not as specified	Construction Manager (Contractor)	Develop inspection and testing plan as per specification requirements
7. Regulatory compliance	The constructed facility does not fully comply with regulatory requirements	Delay in approval by regulatory authority(ies)	Construction Manager (Contractor)	Obtain regulatory approval of all the completed/commissioned works as per regulatory requirements

(Continued)

TABLE 4.11
(Continued)

Serial Number	Activities	Major Elements	Potential Risk	Probable Effect on Project	Ownership of Risk	Risk Treatment/ Control Measures
7	7.1 Conflict Resolution	1. Conflict management process is not followed 2. The training in communication and interpersonal skills is not given to cross-functional team members		1. Delay in resolving the conflict 2. Personality and interpersonal issues may draw conflict	Project Owner/R.E./ Construction Manager	1. Establish conflict management process 2. Resolve conflict by searching for alternate and amicable solution
	7.2 Performance Analysis	Project team members' performance track is not maintained to ensure project performance optimization		Project performance will be affected	R.E. Construction Manager (Contractor)	1. Establish ground rules 2. Coordinate with all team members to understand their issues 3. Create shared vision among team members 4. Track the team performance 5. Manage conflict
	7.3 Material Management	Material management process for project is not followed		1. Delay in getting material as scheduled 2. Delay in project execution	Construction Manager (Contractor)	Establish material management process (refer Figure 3.46 in Chapter 3)

8 Meetings

8.1

			Responsibility	Action
1. Progress meetings	Progress meetings are not conducted on regular basis as per specified frequency in the contract documents	Stakeholders not aware of project progress and exchange information among team members	R.E.	Conduct progress meetings as per contract documents with specified and agreed up frequency 2. Distribute minutes of meeting to the concerned stakeholders
2. Coordination meetings	Coordination meetings are not conducted	Delay in resolving coordination problems among various trades	R.E.	Conduct coordination meeting as needed to resolve the problems, as per agreed-upon frequency
3. Safety meetings	Safety meetings are not conducted	Stakeholders may not know about related health, site safety, and environmental issues	R.E./Safety officer	Conduct safety meetings as per contract requirements
4. Quality meetings	Quality meetings are not conducted	Stakeholders may not know the quality issues at site and how to improve the construction process to avoid/reduce rejection of works	R.E./Quality Manager	Conduct quality meetings as per contract requirements, as per agreed-upon frequency

8.2 Submittal Control

			Responsibility	Action
1. Drawings	Log for submittal of shop drawings is not maintained and updated	Stakeholders are not aware of updated information about shop drawings	R.E./Document Controller Construction Manager (Contractor)	Maintain shop drawing submittal log and update the information to relevant stakeholders Maintain shop drawing submittal log and update the information
2. Material	Log for submittal of material is not maintained and updates	Stakeholders are not aware of updated information about materials	R.E./Document Controller Construction Manager (Contractor)	Maintain material submittal log and update the information Maintain material submittal log and update the information

(Continued)

TABLE 4.11
(Continued)

Serial Number	Activities	Major Elements	Potential Risk	Probable Effect on Project	Ownership of Risk	Risk Treatment/ Control Measures
	8.3	Documents Control				
		1. Correspondence	1. Communication matrix is not established 2. Proper documentation is not used for communication with external stakeholders	1. Smooth flow of communication, roles, and responsibilities 2. Stakeholders not aware of project progress, information	R.E.	1. Maintain logs for all incoming and outgoing documents 2. Distribute the documents among relevant team members, stakeholders
9	9.1	Monitor & Control Risk				
		1. Scope change risk	1. The source of risk is not identified 2. Identified risk is not registered 3. Timely implementation of risk response to the identified risk is not done	Probability and scope of positive events to decrease the probability and impacts of scope change that are averse to the project is not recorded	R.E. Construction Manager (Contractor)	1. Identify occurrence of risk in the project scope and register the same for treatment 2. Determine the action, if any, to be taken on identified and assessed risk 1. Identify occurrence of risk in the project scope and register the same for treatment 2. Determine the action, if any, to be taken on identified and assessed risk

	2. Schedule change risk	1. The source of risk is not identified 2. Identified risk is not registered 3. Timely implementation of risk response to the identified risk is not done	Probability and scope of positive events to decrease the probability and impacts of schedule change that are adverse to the project is not recorded	R.E./PCM	1. Identify occurrence of risk in the project schedule and register the same for treatment 2. Determine the action, if any, to be taken on identified and assessed risk
	3. Cost change risk	1. The source of risk is not identified 2. Identified risk is not registered 3. Timely implementation of risk response to the identified risk is not done	Probability and scope of positive events to decrease the probability and impacts of cost change that are adverse to the project is not recorded	R.E./PCM	1. Identify occurrence of risk in the project cost and register the same for treatment 2. Determine the action, if any, to be taken on identified and assessed risk 3. Implement the risk response without any delay
	4. Mitigate risk	Early actions are not taken to reduce/mitigate the probability of risk	Delay in positive impact for the benefit of the project to minimize/decrease or eliminate the impact of events adverse to the project	R.E./PCM	Determine the action to be taken on identified, assessed risk and implement the response for impact (consequences) on the project
	5. Risk audit	Risk audit is not performed to risk management system	Affect improvement in the quality of project	R.E.	Perform risk audit as per the ISO requirements
10	10.1 Inspection	Inspection of works is not carried out regularly to ensure full compliance with the contract documents	Compliance with the specifications	Construction Manager (Contractor)	Perform inspection of construction works on regular basis throughout the project
	10.2 Checklists	Daily checklist status report is not maintained	Project inspection status not known	R.E.	Maintain a log for checklist status on daily basis

(Continued)

TABLE 4.11
(Continued)

Serial Number	Activities	Major Elements	Potential Risk	Probable Effect on Project	Ownership of Risk	Risk Treatment/ Control Measures
	10.3 Handling of Claims, Disputes					
			1. Unable to resolve claims and disputes	Delay in project closeout	Project Owner R.E.	Resolve claim by applying the following methods:
			2. Delay in resolving the claims, disputes		Construction Manager (Contractor)	1. Negotiation 2. Mediation 3. Arbitration 4. Litigation
11	11.1 Prevention Measures					
		1. Accidents avoidance/ mitigation	Safety and accident prevention program is not followed	Occurrence of more accidents at site	Construction Manager (Contractor)/ Safety Officer	1. Take precautionary measures to avoid and mitigate accidents at site 2. Make work-level site inspection on regular basis
		2. Firefighting system	Firefighting system is not working/operational	Danger in site safety	Construction Manager (Contractor)/ Safety Officer	Check regularly that installed firefighting system at site is operative and functional
		3. Loss prevention measures	Fundamental concepts and methodology of the program to identify all loss exposures, evaluate the risk of each exposure, plan how to handle each risk, and manage according to plan, is not established and followed	1. Relatively high number of injuries and accidents at construction site 2. Strategies to reduce and prevent accidents are not in place	Construction Manager (Contractor)/ Safety Officer	1. Perform loss prevention compliance reviews of the project site 2. Use PPE (Personal Protective Equipment)

11.2 Application of Codes and Standards

12	Applicable safety codes and standards are not followed	Likely to affect prevention of accidents, injury, occupational illness, and property damages	Construction Manager (Contractor)/ Safety Officer	Follow applicable safety codes and standards and regulatory requirements
12.1 Financial Control				
1. Payments to project team members	Delay in payments to project team members	Statutory action for delay in salaries	Project Owner	1. Establish forecast of cash flow for project team members and ensure that adequate funds are available to meet the payment dues 2. Set contingency plan
2. Payments to Contractor(s)/ Subcontractor(s)	1. Cost-loaded schedule for determining the contract payments during the project against the approved progress of works by contractor, subcontractor is not followed 2. The funds to pay progress payments on regular basis are not arranged	1. Affect progress of work execution by the contractor, subcontractor 2. Default in payment toward approved work progress payments	Project Owner	1. Plan the project expenses payments as per approved cost-loading curve 2. Establish forecast of cash flow and ensure sufficient funds are available to make progress in payment related to the approved works
3. Material purchases	Payments toward material purchases is not paid as per the terms in purchase order	Affect getting material as per required schedule	Construction Manager (Contractor)	1. Plan project expenses payments as per approved schedule 2. Allocate sufficient funds toward purchase of material as per purchase contract with the buyers of material
4. Variation order payment	Approved payment toward variation order is not in time	Likely a claim by contractor for payment delay	Project Owner	Make payment toward approved variation orders
5. Insurance and Bonds	Delay in the payment of insurance and bonds	Affect insurance coverage for site works, resources	Construction Manager (Contractor)	Pay insurance and bonds whenever due

(Continued)

TABLE 4.11
(Continued)

Serial Number	Activities	Major Elements	Potential Risk	Probable Effect on Project	Ownership of Risk	Risk Treatment/ Control Measures
	12.2 Cash Flow	S-Curve is not used to track the progress of the project over time and is not updated		Tracking the appropriate level of funding requirements as the project progresses is difficult to manage	R.E./PCM	1. Use S-Curve to predict the amount that will be spent over the established project schedule (time) 2. Update S-Curve for approved changes
13	13.1 Claim Prevention	1. Proper design review	Design review process is not followed	Design errors resulting in variation claim by the contractor	Project Manager (Design)	Designer to review design at every design phase
		2. Unambiguous contract documents' language	Contract documents are not reviewed for unambiguousness and clearly written to enable all team members to understand	Conflict in specifications	Project Manager (Design)	Specify the scope of work in the contract documents that are clearly written in simple language, unambiguous, and convenient to be understood by all the concerned parties for successful completion of the project
		3. Practical schedule	Impractical planning and schedule	Affect project completion time	Project Manager (Design)	Develop the schedule to meet the project objectives and the scope of works, based on timing of each activity, accepted by the stakeholders for tracking and measuring the project performance

	Risk	Effect	Responsible	Action
4. Qualified Contractor(s)	Contractor has lack of knowledge and experience on similar type of projects	1. Poor workmanship 2. Delay in completion of project in accordance with the contract documents	Project Owner	1. Select contractor having expertise in similar type, size of projects 2. Monitor works regularly
5. Competent project team members	1. Incompetent project team members 2. Failure of team members to perform as expected	Project quality not as specified	Project Manager (Supervision)	Select competent team members having experience to achieve project performances (refer Figure 4.4)
			Construction Manager (Contractor)	Select competent team members having experience to achieve project performances (refer Figure 4.4)
6. RFI review procedure	Delay in processing and response to RFI	Delay in the execution of the relevant activity(ies)	R.E.	Follow RFI review procedure (refer Figure 3.21 and Figure 3.22)
7. Negotiations	Negotiations failed to resolve the issue	Delay in execution of particular activity(ies)	All the concerned parties	Amicably resolve the issue as per condition of contract
8. Appropriate project delivery system	1. Project delivery system is not selected considering size, type, and complexity of project 2. Project delivery system is not selected considering level of owner's expertise	1. Delay in developing successful facility/project 2. Successful completion of the project as per owner's requirements	Project Owner (Consultant)	Select project delivery system considering: 1. Size and complexity of the project 2. Type of project 3. Owner's level of expertise in similar type of projects 4. Owner's interest to exert the influence/control over the project

4.5.3 Risk Monitoring in Activities Related to Quality of Monitoring and Control Process Group

Table 4.12 is an example process of risk monitoring of Monitoring and Control Process Group activity.

4.6 RISK MANAGEMENT IN QUALITY OF CLOSING PROCESS GROUP

Figure 4.13 illustrates a flowchart for risk management in quality of Closing Process Group.

4.6.1 Risk in Activities Related to Quality of Closing Process Group

Table 3.5 discussed in Chapter 3, Section 3.2.2, illustrates major activities related to Closing Process group.

Table 4.13 lists potential risks in the activities related to Closing Process Group.

TABLE 4.12
Risk Monitoring of Monitoring and Control Process Group Activity

Serial Number	Description	Action			
1	Risk ID	Monitoring and Control Process Group—1.1 (4)			
2	Description of risk	1. Project performance (forecasted schedule) 1.1 Current situation of all activities/tasks, milestones, sequencing, resources, duration, constraints, and project update are not monitored to establish and control forecasted schedule			
3	Response	1.1 Monitor and control the schedule reflecting the current situation of all activities/tasks, milestones, sequencing, resources, duration, constraints, and project update			
4	Strategy of response	Avoidance	Transfer	Mitigation X	Acceptance
5	Monitoring and control	Risk owner	Review date	Critical issue	
		R.E.	D/M/Y	Monitoring of project progress and project status and recognizing project obstacles	
6	Estimated impact on the project	Project monitoring			
7	Actual impact on the project	Disruption in execution of works smoothly			
8	Revised response	Update project schedule regularly			
9	Record update date	D/M/Y			
10	Communication to stakeholders	Information sent to all stakeholders on D/M/Y			

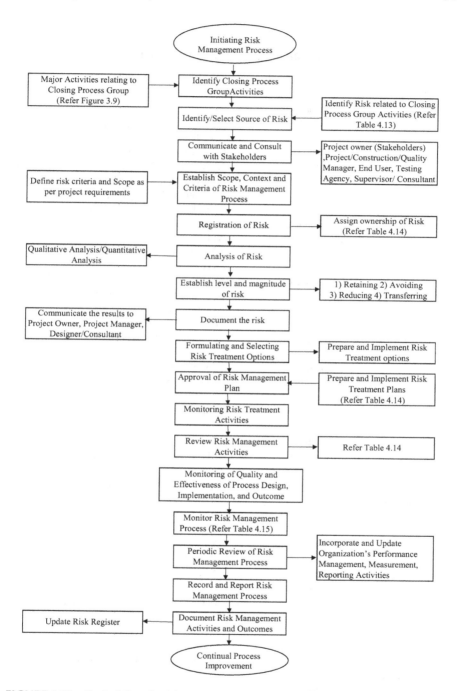

FIGURE 4.13 Typical flowchart for risk management in quality of closing process group.

TABLE 4.13

Potential Risks in Activities Related to Closing Process Group

Serial Number	Activities	Major Elements	Potential Risk
1	1.1	Close Project or Phase	
		1. Testing and commissioning	Testing and commissioning plan is not developed and followed
		2. Authorities' approvals	Delay in obtaining authorities' approvals
		3. Punch list/snag list	1. All the outstanding defects and works are not listed and documented
			2. Contractor not rectifying the punch list items within the agreed upon duration
		4. Handover of project/facility	Delay in handover of facility
		5. As-built drawings	As-built drawings are not reflecting the record drawings indicating the actual installation/ executed works
		6. Technical manuals	Technical manuals are incomplete and are not consisting of related information about the equipment, systems, and other details specified in the contract documents
		7. Spare parts	Spare parts are not as per the listed items in the contract documents
		8. Lesson learned	Lesson learned are not listed and documented for future references to improve the process
2	2.1	Close Project Team	
		1. Demobilization	Delay in demobilization
		2. New assignment	New assignment not existing
	2.2	Material and Equipment	
		1. Excess material removal/ disposal	Delay in removal/disposal of excess material
		2. Equipment removal	Delay in removal of equipment
3	3.1	Close Contract	
		1. Project acceptance/ takeover	Delay in acceptance/takeover of project
		2. Issuance of substantial completion certificate	Delay in issuance of substantial completion certificate
		3. Occupancy	Move-in plan is not ready
4	4.1	Financial Administration and Records	
		1. Payments to all contractors, subcontractors and other team members	All the payments to contractors, subcontractors, and other team members not released
		2. Bank guarantees/ warranties	Bank guarantees/warranties not received as specified in the contract documents
5	5.1	Claim Resolution	
		1. Settlement of claims	Project-related claims are not amicably resolved and settled

4.6.2 Risk Effect and Risk Treatment in Activities Related to Quality of Closing Process Group

Table 4.14 lists potential risks, probable effect on project, and risk treatment in activities related to closing process group.

TABLE 4.14
Potential Risks, Probable Effect, and Risk Treatment in Activities Related to Closing Process Group

Serial Number	Activities	Major Elements	Potential Risk	Probable Effect on Project	Ownership of Risk	Control Measures/ Risk Treatment
1	1.1 Close Project or Phase					
		1. Testing and commissioning	Testing and commissioning plan is not developed and followed	1. Systematic, orderly completion of testing and commissioning process 2. Proper functioning of the systems to the specified standards	Construction Manager (Contractor)	Develop testing, commissioning, and handover plan based on contract requirements to ensure that each individual work/system is fully functional and operational as per specified requirements
		2. Authorities' approvals	Delay in obtaining authorities' approvals	Delay in move-in plan/occupation	Construction Manager (Contractor)	Obtain regulatory approval from the concerned authorities
		3. Punch list/snag list	1. All the outstanding defects and works are not listed and documented 2. Contractor not rectifying the punch list items within the agreed-upon duration	1. Incomplete information to owner/end user 2. Owner will have to rectify on its own	R.E.	1. Prepare punch list (snag list) listing all the items requiring completion or correction 2. Rectify/repair the items as per agreed-upon terms and conditions
		4. Handover of project/ facility	Delay in handover of facility	Delay in occupancy/start-up of the facility	Construction Manager (Contractor)	Complete testing, commissioning of all the works as per specification requirements per approved schedule to enable handover the facility to the project owner/end user

(Continued)

TABLE 4.14
(Continued)

Serial Number	Activities	Major Elements	Potential Risk	Probable Effect on Project	Ownership of Risk	Control Measures/ Risk Treatment
5.		As-built drawings	As-built drawings are not reflecting the record drawings indicating the actual installation/executed works	Incomplete information about as-built/actual installation facility/ project	Construction Manager (Contractor)	Prepare as-built drawings based on actual installed works and record drawings
6.		Technical manuals	Technical manuals are incomplete and are not consisting of related information about the equipment, systems, and other details specified in the contract documents	Difficulties in follow- up upon breakdown of system, equipment	Construction Manager (Contractor)	Prepare technical manual consisting of: 1. Source information 2. Operating procedures 3. Manufacturer's maintenance documentation 4. Maintenance procedure 5. Maintenance and service schedules 6. Spare parts' list and source information 7. Maintenance service contract 8. Warranties and guarantees
7.		Spare parts	Spare parts are not as per the listed items in the contract documents	Owner will have difficulty in getting the parts in case of urgency	Construction Manager (Contractor)	Handover spare parts as per contract requirements properly listing and labelling all the spare parts
8.		Lesson learned	Lessons learned are not listed and documented for future references to improve the process	Lack of information for future reference to improvement of the processes and organizational performance	R.E.	Prepare lesson learned and document the same for future references

Phase	Step	Activity	Risk	Responsible	Response/Action
2	2.1 Close Project Team			Construction Manager (Contractor)	Prepare lesson learned and document the same for future references
		1. Demobilization	Delay in demobilization	Project Manager (Supervision)	Demobilize the team members by terminating their services or engaging on another project
			Additional expenses to project owner	Project Manager (Contractor office)	Demobilize the team members by terminating their services or engaging on another project
		2. New assignment	New assignment not existing	Project Manager (Supervision)	Assign the demobilized team members with new assignment if there are opportunities available
			Loss of expertise due to termination of services	Construction Manager (Contractor)	Assign the demobilized team members with new assignment if there are opportunities available
	2.2 Material and Equipment	1. Excess material removal/disposal	Delay in removal/disposal of excess material	Construction Manager (Contractor)	Remove the excess material from the site
			Delay in site clearance		
		2. Equipment removal	Delay in removal of equipment	Construction Manager (Contractor)	Remove the excess equipment from the site
			Delay in site clearance		
3	3.1 Close Contract	1. Project acceptance/takeover	Delay in acceptance/takeover of project	Project Owner	Takeover the project as per agreed-upon schedule
			Delay in project closeout		
		2. Issuance of substantial completion certificate	Delay in issuance of substantial completion certificate	Project Owner	Issue substantial completion certificate once it is established that project work is complete in accordance with the contract documents
			Delay in the settlement of claims		

(Continued)

**TABLE 4.14
(Continued)**

Serial Number	Activities	Major Elements	Potential Risk	Probable Effect on Project	Ownership of Risk	Control Measures/ Risk Treatment
		3. Occupancy	Move-in plan is not ready	Occupancy in an unorganized way	Project Owner/R.E.	Develop move-in plan as approved by the owner/end user
4	4.1	Financial Administration and Records				
		1. Payments to all contractors, subcontractors, and other team members	All the payments to contractors, subcontractors, and other team members not released	Delay in project closeout	Project Owner	Settle all the approved payments to the contractor, sub- contractor, and other team members
		2. Bank guarantees/ warranties	Bank guarantees/ warranties not received as specified in the contract documents	Delay in closing project accounts	Construction Manager (Contractor)	Submit bank guarantees, warranties as per contract documents
5	5.1	Claim Resolution				
		1. Settlement of claims	Project-related claims are not amicably resolves and settled	Delay in closing of project accounts	Project Owner/R.E./ CM	Amicably settle and resolve all the claims

4.6.3   Risk Monitoring in Activities Related to Quality of Closing Process Group

Table 4.15 is an example process of risk monitoring of Closing Process Group activity.

TABLE 4.15
Risk Monitoring of Closing Process Group Activity

Serial Number	Description	Action			
1	Risk ID	Closing Process Group—1.1 (1)			
2	Description of risk	1. Close project 1.1 Testing and commissioning plan is not developed and followed			
3	Response	1.1 Develop testing, commissioning, and handover plan based on contract requirements to ensure that each individual work/system is fully functional and operational as per specified requirements			
4	Strategy of response	Avoidance X	Transfer	Mitigation	Acceptance
5	Monitoring and control	Risk owner	Review date	Critical issue	
		Construction Manager (Contractor)	D/M/Y	Performing testing and commissioning activities orderly and systematically	
6	Estimated impact on the project	Delay in handover of project			
7	Actual impact on the project	Handing over/taking over of the project			
8	Revised response	Develop testing, commissioning, and handover plan			
9	Record update date	D/M/Y			
10	Communication to stakeholders	Information sent to all stakeholders on D/M/Y			

5 Risk Management in Quality of Project Life Cycle Phases

5.1 PROJECT LIFE CYCLE PHASES

Construction is translating Owner's goals and objectives, by the contractor, to build the facility as stipulated in the contract documents, plans, specifications within budget and on schedule. Construction projects are mainly capital investment projects. They are customized and non-repetitive in nature. Construction projects have become more complex and technical, and the relationships and the contractual grouping of those who are involved are also more complex and contractually varied.

There are several types of projects. Figure 5.1 illustrates types of construction projects.

Most construction projects are custom-oriented having a specific need and a customized design. It is always the owner's desire that his project should be unique and better. Further, it is the owner's goal and objective that the facility is completed on time. Expected time schedule is important for both finance and acquisition of the facility by the owner/end user.

The system life cycle is fundamental to the application of systems engineering. Systems engineering approach to construction projects helps understand the entire process of project management and to manage and control its activities at different levels of various phases to ensure timely completion of the project with economical use of resources to make the construction project most qualitative, competitive, and economical.

Systems engineering starts from the complexity of the large-scale problem as a whole and moves toward structural analysis and partitioning process until the questions of interest are answered. This process of decomposition is called a Work Breakdown Structure (WBS). The WBS is a hierarchical representation of system levels. Being a family tree, the WBS consists of a number of levels, starting with the complete system at level 1 at the top and progressing downward through as many levels as necessary to obtain elements that can be conveniently managed.

Benefits of systems engineering applications are:

- Reduction in cost of system design and development, production/construction, system operation and support, system retirement, and material disposal
- Reduction in system acquisition time
- More visibility and reduction in the risks associated with the design decision-making process

DOI: 10.1201/9781003245612-5

1	**Process Type Projects**		
1.1	Liquid chemical plants		
1.2	Liquid/solid plants		
1.3	Solid process plants		
1.4	Petrochemical plants		
1.5	Petroleum refineries		
2	**Non-Process Type Projects**		
2.1	Power plants		
2.2	Manufacturing plants		
2.3	Support facilities		
2.4	Miscellaneous (R&D) projects		
2.5	Civil construction projects	Residential construction	Family homes, multiunit town houses, garden, apartments, condominiums, high-rise apartments, villas.
2.6	Commercial A/E projects	Building construction (institutional and commercial)	Schools, universities, hospitals, commercial office complexes, shopping malls, banks, theaters, stadiums, government buildings, warehouses, recreation centers, Amusement parks, holiday resorts, neighborhood centers.
		Industrial construction	Petroleum refineries, petroleum plants, power plants, heavy manufacturing plants, steel mills, chemical processing plants.
		Heavy engineering	Dams, tunnels, bridges, highways, railways, airports, urban rapid transit system, ports, harbors, power lines and communication network.
		Environmental	Water treatment and clean water distribution, sanitary and sewage system, waste management.

Categories of Civil construction projects and Commercial A/E projects

FIGURE 5.1 Types of construction projects.

Source: Abdul Razzak Rumane. (2013). *Quality Tools for Managing Construction Projects.* Reprinted with permission of Taylor & Francis Group.

Though it is difficult to generalize project life cycle to system life cycle, however, considering that there are innumerable processes that make up the construction process, the technologies and processes, as applied to systems engineering, can also be applied to construction projects. The number of phases shall depend on the complexity of the project. Duration of each phase may vary from project to project. The project life cycle phases of non-process type (A/E type) of projects differ from that of process type of projects.

Each type of construction projects has many varying risks. Risk management throughout the life cycle of the project is important and essential to prevent unwanted consequences and effects on the project. Construction projects have the involvement of many stakeholders such as project owners, developers, design firms (consultants), contractors, banks, and financial institutions funding the project who are affected by the risk. Each of these parties have an involvement with a certain portion of overall construction project risk; however, the owner has a greater share of risks as the owner is involved from the inception until completion of project and beyond. The owner must take initiatives to develop risk consciousness and awareness among all the parties emphasizing upon the importance of explicit consideration of risk at each stage of the project as the owner is ultimately responsible for overall project const. Traditionally,

1. Owner/client is responsible for the investment/finance risk.
2. Designer (Consultant) is responsible for design risk.
3. Contractors and subcontractors are responsible for construction risk.

Construction projects are characterized as being very complex projects, where uncertainty comes from various sources. Construction projects involve a cross section of many different participants. They have varying project expectations. Those both influence and depend on each other in addition to the "other players" involved in the construction process. The relationships and the contractual groupings of those who are involved are also more complex and contractually varied. Construction projects often require a large quantity of materials and physical tools to move or modify these materials. Most items used in construction projects are normally produced by other construction-related industries/manufacturers. Therefore, risk in construction projects is multifaceted. Construction projects inherently contain a high degree of risk in their projection of cost and time as each is unique. No construction project is without any risk. Risk management in construction projects is mainly focused on delivering the project with:

1. What was originally accepted (as per defined scope)
2. Agreed-upon time (as per schedule without any delay)
3. Agreed-upon budget (no overruns to accepted Cost)

Risk management is an ongoing process. In order to reduce the overall risk in construction projects, the risk assessment (identification, analysis, and evaluation) process must start as early as possible to maximize project benefits. There are number of risks which can be identified at each stage of the project. Early risk identification can lead to better estimation of the cost in the project budget, whether through contingencies, contractual, or insurance. Risk identification is the most important function in construction projects.

Risk factors in construction projects can be categorized into a number of ways according to the level of details or selected viewpoints. These are categorized and based on various risks factors and source of risk. Contractor has to identify related risks having effects on the construction, analyze these risks, evaluate the effects on the contract, and evolve the strategy to counter these risks, before bidding for a construction contract. Construction project risks mainly relate to the following:

- Scope and change management
- Schedule/time management
- Budget/cost management
- Quality management
- Resources and manpower management
- Communication management
- Procurement/contract management
- Health, safety, and environmental management

Following sections discuss the identification of potential risks in the major activities/ elements of project life cycle phases, probable effects on the project, ownership of the risks, and risk treatment, control measures to avoid and mitigate occurrence of risk(s) in the project.

5.2   A/E TYPE PROJECT PHASES

Figure 5.2 illustrates Construction Project Life cycle phases (A/E type of projects).

Table 5.1 illustrates the subdivided activities/components of the Construction Project Life Cycle phases.

These activities may not be strictly sequential; however, the breakdown allows the implementation of project management functions more effectively at different stages.

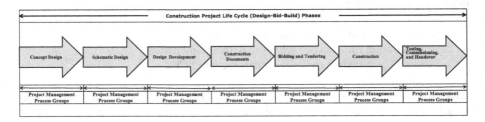

FIGURE 5.2   Construction project life cycle (Design-Bid-Build) phases.

TABLE 5.1

Construction Project Life Cycle Phases (Non-process Type of Project)

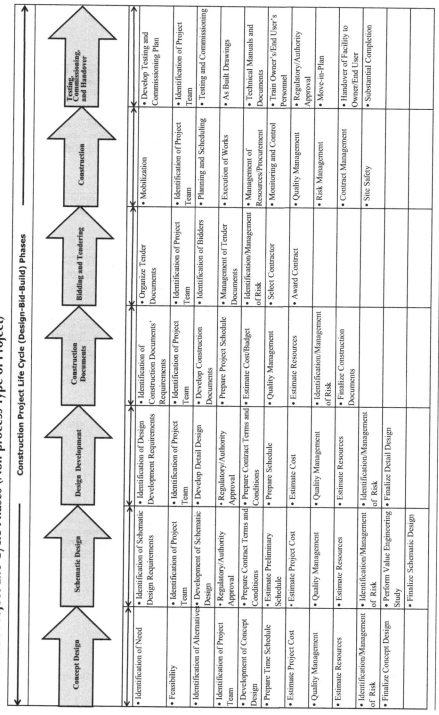

Concept Design	Schematic Design	Design Development	Construction Documents	Bidding and Tendering	Construction	Testing, Commissioning, and Handover
• Identification of Need	• Identification of Schematic Design Requirements	• Identification of Design Development Requirements	• Identification of Construction Documents' Requirements	• Organize Tender Documents	• Mobilization	• Develop Testing and Commissioning Plan
• Feasibility	• Identification of Project Team	• Identification of Project Team	• Identification of Project Team	• Identification of Project Team	• Identification of Project Team	• Identification of Project Team
• Identification of Alternatives	• Development of Schematic Design	• Develop Detail Design	• Develop Construction Documents	• Identification of Bidders	• Planning and Scheduling	• Testing and Commissioning
• Identification of Project Team	• Regulatory/Authority Approval	• Regulatory/Authority Approval	• Prepare Project Schedule	• Management of Tender Documents	• Execution of Works	• As Built Drawings
• Development of Concept Design	• Prepare Contract Terms and Conditions	• Prepare Contract Terms and Conditions	• Estimate Cost/Budget	• Identification/Management of Risk	• Management of Resources/Procurement	• Technical Manuals and Documents
• Prepare Time Schedule	• Estimate Preliminary Schedule	• Prepare Schedule	• Quality Management	• Select Contractor	• Monitoring and Control	• Train Owner's/End User's Personnel
• Estimate Project Cost	• Estimate Project Cost	• Estimate Cost	• Estimate Resources	• Award Contract	• Quality Management	• Regulatory/Authority Approval
• Quality Management	• Quality Management	• Quality Management	• Identification/Management of Risk		• Risk Management	• Move-in-Plan
• Estimate Resources	• Estimate Resources	• Estimate Resources	• Finalize Construction Documents		• Contract Management	• Handover of Facility to Owner/End User
• Identification/Management of Risk	• Identification/Management of Risk	• Identification/Management of Risk			• Site Safety	• Substantial Completion
• Finalize Concept Design	• Perform Value Engineering Study	• Finalize Detail Design				
	• Finalize Schematic Design					

Construction Project Life Cycle (Design-Bid-Build) Phases

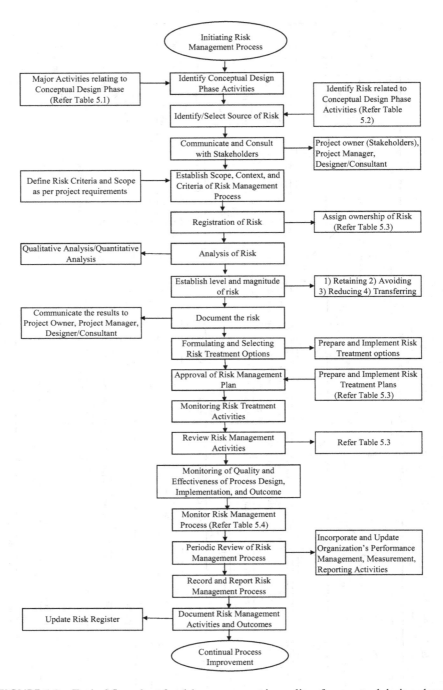

FIGURE 5.3 Typical flow chart for risk management in quality of conceptual design phase.

5.3   RISK MANAGEMENT IN QUALITY OF CONCEPTUAL DESIGN PHASE

Figure 5.3 illustrates a typical flow chart for risk management in quality of Conceptual Design phase.

5.3.1   RISK IN ACTIVITIES RELATED TO QUALITY OF CONCEPTUAL DESIGN PHASE

Table 5.2 lists potential risks in the activities related to Conceptual Design Phase.

5.3.2   RISK EFFECT AND RISK TREATMENT IN ACTIVITIES RELATED TO QUALITY OF CONCEPTUAL DESIGN PHASE

Table 5.3 lists potential risks, probable effect on project, and risk treatment in activities related to conceptual design.

TABLE 5.2
Potential Risks in Activities Related to Conceptual Design Phase

Serial Number	Phase	Major Elements/Activities	Potential Risk
1	1.1	Conceptual Design Phase	
		1. Identification of need	Need is not established based on project goals and objectives to meet owner's requirements
		2. Feasibility	1. Feasibility study not done properly
			2. Incomplete/improper feasibility study
			3. Regulations and policy aspects not considered
		3. Identification of alternatives	Identified alternative is not suitable for further development
		4. Identification of project team	1. Incompetent team members
			2. Lack of knowledge about the project
			3. Team members not performing as expected
		5. Development of concept design	Designer has not considered following major points:
			1. Project goals
			2. Usability
			3. Constructability
			4. Sustainability
			5. Availability of resources
		6. Prepare time schedule	Expected time schedule for completion of the project/facility is not properly worked out
		7. Estimate project cost	Source of funding from the beginning through completion of project is not considered
		8. Quality management	Quality criteria for the project are not established
		9. Estimate resources	Designer has not properly established the resources required to complete the project
		10. Identify/manage risk	Probable risks are not identified, and response is evolved
		11. Finalize concept design	Concept design is not reviewed before finalization

TABLE 5.3

Potential Risks, Probable Effect on Project, and Risk Treatment in Activities Related to Conceptual Design Phase

Serial Number	Phase	Major Elements/ Activities	Potential Risk	Probable Effect on Project	Ownership of Risk	Risk Treatment/Control Measures
1	1.1 Conceptual Design Phase	1. Identification of need	Need is not established based on project goals and objectives to meet owner's requirements	Designer may not be able to ensure that design will meet owner's needs	Project Owner (Consultant)	1. Establish the need based on real (perceived) requirements 2. Define the need by describing the minimum requirements of quality, performance, project completion date, and approved budget
		2. Feasibility	1. Feasibility study not done properly 2. Incomplete/improper feasibility study 3. Regulations and policy aspects not considered	1. Owner will not have sufficient information to enable proceeding or aborting the project 2. Insufficient information that the proposed facility is suitable for intended use by the owner/end user	Project Owner (Consultant)	1. Conduct feasibility study taking into account various factors such as economic, technical, and environmental constraints 2. Analyze feasibility study for potential impact of the identified need of the proposed project 2. Define clearly the viability of feasibility study for of the project that will produce the best or most profitable results 3. Establish broad objectives of the project in the feasibility study 4. Review feasibility study for technical/financial viability of the project to give sufficient information to the client to proceed or abort the project

Activity	Cause / Problem	Effect	Responsibility	Action / Measures
3. Identification of alternatives	Identified alternative is not suitable for further development	Facility is not based on a predetermined set of performance measures to meet owner's requirements	Project Manager (Design)	Identify and develop alternatives based on predetermined set of performance measures to meet the owner's requirements
4. Identification of project team	1. Incompetent team members 2. Lack of knowledge about the project 3. Team members not performing as expected	1. The team members may not be able to perform in achieving the project as per owner's requirements 2. Design deliverables not achieved as per TOR 3. Affect achieving successful completion of project	Project Owner	Approve qualified and knowledgeable team members to work on the project as team members to perform achieving successful project
			Project Manager (Design)	1. Select team members having project-related qualification, experience, and knowledge 2. Select competent candidate 3. Provide training
5. Development of Concept Design	Designer has not considered following major points: 1. Project goals 2. Usability 3. Constructability 4. Sustainability 5. Availability of resources	Owner may not be able to achieve a project that meets the owner's need for intended use	Project Manager (Design)	Develop concept design based on preferred alternative, TOR, and owner's preferred requirements suitable for further development to meet: 1. Project goals 2. Usability 3. Constructability 4. Sustainability 5. Resource availability 6. Environmental compatibility 7. Energy conservation 8. Cost-effectiveness over life cycle
6. Prepare time schedule	Expected time schedule for completion of the project/facility is not properly worked out	Finances and acquisition of the facility by the owner/end user will be affected	Project Manager (Design)	Prepare the expected time schedule to meet owner's need for successful completion of the project/facility

(Continued)

TABLE 5.3
(Continued)

Serial Number	Phase	Major Elements/ Activities	Potential Risk	Probable Effect on Project	Ownership of Risk	Risk Treatment/Control Measures
7.		Estimate project cost	Source of funding from the beginning through completion of project is not considered	Completion of project/facility within approved schedule	Project Manager (Design)	Develop estimated cost to meet owner's budgetary cost and funding capability for successful completion of the project/facility and does not exceed the estimated cost during the feasibility study
8.		Quality management	Quality criteria for the project are not established	Project quality not meeting owner's requirements	Project Manager (Design)	Establish the quality criteria, standards, and codes to ensure technical and functional capability and usage of the project/facility that meets owner's requirements
9.		Estimate resources	Designer has not properly established the resources required to complete the project	1. Estimated project finances will be affected 2. Delay in completion of projects	Project Manager (Design)	Estimate the resources that are sufficient for successful completion of project
10.		Identify/manage risk	Probable risks are not identified, and response is evolved	Successful completion of project	Project Manager (Design)	Identify all the probable risks to avoid disruption of project
11.		Finalize concept design	Concept design is not reviewed before finalization	Design errors, mistakes, omissions	Project Manager (Design)	Review the concept design to minimize errors and mistakes and to ensure it meets the owner's requirements
					Project Owner (Project Manager)	Review the submitted concept design for compliance with owner's requirements

5.3.3 RISK MONITORING IN ACTIVITIES RELATED TO QUALITY OF CONCEPTUAL DESIGN PHASE

Table 5.4 is an example process of risk monitoring of Conceptual Design phase activity.

5.4 RISK MANAGEMENT IN QUALITY OF SCHEMATIC DESIGN PHASE

Figure 5.4 illustrates a typical flowchart for risk management in quality of Schematic Design phase.

TABLE 5.4
Risk Monitoring of Conceptual Design Phase Activity

Serial Number	Description	Action			
1	Risk ID	Development of concept design—1.1 (5)			
2	Description of risk	1. Development of concept design			
		Designer has not considered the following major points:			
		1.1. Project goals			
		1.2. Usability			
		1.3. Constructability			
		1.4. Sustainability			
		1.5. Availability of resources			
3	Response	Develop concept design based on preferred alternative, TOR, and owner's preferred requirements suitable for further development			
4	Strategy of response	Avoidance	Transfer	Mitigation X	Acceptance
5	Monitoring and control	Risk owner	Review date	Critical Issue	
		Project Manager (Design)	D/M/Y	Concept design not meeting owner's need for intended use	
6	Estimated impact on the project	Errors and mistakes in project design			
7	Actual impact on the project	Project design			
8	Revised Response	Develop concept design taking into consideration all the requirements meeting owner's goals and objectives			
9	Record update date	D/M/Y			
10	Communication to stakeholders	Information sent to all stakeholders on D/M/Y			

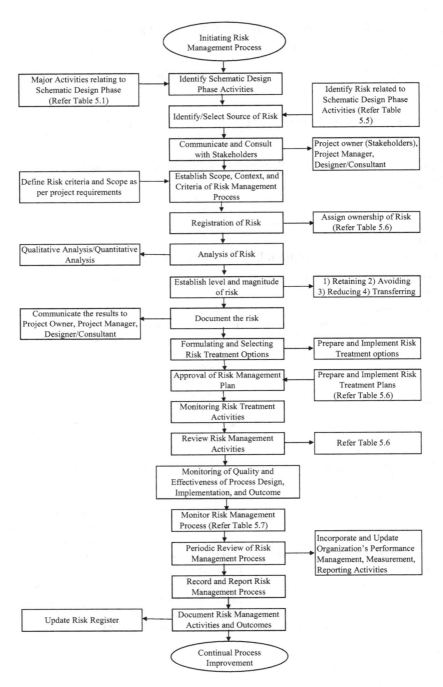

FIGURE 5.4 A typical flowchart for risk management in quality of schematic design phase.

5.4.1 RISK IN ACTIVITIES RELATED TO QUALITY OF SCHEMATIC DESIGN PHASE

Table 5.5 lists potential risks in the activities related to Schematic Design phase.

5.4.2 RISK EFFECT AND RISK TREATMENT IN ACTIVITIES RELATED TO QUALITY OF SCHEMATIC DESIGN PHASE

Table 5.6 lists potential risks, probable effect on project, and risk treatment in activities related to Schematic Design phase.

TABLE 5.5
Potential Risks in Activities Related to Schematic Design Phase

Serial Number	Phase	Major Elements/ Activities	Potential Risk
1	1.1	Schematic Design Phase	
		1. Identification of schematic design requirements	1. TOR requirements are not properly collected 2. Concept design deliverables and review comments are not considered 3. The related project data and information collected are incomplete and are likely to be incorrect and wrongly estimated
		2. Identification of project team	Design team members are not selected based on the organizational structure and suitable skill required to perform the job
		3. Development of schematic design	Schematic design scope of work is incomplete
		4. Regulatory/ authorities' approval	Regulatory authorities' requirements are not taken into consideration
		5. Contract terms and conditions	1. Designer has not developed contract documents that meet the owner's need, specifying the required level of quality, schedule, and budget 2. Inadequate and ambiguous specifications
		6. Estimate preliminary schedule	Schedule is not prepared on the basis of logic, critical path method (CPM), and set of milestones
		7. Estimate project cost	Cost estimate is not based taking into consideration schematic design, cost of activities, and resources
		8. Quality management	Quality management plan is not developed
		9. Estimate resources	Resources are not estimated correctly
		10. Identification/ management of risk	Probable risks are not identified, and risk responses are developed
		11. Perform value engineering study	Value engineering study is not conducted
		12. Finalize schematic design	Schematic design is not reviewed

TABLE 5.6

Potential Risks, Probable Effect on Project, and Risk Treatment in Activities Related to Schematic Design Phase.

Serial Number	Phase	Major Elements/ Activities	Potential Risk	Probable Effect on Project	Ownership of Risk	Risk Treatment/ Control Measure
1	1.1 Schematic Design Phase					
		1. Identification of schematic design requirements	1. TOR requirements are not properly collected 2. Concept design deliverables and review comments are not considered 3. The related project data and information collected are incomplete and are likely to be incorrect and wrongly estimated	Project is not planned to a level where sufficient details are available for initial development of schedule, cost, and contract documents	Project Manager (Design)	Collect the following items to develop schematic design: 1. TOR requirements, 2. Comments made by the project owner on concept design. 3. Relevant data and information 4. Regulatory requirements 5. Energy conservation requirements
		2. Identification of project team	Design team members are not selected based on the organizational structure and suitable skill required to perform the job	Project team members may not be able to perform in achieving project delivery requirements successfully	Project Owner	Approve the qualified team members as per contract requirements
					Project Manager (Design)	1. Select team members as per organizational structure, suitable skills to perform the job 2. Select competent candidate. (refer Figure 4.4 in Chapter 4) 3. Provide training

3. Development of schematic design	Schematic design scope of work is incomplete	Lack of sufficient information to identify the works to be performed and to allow detail design to proceed without significant changes that may adversely affect the project	Project Manager (Design)	Develop schematic design scope considering the following items: 1. Gathering comments on concept design 2. Collecting owner's preferred requirements 3. Collecting regulatory requirements 4. TOR requirements 5. Reliability 6. Approved preliminary schedule 7. Approved preliminary cost 8. Codes and standards 9. Environmental issues
4. Regulatory/ authorities' approval	Regulatory authorities' requirements are not taken into consideration	Project may not be in compliance with latest regulations, codes, and licensing procedures	Project Owner/ Project Manager (Design)	Collect regulatory requirements and incorporate the same in the schematic design
5. Contract terms and conditions	1. Designer has not developed contract documents that meet the owner's need, specifying the required level of quality, schedule, and budget	1. Project may not fully meet owner's needs and requirements 2. Conflict during the execution of works	Project Manager (Design)	1. Prepare the documents based on the contracting arrangements with which the owner would like to handle the project 2. Prepare the documents in line with model contract documents

(Continued)

TABLE 5.6
(Continued)

Serial Number	Phase	Major Elements/ Activities	Potential Risk	Probable Effect on Project	Ownership of Risk	Risk Treatment/ Control Measure
			2. Inadequate and ambiguous specifications			3. Develop specifications taking into consideration all the requirements for the adequacy of the project performance and clearly written without any ambiguity for successful completion of project
		6. Estimate preliminary schedule	Schedule is not prepared on the basis of logic, critical path method (CPM), and set of milestones	1. Delay in project 2. Project completion time	Project Manager (Design)	Prepare the schedule on the basis of logic, critical path method (CPM) using top-down planning using key events and milestones
		7. Estimate project cost	Cost estimate is not based taking into consideration schematic design, cost of activities, and resources	Time- phased plan for project financing, expected expenses	Project Manager (Design)	Estimate and determine the cost by estimating the cost of activities and resources, which is related to schedule of the project
		8. Quality management	Quality management plan is not developed	Design errors, omissions	Project Manager (Design)	Designer to develop quality management plan to minimize design errors and omissions

9. Estimate resources	Resources are not estimated correctly	Successful completion of project as per owner's requirements	Project Manager (Design)	Estimate resources considering the activities and works to be performed during the construction, testing, and commissioning of the project
10. Identification/ management of risk	Probable risks are not identified, and risk responses are developed	Project quality	Project Manager (Design)	Identify probable risks to ensure minimize the design errors and omissions
11. Perform value engineering study	Value engineering study is not conducted	Project may not be of optimizing value to achieve better performance while maintain all functional requirements	Project Manager (Design)	Perform value engineering study to achieve optimizing design, construction, operations, and the comments are incorporated in the design
12. Finalize schematic design	Schematic design is not reviewed	Design errors	Project Manager (Design)	Review schematic design before submitting to the project owner in order to minimize errors and mistakes
			Project Owner (Project Manager)	Review submitted schematic design to ensure compliance with owner's requirements

TABLE 5.7

Risk Monitoring of Schematic Design Phase Activity

Serial Number	Description	Action			
1	Risk ID	Schematic design—1.1 (1)			
2	Description of risk	1. Identification of schematic design requirements			
		1.1. TOR requirements are not properly collected			
		1.2. Concept design deliverables and review comments are not considered			
		1.3. The related project data and information collected are incomplete and are likely to be incorrect and wrongly estimated			
3	Response	Collect the following items to develop schematic design			
		1.1. TOR requirements			
		1.2. comments made by the project owner on concept design,			
		1.3. relevant data and information, and			
		1.4. regulatory requirements			
4	Strategy of response	Avoidance	Transfer	Mitigation X	Acceptance
5	Monitoring and control	Risk owner	Review date	Critical issue	
		Project Manager (Design)	D/M/Y	Schematic design is not developed to a level where sufficient details are available for initial development of schedule, cost, and contract documents	
6	Estimated impact on the project	Errors and mistakes in schematic design			
7	Actual impact on the project	Project design			
8	Revised response	Develop schematic design taking into consideration all the requirements meeting owner's goals and objectives			
9	Record update date	D/M/Y			
10	Communication to stakeholders	Information sent to all stakeholders on D/M/Y			

5.4.3 RISK MONITORING IN ACTIVITIES RELATED TO QUALITY OF SCHEMATIC DESIGN PHASE

Table 5.7 is an example process of risk monitoring of Schematic Design phase activity.

5.5 RISK MANAGEMENT IN QUALITY OF DESIGN DEVELOPMENT PHASE

Figure 5.5 illustrates a typical flowchart for risk management in quality of Design Development phase.

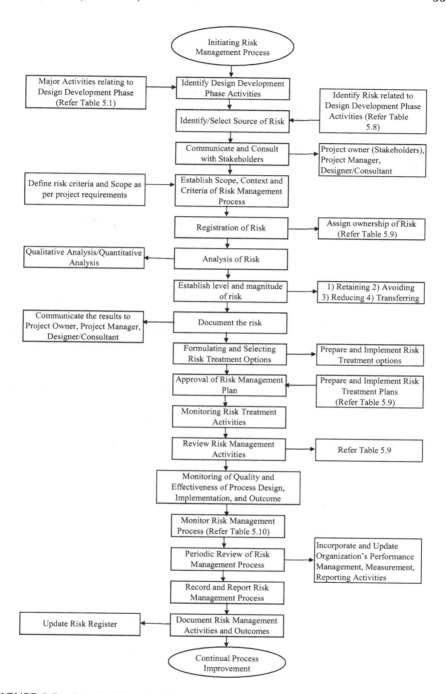

FIGURE 5.5 A typical flowchart for risk management in quality of design development phase.

5.5.1 Risk in Activities Related to Quality of Design Development Phase

Table 5.8 lists potential risks in the activities related to Design Development phase.

5.5.2 Risk Effect and Risk Treatment in Activities Related to Quality of Design Development Phase

Table 5.9 lists potential risks, probable effect on project, and risk treatment in activities related to Design Development phase.

Figure 5.6 illustrates detail design review steps.

TABLE 5.8
Potential Risks in Activities Related to Design Development Phase

Serial Number	Phase	Major Elements/ Activities	Potential Risk
1	1.1 Design Development Phase		
		1. Identification of design development requirements	1. Schematic design deliverables and review comments are not taken into consideration while preparing the detail design 2. TOR requirements are not taken into consideration 3. Environmental considerations are not taken into account
		2. Identification of project team	The members are not selected based on the organizational structure and suitable skills required to perform the job
		3. Development of detail design of works	1. The designer has not prepared comprehensive design of the works with a detailed work breakdown structure and work packages. 2. Detail design scope of work is not properly established and is incomplete
		4. Regulatory/authorities' approval	Regulatory authorities' requirements are not taken into consideration
		5. Prepare contract documents and specifications	1. Incomplete contract documents 2. Conflict/errors in the documents
		6. Prepare schedule	1. Schedule is not based on the detailed engineering and design drawings and contract documents 2. Project schedule is not as per detailed data and project assumptions
		7. Estimate cost	Errors in detailed cost estimation
		8. Quality management	Quality management plan is not established
		9. Estimate resources	Resource estimation is not done based on detailed information about resources
		10. Identification/ management of risk	Potential risks are not identified, and response is evolved
		11. Finalize detail design	Detail design is not reviewed prior to next stage

TABLE 5.9

Potential Risks, Probable Effect on Project, and Risk Treatment in Activities Related to Design Development Phase

Serial Number	Phase	Major Elements/ Activities	Potential Risk	Probable Effect on Project	Ownership of Risk	Risk Treatment/ Control Measures
1	1.1 Design Development Phase	1. Identification of design development requirements	1. Schematic design deliverables and review comments are not taken into consideration while preparing the detail design 2. TOR requirements are not taken into consideration 3. Environmental considerations are not taken into account	Suggested changes in schematic are not considered or revaluated to ensure that the changes will not detract from meeting the project design goals/objectives	Project Manager (Design)	Develop detail design considering the following points: 1. Review comments on schematic design 2. Regulatory authorities' comments on schematic design 3. TOR requirements 4. Establishing and integrating in the design the size, shape, levels, performance characteristics, technical details, and requirements of all the individual components 5. Interdisciplinary coordination 6. Project schedule 7. Project cost 8. Project specifications 9. Availability of resources 10. Project documents 11. Relevant codes and standards 12. Environmental considerations
		2. Identification of project team	The members are not selected based on the organizational structure and suitable skills required to perform the job.	The team members may not be able to perform achieving project deliverables and develop detail design	Project Owner (Project Manager)	Approve team members having suitable skills and knowledge of similar size and types of projects

(Continued)

TABLE 5.9
(Continued)

Serial Number	Phase	Major Elements/ Activities	Potential Risk	Probable Effect on Project	Ownership of Risk	Risk Treatment/ Control Measures
					Project Manager (Design)	1. Select team members having skills and experience in similar size and types of projects
						2. Select team having adequate knowledge of technical conditions about the project
3.		Development of detail design of works	1. The designer has not prepared comprehensive design of the works with a detailed work breakdown structure and work packages.	1. The detail design may not meet owner's requirement/TOR	Project Manager (Design)	1. Develop detail design meeting owner's requirements
				2. The developed design is not accurate to the possible extent, free of errors, and with minimum omissions		2. Develop detail design to meet TOR requirements
			2. Detail design scope of work is not properly established and is incomplete			3. Develop detail design drawings for all the relevant trades
						4. Develop detail design to match with the specifications and contract documents
						5. Constructability
						6. Energy conservation
						7. Environmental compatibility
4.		Regulatory/ authorities' approval	Regulatory authorities' requirements are not taken into consideration	1. Impact on precontract planning of the project	Project Manager (Design)	1. Take into consideration the regulatory requirements and compatibility with local codes and regulations while developing the detail design
				2. Delay in getting/ connecting utility services from authorities		2. Take approval of regulatory agencies
5.		Prepare contract documents and specifications	1. Incomplete contract documents	1. Project may not meet the owner's requirements	Project Manager (Design)	1. Prepare contract documents taking into consideration detail design and technical specifications and owner's requirements
			2. Conflict/errors in the documents	2. Resolving issues will delay in execution of		2. Review contract documents to minimize errors

#	Activity	Possible problems	Effect	Responsibility	Recommendations
6.	Prepare schedule	1. Schedule is not based on the detailed engineering and design drawings and contract documents 2. Project schedule is not as per detailed data and project assumptions	1. Delay in completion of project 2. Affect project progress measurement and project control	Project Manager (Design)	1. Prepare schedule based on detail drawings and contract documents and resource management 2. Develop schedule considering practical aspects, project assumptions, and detail data
7.	Estimate cost	Errors in detailed cost estimation	1. Owner's capability of financing the project 2. Time-phased plan summarizing expected expenditure	Project Manager (Design)	1. Estimate the project cost based on work packages and BOQ to meet the owner's capability of financing the project 2. Estimate time-phased expenses toward the project
8.	Quality management	Quality management plan is not established	Project quality	Project Manager (Design)	Develop quality management plan to ensure that the design fully meet owner's objectives/goals
9.	Estimate resources	Resource estimation is not done based on detailed information about resources	Completion of project as per owner's requirements within schedule	Project Manager (Design)	Estimate the resources taking into consideration resource calendar to meet the requirements to complete the project
10.	Identification/ management of risk	Potential risks are not identified, and response is evolved	Project design and project execution	Project Manager (Design)	Identify probable risks to ensure design is error/defect free
11.	Finalize detail design	Detail design is not reviewed prior to next stage	1. Design errors, omissions 2. Errors in contract documents	Project Manager (Design) Project Owner (Project Manager)	Review detail design before submitting to the project owner in order to minimize errors and mistakes (refer Figure 5.4 for design review steps) Review submitted detail design to ensure compliance with owner's requirements

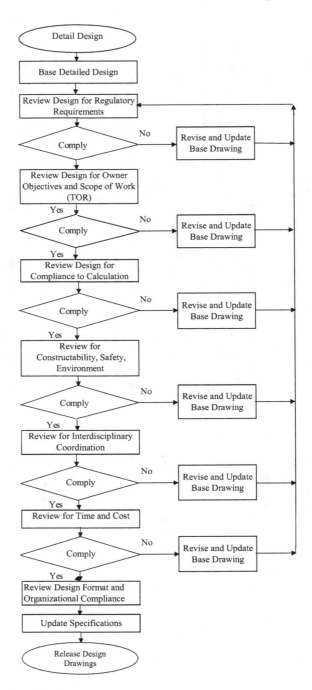

FIGURE 5.6   Detail design review steps.

Source: Abdul Razzak Rumane. (2013). Quality Tools for Managing Construction Projects. Reprinted with permission of Taylor & Francis Group.

5.5.3 RISK MONITORING IN ACTIVITIES RELATED TO QUALITY OF DESIGN DEVELOPMENT PHASE

Table 5.10 is an example process of risk monitoring of Design Development phase activity.

TABLE 5.10
Risk Monitoring of Design Development Phase Activity

Serial Number	Description	Action			
1	Risk ID	Design development—1.1 (3)			
2	Description of risk	1. Development of detail design			
		1.1. The designer has not prepared comprehensive design of the works with a detailed work breakdown structure and work packages			
		1.2. Detail design scope of work is not properly established and is incomplete			
3	Response	1.1. Develop detail design meeting owner's requirements			
		1.2. Develop detail design to meet TOR requirements			
		1.3. Develop detail design drawings for all the relevant trades			
		1.4. Develop detail design to match with the specifications and contract documents			
4	Strategy of response	Avoidance	Transfer	Mitigation X	Acceptance
5	Monitoring and control	Risk owner Project Manager (Design)	Review date D/M/Y	Critical issue 1. The detail design may not meet owner's requirement/TOR 2. The developed design is not accurate to the possible extent, free of errors, and with minimum omissions	
6	Estimated impact on the project	Errors and mistakes in detail design			
7	Actual impact on the project	Project design			
8	Revised response	Develop detail design taking into consideration all the requirements meeting owner's goals and objectives			
9	Record update date	D/M/Y			
10	Communication to stakeholders	Information sent to all stakeholders on D/M/Y			

5.6   RISK MANAGEMENT IN QUALITY OF CONSTRUCTION DOCUMENTS PHASE

Figure 5.7 illustrates a typical flow chart for risk management in quality of Construction Documents phase.

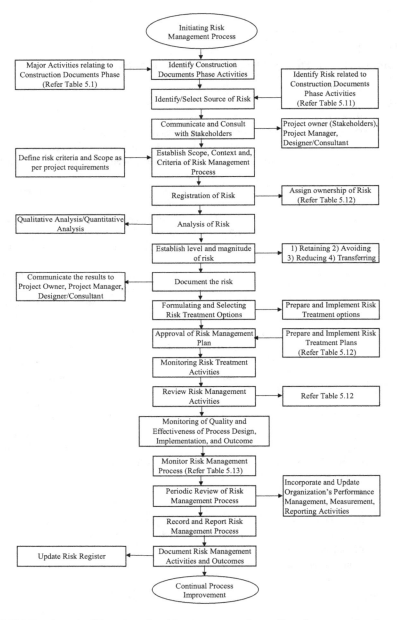

FIGURE 5.7   A typical flowchart for risk management in quality of construction documents phase.

5.6.1 RISK IN ACTIVITIES RELATED TO QUALITY OF CONSTRUCTION DOCUMENTS PHASE

Table 5.11 lists potential risks in the activities related to Construction Documents phase.

5.6.2 RISK EFFECT AND RISK TREATMENT IN ACTIVITIES RELATED TO QUALITY OF CONSTRUCTION DOCUMENTS PHASE

Table 5.12 lists potential risks, probable effect on project, and risk treatment in activities related to Construction Documents phase.

TABLE 5.11
Potential Risks in Activities Related to Construction Documents Phase

Serial Number	Phase	Major Elements/Activities	Potential Risk
1	1.1 Construction documents phase		
		1. Identification of construction document requirements	1. Construction documents are not prepared gathering the comments on the detail design 2. Construction documents are not prepared collecting TOR requirements 3. Scope of work to produce construction documents is nor properly established
		2. Identification of project team	Project team does not have members having direct involvement in the construction documents phase
		3. Develop construction documents	Construction documents do not consist of the following: 1. Working drawings 2. Technical specifications 3. Documents
		4. Prepare project schedule	Project schedule is not as per detail data and project assumptions
		5. Estimate cost/budget	Errors in definitive cost estimation
		6. Quality management	Quality management plan is not established
		7. Estimate resources	Accurate and exact resource estimation is not done
		8. Identification/management of risk	Potential risks are not identified, and response is evolved
		9. Finalize construction documents	Documents are not reviewed prior to submission for the next stage

TABLE 5.12
Potential Risks, Probable Effect on Project, and Risk Treatment in Activities Related to Construction Documents Phase

Serial Number	Phase	Major Elements/ Activities	Potential Risk	Probable Effect on Project	Ownership of Risk	Risk Treatment/ Control Measures
1	1.1 Construction Documents Phase					
		1. Identification of construction document requirements	1. Construction documents are not prepared gathering the comments on the detail design	1. Bidding price will be different than definitive cost	Project Manager (Design)	1. Incorporate the comments collected on the detail design
			2. Construction documents are not prepared collecting TOR requirements	2. Quality of construction will not be as per owner's requirements		2. Prepare construction documents by taking into consideration TOR requirements, regulatory requirements, environmental requirements, and owner's requirements
			3. Scope of work to produce construction documents is not properly established	3. Errors in compilation of construction documents		3. Prepare construction documents based on the scope of work to meet owner's requirements
		2. Identification of project team	Project team does not have members having direct involvement in the construction documents phase	The team members will not be able to perform properly the coordination and assembling of construction documents	Project Manager (Design)	1. Identify and select the team members consisting of stakeholders having direct involvement in preparation of construction documents
						2. Select the team members have skills and knowledge on the preparation of construction documents for coordinating and assemble all the required documents

3. Develop construction documents	Construction documents does not consist of the following: 1. Working drawings 2. Technical specifications 3. Documents	Incomplete tendering documents	Project Manager (Design)	Organize the construction documents consisting of the following items: 1. Working drawings 2. Technical specifications 3. BOQ 4. Project schedule 5. Existing site conditions/site plans/site surveys 6. Design calculations
4. Prepare project schedule	Project schedule is not as per detail data and project assumptions	Errors in construction schedule	Project Manager (Design)	Review and update construction schedule based on the construction documents
5. Estimate cost/ budget	Errors in definitive cost estimation	Estimated definitive cost may change budget amount that may exceed owner's capability of financing the project	Project Manager (Design)	Develop cost estimate based on detail costing methodology based on project activities and BOQ
6. Quality management	Quality management plan is not established	Errors and omission in assembling of construction documents	Project Manager (Design)	Establish quality management plan to ensure that working drawings are suitable for construction, specifications are comprehensively and correctly described and coordinated
7. Estimate resources	Accurate and exact resource estimation is not done	The execution of construction project activities	Project Manager (Design)	Estimate the resources correctly based on the available details
8. Identification/ management of risk	Potential risks are not identified, and response is evolved	1. Incomplete construction documents 2. Conflict between working drawings and specification	Project Manager (Design)	Identify the probable risks and treatment plan to minimize errors and omissions
9. Finalize construction documents	Documents are not reviewed prior to submission for next stage	Final construction documents package not as per requirements for tendering purpose	Project Manager (Design)	Review construction documents for constructability of design and for tendering purpose
			Project Owner (Project Manager)	Review construction documents to ensure compliance with owner's requirements

5.6.3   Risk Monitoring in Activities Related to Quality of Construction Documents Phase Activity

Table 5.13 is an example process of risk monitoring of Construction Documents phase activity.

5.7   RISK MANAGEMENT IN QUALITY OF BIDDING AND TENDERING PHASE

Figure 5.8 illustrates a typical flowchart for risk management in Quality of Bidding and Tendering Phase.

TABLE 5.13
Risk Monitoring of Construction Documents Phase Activity

Serial Number	Description	Action			
1	Risk ID	Construction Documents—1.1 (1)			
2	Description of Risk	1. Identification of construction documents requirements			
		1.1. Construction documents are not prepared gathering the comments on the detail design			
		1.2. Construction documents are not prepared collecting TOR requirements			
		1.3. Scope of work to produce construction documents is not properly established			
3	Response	1.1. Incorporate the comments collected on the detail design			
		1.2. Prepare construction documents by taking into consideration TOR requirements, regulatory requirements, environmental requirements, and owner's requirements			
4	Strategy of response	Avoidance	Transfer	Mitigation X	Acceptance
5	Monitoring and control	Risk owner	Review date	Critical issue	
		Project Manager (Design)	D/M/Y	Construction documents not properly developed for bidding and tendering having effect on bid price	
6	Estimated impact on the project	Different bidding price			
7	Actual impact on the project	Construction documents are not properly complied			
8	Revised response	Develop construction documents taking into consideration all the requirements as per procurement strategy			
9	Record update date	D/M/Y			
10	Communication to stakeholders	Information sent to all stakeholders on D/M/Y			

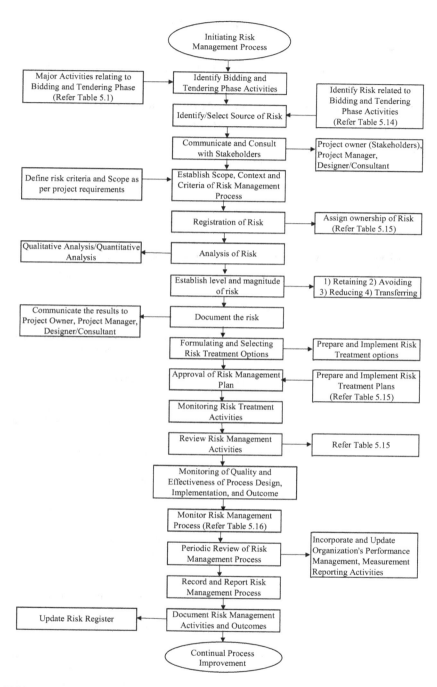

FIGURE 5.8 A typical flowchart for risk management in quality of bidding and tendering phase.

5.7.1 RISK IN ACTIVITIES RELATED TO QUALITY OF BIDDING AND TENDERING PHASE

Table 5.14 lists potential risks in the activities related to Bidding and Tendering Phase.

5.7.2 RISK EFFECT AND RISK TREATMENT IN ACTIVITIES RELATED TO QUALITY OF BIDDING AND TENDERING PHASE

Table 5.15 lists potential risks, probable effect on project, and risk treatment in activities related to Bidding and Tendering Phase.

TABLE 5.14
Potential Risks in Activities related to Bidding and Tendering Phase

Serial Number	Phase	Major Elements/ Activities	Potential Risk
1	1.1 Bidding and Tendering Phase		
		1. Organize tendering documents	Bid documents are not prepared as per procurement method and contract strategy adopted at the early stage of project
		2. Identification of project team	Identified project team members may not have direct involvement in the bidding and tendering phase
		3. Identification of bidders	Short listing to identify bidders is not done with proper system
		4. Management of tender documents	1. Tender documents do not consist of all the documents specified in tendering procedures
			2. Errors in construction documents
		5. Identification/ management of risk	Risks are not identified, and risk response evolved
			BOQ not matching with construction/design drawings
		6. Select contractor	Bid evaluation
			Contractor is not selected as per the procurement strategy adopted by the owner
			Successful bidder fails to submit performance bond
		7. Award contract	1. Delay in submitting performance bond by the contractor
			2. Delay in awarding the contract

TABLE 5.15

Potential Risks, Probable Effect on Project, and Risk Treatment in Activities Related to Bidding and Tendering Phase

Serial Number	Phase Major Elements/ Activities	Potential Risk	Probable Effect on Project	Ownership of Risk	Risk Treatment/Control Measures
1	1.1 Bidding and Tendering Phase				
	1. Organize tendering documents	Bid documents are not prepared as per procurement method and contract strategy adopted at the early stage of project	1. Delay to organize the documents 2. Bid price will be different from the estimated budget	Project Manager (Design)	Prepare and organize the tender documents as per the procurement method and contract strategy
				Tendering Team Manager	Review and update the tender documents submitted by the designer and insert the necessary owner-related information in the tender documents
	2. Identification of project team	Identified project team members may not have direct involvement in the bidding and tendering phase	Team members will not be able to follow and perform bidding and tendering process	Tendering Team Manager	Select team members having skills and knowledge to perform bidding and tendering activities and having experience of the bidding and tendering process
	3. Identification of bidders	Short listing to identify bidders is not done with proper system	Selected bidders are incompetent and are not qualified	Tendering Team Manager	Identify and short list bidders properly as per the procurement strategy with announcing to submit response to pre-qualification questionnaires and evaluating their response

(Continued)

TABLE 5.15 (Continued)

Serial Number	Phase	Major Elements/ Activities	Potential Risk	Probable Effect on Project	Ownership of Risk	Risk Treatment/Control Measures

4.		Management of Tender Documents	1. Tender documents do not consist of all the documents specified in tendering procedures
			2. Errors in construction documents
				Amendment to the tendering documents resulting in delay in bidding and tendering process	Tendering Team Manager	1. Verify for inclusion of all documents for tendering
						2. Review documents for error and omissions
						3. Manage the tendering process by following activities as listed below:
						  1. Bid notification
						  2. Distribution of tender documents
						  3. Pre-bid meeting
						  4. Issuing addendum, if any
						  5. Bid submission
						  6. Bid evaluation

5.		Identification/ management of Risk	Risks are not identified, and risk response evolved	Bid value exceeding the estimated definitive cost	Tendering Team Manager	Identify risks associated with bidding and tendering to ensure qualified bidders are taking parts and proper evaluation is made on submitted tenders
			BOQ not matching with construction/design drawings	Errors in calculating bid value	Bidder	Verify quantities in BOQ for correctness with contract documents and drawings
6.		Select contractor	Bid evaluation	Selection of qualified contractor	Tendering Team Manager Project Manager (Design	Review and evaluate the bids with proper checklist and evaluation process

Phase	Cause	Effect	Responsibility	Action
	Contractor is not selected as per the procurement strategy adopted by the owner	1. Unqualified contractor 2. Quality of construction	Tendering Team Manager Project Owner	Select the contractor based on the procurement strategy adopted by the owner / Award the contract to qualified contractor
	Successful bidder fails to submit performance bond	Delay in the selection of contractor	Tendering Team Manager	Ensure that successful bidder follows bidding and tending conditions
7. Award contract	1. Delay in submitting performance bond by the contractor 2. Delay in awarding the contract	1. Delay in award of the project 2. Delay in start of the project	Project Owner	Follow the contract award process in all the stages of bidding and tendering phase (refer Figure 5.8)

Figure 5.9 illustrates logic flow process for Bidding and Tendering phase.

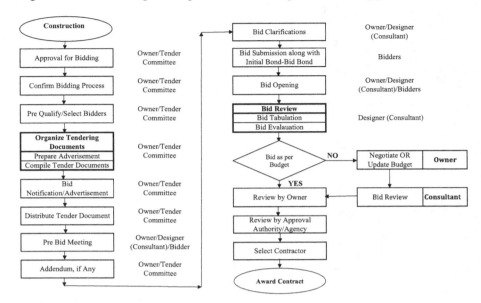

FIGURE 5.9   Logic flow process for bidding and tendering phase.

5.7.3   RISK MONITORING IN ACTIVITIES RELATED TO QUALITY OF BIDDING AND TENDERING PHASE ACTIVITY

Table 5.16 is an example process of risk monitoring of Bidding and Tendering phase activity.

TABLE 5.16
Risk Monitoring of Bidding and Tendering Phase Activity

Serial Number	Description	Action			
1	Risk ID	Bidding and Tendering—1.1 (1)			
2	Description of risk	1. Identification of bidders 1.1. Short listing to identify bidders is not done with proper system			
3	Response	1.1 Identify and short list bidders properly as per the procurement strategy with announcing to submit response to pre-qualification questionnaires and evaluating their response			
4	Strategy of response	Avoidance	Transfer	Mitigation X	Acceptance

Serial Number	Description	Action		
5	Monitoring and control	Risk owner	Review date	Critical issue
		Tendering Team Manager	D/M/Y	Incompetent and unqualified selected bidders
6	Estimated impact on the project	Incompetent contractor		
7	Actual impact on the project	Project quality		
8	Revised response	Identify and short list bidders properly		
9	Record update date	D/M/Y		
10	Communication to stakeholders	Information sent to all stakeholders on D/M/Y		

5.8   RISK MANAGEMENT IN QUALITY OF CONSTRUCTION PHASE

Figure 5.10 illustrates a typical flowchart for risk management in Quality of Construction phase.

5.8.1   RISK IN ACTIVITIES RELATED TO QUALITY OF CONSTRUCTION PHASE

Table 5.17 lists potential risks in the activities related to Construction phase.

5.8.2   RISK EFFECT AND RISK TREATMENT IN ACTIVITIES RELATED TO QUALITY OF CONSTRUCTION PHASE

Table 5.18 lists potential risks, probable effect on project, and risk treatment in activities related to Construction phase.

5.8.3   RISK MONITORING IN ACTIVITIES RELATED TO QUALITY OF CONSTRUCTION PHASE ACTIVITY

Table 5.19 is an example process of risk monitoring of Construction phase activity.

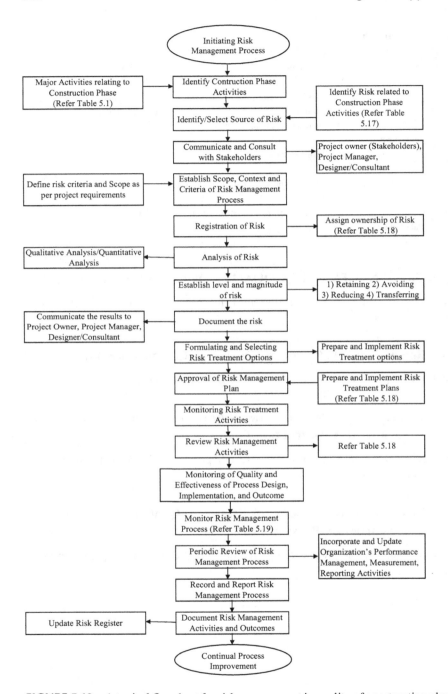

FIGURE 5.10 A typical flowchart for risk management in quality of construction phase.

TABLE 5.17
Potential Risks in Activities Related to Construction Phase

Serial Number	Phase	Major Elements/Activities	Potential Risk
1	1.1	Construction Phase	
		1. Mobilization	Delay in mobilization
		2. Identification of project team	1. Incompetent subcontractor 2. Team members not performing as expected
		3. Planning and scheduling	1. Inappropriate schedule/ plan 2. Contractor's construction schedule (CCS) is not in an organized and structured manner 3. Continuous (uninterrupted) flow of work is not considered in CCS
		4. Execution of works	1. Different site conditions to the information provided 2. Work is not executed as per approved shop drawings 3. Approved material is not installed
		5. Management of resources/ procurement	1. Availability of resources 2. Delay in approval and procurement of material
		6. Monitoring and control	Works not properly monitored as planned and status updated
		7. Quality management	Contractor's quality control plan not approved
		8. Risk management	Potential risks are not identified, and risk response is evolved
		9. Contract management	Contract management plan is not established
		10. Site safety	1. Site safety is not monitored 2. Hazardous areas are not identified 3. Safety awareness plan is not in place
		11. Inspection	1. Most of the executed/ installed works submitted for approval through checklists are rejected 2. Sequence for approval of executed works is not followed, hence executed works are rejected 3. All the executed works are not inspected on a regular basis

TABLE 5.18
Potential Risks, Probable Effect on Project, and Risk Treatment in Activities Related to Construction Phase

Serial Number	Phase	Major Elements/ Activities	Potential Risk	Probable Effect on Project	Ownership of Risk	Risk Treatment/ Control Measures
1	1.1 Construction Phase	1. Mobilization	Delay in mobilization	Delay in start of project execution	Construction Manager (Contractor)	1. Obtain necessary permits from the relevant authorities immediately after signing of contract 2. Prepare plan for mobilization activities immediately after signing of contract
					Project Owner	Transfer the project site without any delay
		2. Identification of project team	1. Incompetent subcontractor 2. Team members not performing as expected	1. Delay in execution of project works 2. Quality of project works	Construction Manager (Contractor)	1. Submit competent and qualified subcontractors for approval to perform the works and who are from the approved list 2. Submit core staff as per the required qualification in the contract documents 3. Select and submit team members having skills and experience to perform the contracted works (refer Figure 4.4 in Chapter 4)
		3. Planning and scheduling	1. Inappropriate schedule/ plan 2. Contractor's construction schedule (CCS) is not in an organized and structured manner 3. Continuous (uninterrupted) flow of work is not considered in CCS	1. Proper monitoring of project activities and progress of works 2. Control and monitor the project for a successful completion of project quality 3. Delay in execution of project activities	Construction Manager (Contractor)	1. Develop the plan showing the period for all sections/divisions of the works and activities specifying the dates for installation/execution of all the activities 2. Prepare the schedule in an organized structure following a logical process to determine what work must be done to achieve project objectives 3. Consider relationships between project activities and their dependency and precedence to ensure smooth flow of works 4. Prepare schedule taking into consideration resource calendar

Phase	Risk	Consequence	Responsible	Activities
4. Execution of works	1. Site conditions different from the information provided 2. Work is not executed as per approved shop drawings 3. Approved material is not installed	1. Delay in start of relevant activity(ies) 2. Quality of executed works not meeting the specification requirements 3. Quality of project	Project Manager (Design)	1. Investigate and study the relevant data about the site conditions prior to starting of relevant activities 2. Execute the works as per approved shop drawings 3. Install approved material, equipment, systems on the project
5. Management of resources/procurement	1. Availability of resources 2. Delay in the approval and procurement of material	1. Delay in execution of project 2. Delay in project	Construction Manager (Contractor)	1. Prepare detailed plan, once the contract is awarded, for all the resources to complete the project 2. Search for approved resources, competent workforce well in advance 3. Submit material as per the specifications for approval as per the schedule
6. Monitoring and control	Works not properly monitored as planned and status updated	Completion of project	R.E. Construction Manager (Contractor)	Monitor and control the project progress and status as per contract documents and approved schedule to ensure smooth flow of construction activities Execute the project activities as per approved schedule and method of installation
7. Quality management	Contractor's quality control plan not approved	1. Delay in execution of works 2. Quality of project	Construction Manager (Contractor)	Prepare and get approval of quality control plan and execute the works activities as per approved shop drawings using approved material following proper sequencing and method statement for installation of works to avoid rejection/rework (refer Figure 3.43)
8. Risk management	Potential risks are not identified, and risk response is evolved	Delay in project	R.E.	Maintain risk register and resolve the risks

(Continued)

TABLE 5.18
(Continued)

Serial Number	Phase	Major Elements/ Activities	Potential Risk	Probable Effect on Project	Ownership of Risk	Risk Treatment/ Control Measures
9.		Contract management	Contract management plan is not established	Management of contract agreements in an organized manner	Construction Manager (Contractor)	Maintain risk register and resolve the risks. Establish contract management plan and execute the work in an organismal method, process, and procedure for successful completion
10.		Site safety	1. Site safety is not monitored 2. Hazardous areas are not identified 3. Safety awareness plan is not in place	1. Unsafe site likely to be prone to more accidents, injuries 2. Dangerous environment at site 3. More accidents, injuries, and occupational illness	Construction Manager (Contractor)/ Safety Officer	1. Follow safety and accident prevention program. 2. Conduct safety meetings 3. Conduct safety drills and safety awareness program
11.		Inspection	1. Most of the executed/ installed works submitted for approval through checklists are rejected 2. Sequence for approval of executed works is not followed, hence executed works are rejected 3. All the executed works are not inspected on regular basis	1. Delay in the approval of executed works 2. Rejection of executed works by the supervisor 3. Quality of executed works	Construction Manager (Contractor); R.E.	1. Perform the inspection of work throughout the execution of project 2. Submit the checklists to the supervision team to inspect the executed/installed works (refer Figure 4.7 in Chapter 4). Monitor regularly the executed/installed works for quality compliance throughout the project

TABLE 5.19
Risk Monitoring of Construction Phase Activity

Serial Number	Description	Action			
1	Risk ID	Construction Phase—1.1 (3)			
2	Description of risk	1. Planning and scheduling 　1.1 Inappropriate schedule/ plan 　1.2. Contractor's construction schedule (CCS) is not in an organized and structured manner 　1.3. Continuous (uninterrupted) flow of work is not considered in CCS			
3	Response	1.1. Develop the plan showing the period for all sections/ divisions of the works and activities specifying the dates for installation/execution of all the activities 1.2. Prepare the schedule in an organized structure following a logical process to determine what work must be done to achieve project objectives 1.3. Consider relationships between project activities and their dependency and precedence to ensure smooth flow of works			
4	Strategy of response	Avoidance	Transfer	Mitigation X	Acceptance
5	Monitoring and control	Risk owner	Review date	Critical issue	
		Construction Manager (Contractor)	D/M/Y	Proper monitoring and control of project activities, progress of works, and project status	
6	Estimated impact on the project	Monitoring of project progress and status			
7	Actual impact on the project	Project monitoring			
8	Revised response	Develop schedule taking into consideration all the required parameters			
9	Record update date	D/M/Y			
10	Communication to stakeholders	Information sent to all stakeholders on D/M/Y			

5.9 RISK MANAGEMENT IN QUALITY OF TESTING, COMMISSIONING, AND HANDOVER PHASE

Figure 5.11 illustrates a typical flowchart for risk management in Quality of Testing, Commissioning, and Handover phase.

5.9.1 RISK IN ACTIVITIES RELATED TO QUALITY OF TESTING, COMMISSIONING, AND HANDOVER PHASE

Table 5.20 lists potential risks in the activities related to Testing, Commissioning, and Handover phase.

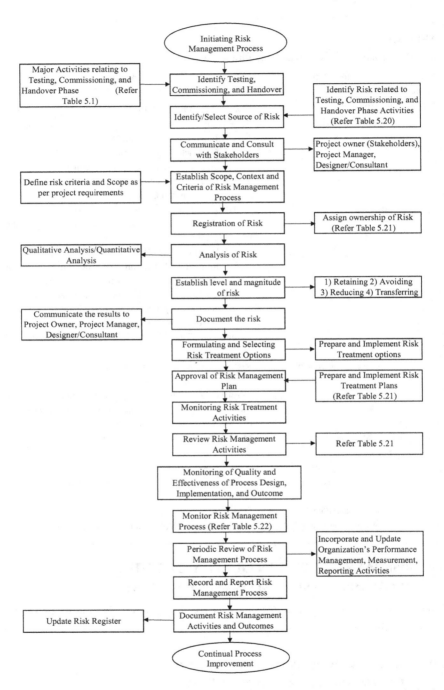

FIGURE 5.11 A typical flowchart for risk management in quality of testing, commissioning, and handover phase.

TABLE 5.20
Potential Risks in Activities related to Testing, Commissioning, and Handover Phase

Serial Number	Phase	Major Elements/Activities	Potential Risk
1	1.1	Testing, Commissioning, and Handover Phase	
		1. Develop testing and commissioning plan	Testing and commissioning plan is not developed and approved
		2. Identification of project team	All the stakeholders to be involved during testing, commissioning, and handover phase are not identified
		3. Testing and commissioning	Contractor has not carried out inspection and tests of certain activities in presence of relevant stakeholders
		4. As-built drawings	All the information from record drawings is not incorporated under as-built drawings
		5. Technical manuals and documents	Technical manuals don't consist of all the documents specified in contract documents
		6. Train owner's/end user's personnel	Training program as per specifications is not established and approved
		7. Regulatory/authorities' approval	Regulatory approvals from concerned authorities are not obtained or delayed
		8. Move-in plan	Move-in plan is not prepared on substantial completion schedule
		9. Handover of facility to owner/end user	Delay in handover/acceptance of the facility
		10. Substantial completion	Delay in issuance of substantial certificate

5.9.2 Risk Effect and Risk Treatment in Activities Related to Quality of Testing, Commissioning, and Handover Phase

Table 5.21 lists potential risks, probable effect on project, and risk treatment in activities related to Testing, Commissioning, and Handover Phase.

5.9.3 Risk Monitoring in Activities Related to Quality of Testing, Commissioning, and Handover Phase Activity

Table 5.22 is an example process of risk monitoring of Testing, Commissioning, And Handover phase activity.

TABLE 5.21

Potential Risks, Probable Effect on Project, and Risk Treatment in Activities Related to Testing, Commissioning, and Handover Phase

Serial Number	Phase	Major Elements/ Activities	Potential Risk	Probable Effect on Project	Ownership of Risk	Risk Treatment/Control Measures
1	1.2	Testing, Commissioning, and Handover Phase				
		1. Develop testing and commissioning plan	Testing and commissioning plan is not developed and approved	Lacking systematic approach that each individual work/ system is fully functional/ operational as specified	Construction Manager (Contractor)	Develop testing and commissioning plan in an orderly manner based on contract documents requirements and approved by R.E.
		2. Identification of project team	All the stakeholders to be involved during testing, commissioning, and handover phase are not identified	The testing and commissioning may not meet the stakeholder's requirements	Project Owner (Project Manager) R.E.	1. Identify and select all the relevant stakeholders that are involved during testing, commissioning, and handover phase 2. Involve manufacturer's representative to test the supplied systems/ equipment Select team members having experience in testing and commissioning of major projects
		3. Testing and commissioning	Contractor has not carried out inspection and tests of certain activities in the presence of relevant stakeholders	Contractor's action to be repeated as per contract documents for the satisfaction of stakeholders	Construction Manager (Contractor)	Invite all the relevant stakeholders to be present to witness the testing and commissioning of systems/equipment as per contract requirements
		4. As-built drawings	All the information from record drawings is not incorporated under as-built drawings	As-built drawings not indicating the actual installation as installed/ executed at site	Construction Manager (Contractor)	Prepare as-built drawings based on record drawings and actual installation/ executed works

5. Technical manuals and documents	Technical manuals don't consist of all the documents specified in contract documents	Incomplete information about installed equipment/ systems	Construction Manager (Contractor)	Submit technical manuals consisting of all the information about installed equipment/systems as per contract documents
6. Train owner's/end user's personnel	Training program as per specifications is not established and approved	Owner's personnel will not be fully aware of maintenance and operation procedures	Construction Manager (Contractor)	Conduct the training to owner's/ user's personnel to familiarize with the installed works and their operations as per contract documents
7. Regulatory/ authorities' approval	Regulatory approvals from concerned authorities are not obtained or delayed	Delay in handover/ takeover of project	Construction Manager (Contractor)	Obtain necessary regulatory approvals from the respective concerned authorities
8. Move-in plan	Move-in plan is not prepared on substantial completion schedule	Delay in occupancy of the facility/project	R.E.	Prepare move-in plan and obtain approval from the project owner
9. Handover of facility to owner/ end user	Delay in handover/ acceptance of the facility	Delay in usage of facility/ project by the owner/end user	Construction Manager (Contractor)	Follow approved schedule to handover the project/facility
10. Substantial completion	Delay in issuance of substantial certificate	Delay in project closeout	Project Owner (Project Manager)	Issue substantial completion certificate once it is established that contractor has completed the works as per contract documents and to the satisfaction of owner

TABLE 5.22
Risk Monitoring of Testing, Commissioning, and Handover Phase Activity

Serial Number	Description	Action			
1	Risk ID	Testing and Commissioning—1.1 (4)			
2	Description of risk	1. As-built drawings 1.1. All the information from record drawings is not incorporated under as-built drawings			
3	Response	1.1. Prepare as-built drawings based on record drawings and actual installation/executed works			
4	Strategy of response	Avoidance	Transfer	Mitigation X	Acceptance
5	Monitoring and control	Risk owner Construction Manager (Contractor)	Review date D/M/Y	Critical issue Problem during maintenance as the references will not be as per actual installation work	
6	Estimated impact on the project	Problem during maintenance			
7	Actual impact on the project	Project maintenance			
8	Revised response	Develop as-built drawings taking into consideration record drawings			
9	Record update date	D/M/Y			
10	Communication to stakeholders	Information sent to all stakeholders on D/M/Y			

5.10 PROCESS TYPE OF PROJECT PHASES

Figure 5.12 illustrates construction Project Life Cycle phases (process type of projects).

Table 5.23 illustrates construction Project Life Cycle phases (process type of projects) with subdivided activities/components.

These activities may not be strictly sequential; however, the breakdown allows implementation of project management functions more effectively at different stages.

5.11 RISK MANAGEMENT IN QUALITY OF STUDY STAGE PHASE

Figure 5.13 illustrates a typical flowchart for risk management in Quality of Feasibility Study Stage phase.

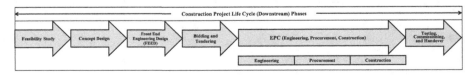

FIGURE 5.12 Construction project life cycle phases (process type projects).

TABLE 5.23
Construction Project Life Cycle Phases (Process Type of Projects)

Construction Project Life Cycle (Downstream) Phases

Feasibility Study	Concept Design	Front End Engineering Design (FEED)	Bidding and Tendering	Engineering	Procurement	Construction	Testing, Commissioning, and Handover
• Project Initiation	• Select Design Team Members	• Identify Design Team Members	• Organize Tender Documents	• Notice to Proceed • Kick off Meeting	• Develop Project Site Facilities	• Mobilization • Temporary Facilities	• Identify Testing, Commissioning and Startup Requirements
• Identification of Need	• Identify Project Stakeholders	• Identify Project Stakeholders	• Identify Project Team/Stakeholders	• Identify Stakeholders • Responsibility Matrix	• Contractor's Design Team • Contractor's Core Staff • Sub contractors + Vendors	• Project Management Consultant • Supervision Consultant	• Identify Stakeholders
• Feasibility Study	• Establish Concept Design Requirements	• Establish FEED Requirements	• Identify Tendering Procedure	• Construction Phase Requirements	• Construction Phase Scope	• Contractor's Construction Schedule	• Develop Scope of Work
• Establish Project Goals and Objectives	• Identify Concept Design Deliverables	• Develop FEED (Technical Study to Select Process)	• Identify Bidders	• Project S-Curve • Contractor's Quality Control Plan	• Stakeholder Mgmt Plan • Resource Mgmt Plan • Communication Mgmt Plan	• Risk Mgmt Plan • Contract Mgmt Plan • HSE Mgmt Plan	• Develop Testing and Commissioning Plan
• Identification of Alternatives/Options	• Develop Concept Design	• Prepare Front End Engineering Design	• Manage Tendering Process	• Identify Detailed Engineering Deliverables	• Identify Procurement Deliverables	• Identify Construction Works Deliverables	• Execute Testing and Commissioning Works
• Analyze and Evaluate Alternatives/Options	• Identify/Collect Regulatory Requirement	• Prepare Project Schedule/Plan	• Distribute Tender Documents	• Develop Detailed Engineering	• Develop Procurement Documents	• Manage Execution/Installation of Works	• Manage Testing and Commissioning Quality
• Select Preferred Alternative	• Prepare Preliminary Schedule	• Evaluate and Estimate Project Cost	• Conduct Pre Bid Meetings	• Perform Interdisciplinary Coordination	• Prepare Material Procurement List	• Manage Scope Change	• As Built Drawings
• Finalize Project Delivery and Contracting System	• Prepare Preliminary Schedule	• Manage FEED Quality	• Submit/Receive Bids	• Develop Project Schedule	• Identify Vendors	• Manage Construction Quality	• Technical Manuals and Documents
• Project Charter	• Estimate Conceptual Project Cost	• Estimate Resources	• Manage Bidding and Tendering Quality	• Estimate Project Cost	• Manage Procurement Quality	• Manage Construction Resources	• Record Books (PRB, CRB, MRB)
	• Establish Project Quality Requirements	• Develop Communication Plan	• Manage Risks	• Manage Detailed Engineering Quality	• Finalize Procurement	• Manage Communication	• Train Owner's/End User's Personnel
	• Manage Concept Design Quality	• Manage Project Risks	• Review Bid Documents	• Estimate Project Resources		• Manage Construction Risk	• Regulatory/Authority Approval
	• Estimate Project Resources	• Manage HSE Requirements	• Evaluate Bids	• Manage Project Risk		• Manage Contract	• Handover of Project to the Owner/End User
	• Identify Project Risks	• Develop FEED Documents	• Select Contractor	• Manage Project HSE Requirements		• Manage HSE Requirements	• Issue Substantial Completion
	• Identify Project HSE Issues and Requirements	• Perform Value Engineering Study	• Award Contract	• Review Detailed Engineering		• Manage Project Finances	• Settle Payments
	• Review Concept Design	• Review Front End Engineering Design		• Finalize Detailed Engineering		• Manage Claims	• Settle Claims
	• Finalize Concept Design	• Finalize Front End Engineering Design				• Monitor and Control Project Works	
		• Develop Tender Documents				• Validate Executed Works	

EPC (Engineering, Procurement, Construction)

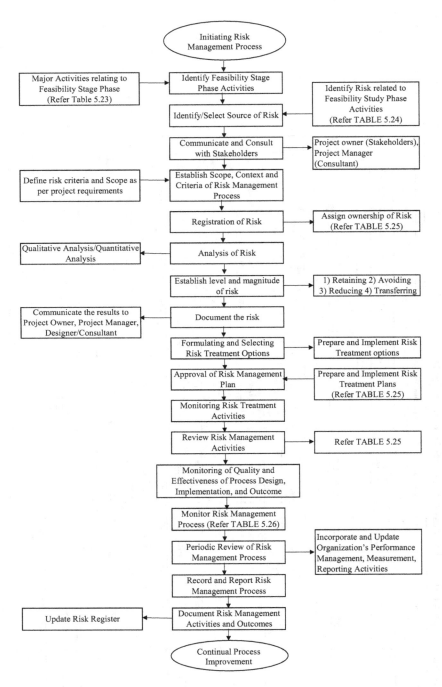

FIGURE 5.13   A typical flowchart for risk management in quality of feasibility study phase.

5.11.1 RISK IN ACTIVITIES RELATED TO QUALITY OF STUDY STAGE PHASE

Table 5.24 lists potential risks in the activities related to Feasibility Study Stage phase.

5.11.2 RISK EFFECT AND RISK TREATMENT IN ACTIVITIES RELATED TO QUALITY OF STUDY STAGE PHASE

Table 5.25 lists potential risks, probable effect on project, and risk treatment in activities related to Feasibility Study Stage phase.

Figure 5.14 illustrates evaluation and analysis of alternatives to select preferred alternative.

5.11.3 RISK MONITORING IN ACTIVITIES RELATED TO QUALITY OF STUDY STAGE PHASE

Table 5.26 is an example process of risk monitoring of Study Stage phase activity.

TABLE 5.24
Potential Risks in Activities Related to Feasibility Study Phase

Serial Number	Phase	Major Elements/ Activities	Potential Risk
1	1.1	Feasibility Study Phase	
		1. Project initiation	Project justification and business case are not established
		2. Identification of need	Project purpose and need are not established
		3. Feasibility study	1. Feasibility study is not performed taking into consideration technical, financial, and environmental impacts on the project
			2. Incomplete/improper feasibility study
			3. Environmental and spatial plan not considered
		4. Establish project goals and objectives	Project goals and objectives as being the most fundamental elements of project planning are not established
		5. Identification of alternatives/options	Identified alternatives/ options is not suitable for further development
		6. Analyze and evaluate alternatives/options	Alternative/options are not properly analyzed and evaluated taking into consideration qualitative and quantitative comparison
		7. Select preferred alternatives/option	Selected alternatives may not satisfy owner's goals and objectives
		8. Finalize project delivery and contracting system	Project delivery system is not finalized and selected taking into consideration size, complexity of the project, innovation, uncertainty, and the degree of involvement of owner
		9. Project charter	Project charter or TOR is not defined clearly

TABLE 5.25

Potential Risks, Probable Effects on Project, and Risk Treatment in Activities Related to Feasibility Study Phase

Serial Number	Phase	Major Elements/Activities	Potential Risk	Probable Effect on Project	Ownership of Risk	Risk Treatment/Control Measures
1	1.1 Feasibility Study Phase	1. Project initiation	Project justification and business case are not established	Enough information not available to proceed or to convince that the project should not proceed and be aborted	Project Owner	Identify and establish business case properly, and performance measures are set to achieve project goals and objectives
		2. Identification of need	Project purpose and need are not established	Requirements of quality, project performance, project completion date, and approved budget for the project are not clear to develop new facility/project	Project Owner	Define the project need very well by the minimum requirements of quality, performance, project completion date based on the available financial resources to develop the project
		3. Feasibility study	1. Feasibility study is not performed taking into consideration technical, financial, and environmental impacts on the project	1. Owner/decision-makers will not be able to take decision that will be in their best interest	Project Owner (Consultant)	1. Conduct feasibility study by specialist consultant in order to assist owner/decision makers in making the decision that will be in the best interest of owner.
			2. Incomplete/improper feasibility study	2. Evaluation of the project need and decide whether to proceed with the project or to stop		2. Perform the study for technical, financial, and economical viability of the project, potential environmental and social impact of the identified need of the project
			3. Environmental and spatial plan not considered			3. Environmental compatibility

4.	Establish project goals and objectives	Project goals and objectives as the most fundamental elements of project planning are not established	Project team member will not know the appropriate boundaries to make decisions about the project and ensure that the project/facility will satisfy the owner's/end user's requirements fulfilling their needs	Project Owner (Consultant)	Prepare and establish project goals and objectives taking into consideration final recommendations/outcome of the feasibility study clearly defining appropriate boundaries that satisfy the owner's requirements
5.	Identification of alternatives/options	Identified alternatives/options are not suitable for further development	Develop alternatives based on predetermined set of performance measures to meet owner's requirements	Project Manager (Consultant/ Design)	Develop and identify several alternative schemes and solutions based on the predetermined set of performance measures to meet the owner's requirements
6.	Analyze and evaluate alternatives/options	Alternative/options are not properly analyzed and evaluated taking into consideration qualitative and quantitative comparison	Alternative/option will not meet predetermined performance and owner's requirements	Project Manager (Consultant/ Design)	Perform qualitative and quantitative comparison, evaluation, and analysis of the identified alternatives by considering the advantage and disadvantage of each item systematically. Refer Figure 5.14.)
7.	Select preferred alternatives/Option	Selected alternatives may not satisfy owner's goals and objectives	Redevelopment and selection of new alternative that satisfy the goals and objectives	Project Manager (Design)	Select preferred alternative that satisfies project goals and objectives and meets the owner's requirements
8.	Finalize project delivery and contracting system	Project delivery system is not finalized and selected taking into consideration size, complexity of the project, innovation, uncertainty, and the degree of involvement of owner	Completion of project successfully to meet owner's objectives and requirements	Project Owner (Consultant)/ PMC	Select the project delivery system to meet the size and complexity suitable for successful completion of process type of project
9.	Project charter	Project charter or TOR is not defined clearly	Project not meeting the objectives and requirements of the owner to achieve qualitative and competitive project	Project Owner (Consultant)/ PMC	1. Prepare accurate and comprehensive TOR clearly documenting the works to be done for the development of the project 2. Define scope of work properly

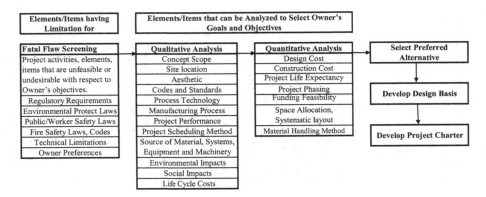

FIGURE 5.14 Evaluation and analysis of alternative methods to select preferred alternative.

TABLE 5.26
Risk Monitoring of Feasibility Study Phase Activity

Serial Number	Description	Action			
1	Risk ID	Feasibility Study—1.1 (3)			
2	Description of risk	1. Feasibility study 1.1. Feasibility study is not performed taking into consideration technical, financial, and environmental impact on the project 1.2. Incomplete/improper feasibility study 1.3. Environmental and spatial plan not considered			
3	Response	1.1. Conduct feasibility study by specialist consultant in order to assist owner/decision-makers in making the decision that will be in the best interest of owner. 1.2. Ensure that consultant performs the study for technical, financial, and economical viability of the project; potential environmental and social impact of the identified need of the project			
4	Strategy of response	Avoidance	Transfer	Mitigation X	Acceptance
5	Monitoring and control	Risk owner	Review date	Critical issue	
		Project Owner/ Consultant	D/M/Y	Owner/decision-makers will not be able to take correct decisions that will be in their best interests whether to proceed with the project or to stop	
6	Estimated impact on the project	Inception of the project			
7	Actual impact on the project	Decision to initiate the project			
8	Revised response	Conduct feasibility study taking into consideration all the factors for appraisal of need to proceed with the project			
9	Record update date	D/M/Y			
10	Communication to stakeholders	Information sent to all stakeholders on D/M/Y			

5.12 RISK MANAGEMENT IN QUALITY OF CONCEPT DESIGN PHASE

Figure 5.15 illustrates a typical flowchart for risk management in quality of Concept Design phase.

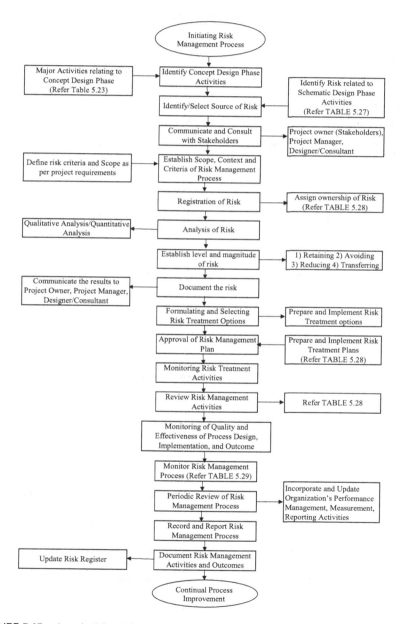

FIGURE 5.15 A typical flowchart for risk management in quality of concept design phase.

5.12.1 Risk in Activities Related to Quality of Concept Design Phase

Table 5.27 lists potential risks in the activities related to Concept Design Phase.

5.12.2 Risk Effect and Risk Treatment in Activities Related to Quality of Concept Design Phase

Table 5.28 lists potential risks, probable effect on project, and risk treatment in activities related to Concept Design Phase.

TABLE 5.27
Potential Risks in Activities Related to Concept Design Phase

Serial Number	Phase	Major Elements/Activities	Potential Risk
1	1.1	Concept Design Phase	
		1. Select design team members	Design team members are not selected as per the procurement strategy and organization structure
		2. Identify project stakeholders	1. All the stakeholders having direct involvement in the project are not identified 2. Responsibilities matrix of stakeholders is not established
		3. Establish concept design phase requirements	Lack of input from owner about the project goals and objectives
		4. Establish concept design deliverables	Concept design deliverables are not established as per TOR requirements
		5. Develop concept design	Concept design is not developed based on requirements listed in project charter or TOR
		6. Identify/collect regulatory requirements	Regulatory requirements are not identified
		7. Prepare preliminary schedule	Errors in estimating the project schedule
		8. Estimate conceptual project cost	Errors in cost estimation
		9. Establish project quality requirements	Project quality management is not established
		10. Manage concept design quality	Quality of concept design in not established
		11. Estimate project resources	Errors in resource estimation
		12. Identify project risks	Potential risks are not identified, and response evolved
		13. Identify project HSE issues and requirements	Environmental issues are not correctly identified
		14. Review concept design	Concept design is not fully reviewed
		15. Finalize concept design	Review comments are not incorporated

TABLE 5.28

Potential Risks, Probable Effect on Project, and Risk Treatment in Activities Related to Concept Design Phase

Serial Number	Phase	Major Elements/Activities	Potential Risk	Probable Effect on Project	Ownership of Risk	Risk Treatment/Control Measures
1	1.1 Concept Design Phase	1. Select design team members	Design team members are not selected as per the procurement strategy and organization structure	Team members will not be able to perform the assigned duties, roles, and responsibilities	Project Owner (PMC)	Approve the team members having skills and knowledge in similar size and type of projects
					Project Manager (Design)	1. Selected the team members have knowledge, experience, and skills 2. Select competent candidate 3. Refer Figure 4.4 in Chapter 4 4. Provide training
		2. Identify project stakeholders	1. All the stakeholders having direct involvement in the project are not identified 2. Responsibilities matrix of stakeholders is not established	1. Relevant stakeholders will not be aware of their roles and responsibilities to monitor and review project progress activities of their interest in the construction project	Project Owner (PMC)	1. Identify and select stakeholders having direct involvements, interest, or impact in the project, and responsibilities matrix is prepared involving the selected stakeholders 2. Establish responsibilities matrix involving stakeholders having interest in the project
		3. Establish concept design phase requirements	Lack of input from owner about the project goals and objectives	Project team members will not be aware of appropriate boundaries to make decisions about the project and to ensure that the project will meet owner's requirements and satisfy owner's needs	Project Manager (Design)	Gather/collect owner's requirements and all the related data and TOR requirements to develop concept design

(Continued)

TABLE 5.28
(Continued)

Serial Number	Phase	Major Elements/Activities	Potential Risk	Probable Effect on Project	Ownership of Risk	Risk Treatment/Control Measures
4.		Establish concept design deliverables	Concept design deliverables are not established as per TOR requirements	Project will not meet owner's requirements	Project Manager (Design)	Establish concept design deliverables as per TOR requirements
5.		Develop concept design	Concept design is not developed based on requirements listed in project charter or TOR	Project design will not meet or satisfy owner's/end user's requirements and fulfil owner's needs	Project Manager (Design)	Develop concept design taking into consideration requirements listed in TOR and related data/information to meet owner's requirements considering the following: 1. Usage 2. Process technology 3. Sustainability 4. Reliability 5. Cost-effectiveness 6. Energy conservation requirements 7. Environmental/HSE issues
6.		Identify/collect regulatory requirements	Regulatory requirements are not identified	Project not complying with the regulations, codes, and licensing procedures	Project Manager (Design)	Identify regulatory requirements, taking into consideration these requirements while developing concept design
7.		Prepare preliminary schedule	Errors in estimating the project schedule	Finances and acquisition of project will be affected	Project Manager (Design)	Prepare preliminary schedule using top-down planning using key events, and the guidelines published by AACE International to prepare preliminary schedule
8.		Estimate conceptual project cost	Errors in cost estimation	Project funding/ arranging the financial requirements	Project Manager (Design)	Estimate conceptual cost based on assumptions and historical data available for similar type and size of projects by maintaining the accuracy

Activity	Risk	Impact	Responsible	Action
9. Establish project quality requirements	Project quality management is not established	Project quality not meeting owner's requirements	Project Manager (Design)	Develop quality management plan by considering the following: 1. Owner's requirements 2. Quality standards and codes to be complied 3. Regulatory requirements 4. Conformance to owner's requirements 5. Conformance to requirements listed under project charter (TOR) 6. Design review procedure 7. Drawings' review procedure 8. Document review procedure 9. Quality management during all the phases of project life cycle
10. Manage concept design quality	Quality of concept design in not established	Design quality will be affected	Project Manager (Design)	Manage the quality criteria and quality of concept design to meet the owner's requirements
11. Estimate project resources	Errors in resource estimation	Execution of project as per schedule will be affected	Project Manager (Design)	Estimate the project resources that are adequate for successful completion of the project
12. Identify project risks	Potential risks are not identified, and response evolved	Affect successful completion of project	Project Manager (Design)	Identify the risks that will affect the successful completion of the project and response plan is established to mitigate/avoid risks
13. Identify project HSE issues and requirements	Environmental issues are not correctly identified	Environmental effect on the project	Project Manager (Design)	Identify HSE issues that affect the environment and consider the same while developing the concept design
14. Review concept design	Concept design is not fully reviewed	Design errors	Project Manager (Design)	Review concept design to avoid mistakes and errors
15. Finalize concept design	Review comments are not incorporated	Design not meeting owner's requirements	Project Manager (Design)	Review concept design for compliance with owner's requirements
			Project Owner (PMC)	Review the submitted concept design to ensure compliance with owner's needs and requirements

5.12.3 Risk Monitoring in Activities Related to Quality of Concept Design Phase

Table 5.29 is an example process of risk monitoring of concept design phase activity.

5.13 RISK MANAGEMENT IN QUALITY OF FRONT-END ENGINEERING DESIGN PHASE

Figure 5.16 illustrates a typical flowchart for risk management in quality of Front-End Engineering Design phase.

TABLE 5.29
Risk Monitoring of Concept Design Phase Activity

Serial Number	Description	Action			
1	Risk ID	Concept Design—1.1 (5)			
2	Description of risk	1. Develop concept design 1.1. Concept design is not developed based on requirements listed in project charter or TOR			
3	Response	1.1. Develop concept design taking into consideration requirements listed in TOR and related data/information to meet owner's requirements			
4	Strategy of response	Avoidance	Transfer	Mitigation X	Acceptance
5	Monitoring and control	Risk owner	Review date	Critical issue	
		Project Manager (Design)	D/M/Y	Concept design not meeting owner's/end user's requirements	
6	Estimated impact on the project	Project design			
7	Actual impact on the project	Design not meeting owner's requirements			
8	Revised response	Develop concept design taking into consideration TOR requirements and related data			
9	Record update date	D/M/Y			
10	Communication to stakeholders	Information sent to all stakeholders on D/M/Y			

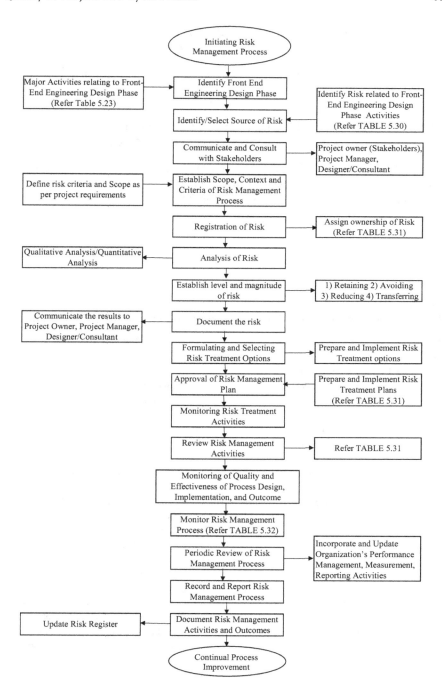

FIGURE 5.16 A typical flowchart for risk management in quality of front-end engineering design phase.

5.13.1 RISK IN ACTIVITIES RELATED TO QUALITY OF FRONT-END ENGINEERING DESIGN PHASE

Table 5.30 lists potential risks in the activities related to Front-End Engineering Design Phase.

5.13.2 RISK EFFECT AND RISK TREATMENT IN ACTIVITIES RELATED TO QUALITY OF FRONT-END ENGINEERING DESIGN PHASE

Table 5.31 lists potential risks, probable effect on project, and risk treatment in activities related to Front-End Engineering Design Phase.

Figure 5.17 illustrates a logic flow chart for the selection of process technology.

TABLE 5.30
Potential Risks in Activities Related to Front-End Engineering Design Phase

Serial Number	Phase	Major Elements/ Activities	Potential Risk
1	1.1 Front-End Engineering Design Phase		
		1. Select project team members	Designer team members are not selected based on the organizational structure and skills in process type of construction projects
		2. Identify project stakeholders	Identifies stakeholders are not directly involved in the project
		3. Establish FEED requirements	The designer has not taken into consideration the concept design deliverables and reviewed the comments
		4. Develop FEED (technical study to select process)	The designer has not identified and collected the following information: • TOR requirements • Concept design comments • Owner's requirements • Regulatory requirements • Energy conservation requirements
		5. Prepare front-end engineering design	Process selection procedure, equipment, machinery, and system are not taken into consideration
		6. Prepare project schedule/ plan	1. The logic of construction program is not set, and execution plan/project schedule is developed 2. Errors in estimating the project schedule
		7. Evaluate and estimate project cost	Errors in cost estimation

Serial Number	Phase	Major Elements/ Activities	Potential Risk
		8. Manage FEED quality	The designer has not established quality management plan
		9. Estimate resources	Error in estimating resources
		10. Develop communication plan	1. Communication plan is not developed 2. Matrix for communications not developed
		11. Manage risks	1. Potential risks are not identified, and response method is evolved 2. Wrong selection of process, materials, and systems
		12. Identify and manage HSE requirements	1. HSE plan is not established 2. HSE issues are not identified
		13. Develop FEED documents	Errors and omissions in the contract documents
		14. Perform value engineering	Value engineering is not performed
		15. Review FEED	1. FEED documents not reviewed 2. FEED documents are updated with value engineering comments 3. A number of FEED drawings are not as per TOR
		16. Finalize FEED	FEED is not reviewed prior to finalization
		17. Develop tender documents	Errors in tender documents

TABLE 5.31

Potential Risks, Probable Effect on Project, and Risk Treatment in Activities Related to Front-End Engineering Design Phase

Serial Number	Phase	Major Elements/ Activities	Potential Risk	Probable Effect on Project	Ownership of Risk	Risk Treatment/Control Measures
1	1.1 Front-End Engineering Design Phase					
		1. Select project team members	Designer team members are not selected based on the organizational structure and skills in process type of construction projects	Team members will not be able to perform achieving successful project as per owner's requirements and specifications	Project Manager (Design)	1. Select the team members based on the organizational structure basis having suitable qualification, knowledge, and skill required to perform the duties and responsibilities achieving successful project (refer Figure 4.4 in Chapter 4) completion 2. Provide training
		2. Identify project stakeholders	Identified stakeholders are not directly involved in the project	Stakeholders will not be able to address the issues and play a vital role in determination, formation, and successful implementation of project process	Project Owner (PMC) Project Manager (Design)	Identify stakeholders having involvement, interest, or impact in the project processes
		3. Establish FEED requirements	The designer has not taken into consideration the concept design deliverables and reviewed the comments	Designer will not be able to develop FEED as per owner's requirements and TOR to ensure that the developed design is error free and has minimum omissions	Project Manager (Design)	Consider concept design requirements, owner's comments, and TOR requirements to develop FEED
		4. Develop FEED (technical study to select process)	The designer has not identified and collected the following information: • TOR requirements • Concept design comments • Owner's requirements • Regulatory requirements • Energy conservation requirements	Selection of process, process technology, and process flow diagram not meeting the specification requirements	Project Manager (Design)	Develop FEED deliverables taking into consideration: 1. TOR requirements 2. Owner's requirements 3. Comments on concept design 4. Related data and technical studies for selection of process technology 5. HSE requirements

5. Prepare front-end engineering design	Process selection procedure, equipment, machinery, and system are not taken into consideration	The process not meeting the owner's objectives, and front-end engineering design	Project Manager (Design)	1. Gather all the related information such as
				  1. TOR requirements
				  2. Concept design comments
				  3. Owner's requirements
				  4. Regulatory requirements
				  5. Project-related data
				  6. Site investigation reports
				  7. Energy conservation requirements
				  8. Technical study and evaluation to select process technology. (Refer Figure 5.17)
				  9. Perform value engineering study

6. Prepare project schedule / plan	1. The logic of construction program is not set, and execution plan/project schedule is developed 2. Errors in estimating the project schedule	1. Identification of relationship and dependency among the project objectives for execution of project are not clearly defined 2. Errors in performing the activity/task during project execution	Project Manager (Design)	1. Prepare schedule by setting the logic program and the project execution plan/project schedule is developed by considering CPM and contract milestones 2. Review and analyze schedule to minimize errors, mistakes in the schedule
7. Evaluate and estimate project cost	Errors in cost estimation	1. Impact on overall success of project 2. Errors in developing time-phased plan summarizing the expected cost toward the construction	Project Manager (Design)	1. Estimate and prepare the Total Investment Cost (TIC) based on FEED-related activities, resources, and related schedule/execution plan 2. Review and evaluate cost estimate for accuracy of TIC
8. Manage FEED quality	The designer has not established quality management plan	1. Design criteria 2. Design errors and omissions	Project Manager (Design)	Establish the quality management plan to minimize design errors and omissions

(Continued)

TABLE 5.31 (Continued)

Serial Number	Phase	Major Elements/ Activities	Potential Risk	Probable Effect on Project	Ownership of Risk	Risk Treatment/Control Measures
9.		Estimate resources	Error in estimating resources	Successful completion of project as per schedule	Project Manager (Design)	Estimate the project resource based on FEED activities and works to be performed during the construction and testing, commissioning phase and prepare manpower histogram
10.		Develop communication plan	1. Communication plan is not developed 2. Matrix for communications not developed	1. Identified stakeholders will not have timely and accurate information	Project Manager (Design)	1. Develop communication plan and prepare responsibilities matrix for a smooth flow of communication to the relevant stakeholders 2. Develop matrix for communication of all the information and distribution of related documents
11.		Manage risks	Potential risks are not identified, and response method is evolved	Potential damage to the functioning of project progress and completion of project	Project Manager (Design)	Identify potential risks and manage to mitigate, avoid errors in the FEED
			Wrong selection of process, materials, and systems	Project functioning and usage as per owner's requirements	Project Manager (Design)	Develop FEED considering the following: 1. Project usage 2. Site investigation 3. Engineering survey 4. Technical studies for process selection 5. Technical and functional capability 6. Operational capability 7. Constructability
12.		Identify and manage HSE requirements	1. HSE plan is not established 2. HSE issues are not identified	Environmental impact on the project	Project Manager (Design)	1. Develop HSE plan 2. Identify issues to manage the environmental impact on the project

Task	Failure mode	Consequence	Responsible	Description
13. Develop FEED documents	Errors and omissions in the contract documents	Successful completion of project meeting owner's requirements and needs	Project Manager (Design)	Develop a complete set of documents having working drawings of all disciplines, site plans, technical specifications, material take off, schedule, and related information for bid
14. Perform value engineering	Value engineering is not performed	Cost optimization of the project	Value Engineering Specialist	Perform value engineering study to achieve optimize design the project functional cost
15. Review FEED	1. FEED documents not reviewed 2. FEED documents are not updated with value engineering comments 3. Number of FEED drawings are not as per TOR	1. Errors in FEED documents 2. Determination of optimized value of the project 3. Tendering procedure will be affected	Project Manager (Design)	1. Review FEED drawings and documents to minimize errors, omissions, and design mistakes 2. Update FEED drawings and documents incorporating value engineering comments 3. Develop FEED drawing to match TOR requirements
16. Finalize FEED	FEED is not reviewed prior to finalization	Errors/mistakes in FEED	Project Manager (Design) Project Manager (PMC)	Review FEED documents prior to submission to the owner/PMC Review the submitted FEED documents to ensure compliance with owner's requirements and needs
17. Develop tender documents	Errors in tender documents	Delay in tendering procedure	Project Manager (Design)	Develop tender documents based on FEED and contract documents

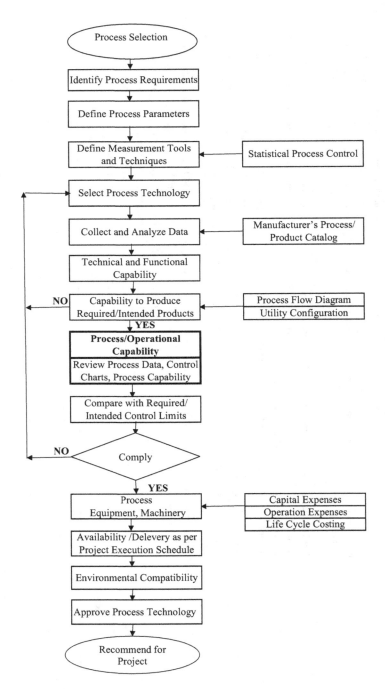

FIGURE 5.17 A logic flowchart for the selection of process technology.

Source: Abdul Razzak Rumane. (2021). Quality Management in Oil and Gas Projects. Reprinted with permission from Taylor & Francis Group Company.

TABLE 5.32
Risk Monitoring of Front-End Engineering Design Phase Activity

Serial Number	Description	Action			
1	Risk ID	FEED—1.1 (5)			
2	Description of risk	1. Prepare Front-End Engineering Design 1.1. Process selection procedure, equipment, machinery, and system are not taken into consideration			
3	Response	1.1. Gather all the related information such as TOR requirements, concept design comments, owner's requirements, project related data, site investigation reports, technical study, and do the evaluation to select process technology.			
4	Strategy of response	Avoidance	Transfer	Mitigation X	Acceptance
5	Monitoring and control	Risk owner	Review date	Critical issue	
		Project Manager (Design)	D/M/Y	The process not meeting the owner's objectives and front-end engineering design	
6	Estimated impact on the project	Project design			
7	Actual impact on the project	Front-end engineering design not meeting owner's requirements			
8	Revised response	Develop FEED taking into consideration TOR requirements and related data			
9	Record update date	D/M/Y			
10	Communication to stakeholders	Information sent to all stakeholders on D/M/Y			

5.13.3 RISK MONITORING IN ACTIVITIES RELATED TO QUALITY OF FRONT-END ENGINEERING DESIGN PHASE

Table 5.32 is an example process of risk monitoring of Front-End Engineering Design phase activity.

5.14 RISK MANAGEMENT IN QUALITY OF BIDDING AND TENDERING PHASE

Figure 5.18 illustrates a typical flowchart for risk management in Quality of Bidding and Tendering phase.

5.14.1 RISK IN ACTIVITIES RELATED TO QUALITY OF BIDDING AND TENDERING PHASE

Table 5.33 lists potential risks in the activities related to Bidding and Tendering phase.

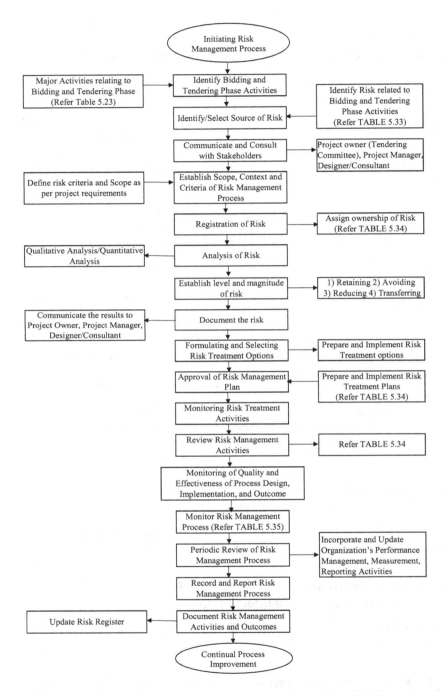

FIGURE 5.18   A typical flow chart for risk management in quality of bidding and tendering phase.

5.14.2   Risk Effect and Risk Treatment in Activities Related to Quality of Bidding and Tendering Phase

Table 5.34 lists potential risks, probable effect on project, and risk treatment in activities related to Bidding and Tendering Phase.

TABLE 5.33

Potential Risks in Activities Related to Bidding and Tendering Phase

Serial Number	Phase	Major Elements/Activities	Potential Risk
1	1.1 Bidding and Tendering Phase		
		1. Organize tender documents	Bid documents are not prepared as per procurement method and contract strategy adopted at the early stage of project
		2. Identify project team	Identified project team members may not have direct involvement in the bidding and tendering phase
		3. Identify tendering procedure	Tendering procedure is not as per procurement strategy
		4. Identify bidders	Short listing to identify bidders is not done with proper system
		5. Manage tendering process	Tendering process is not followed considering the following activities: • Bid notification • Distribution of tender documents • Pre-tender meeting • Issuing addendum if any • Bid submission
		6. Distribute tendering documents	Tender documents are not distributed to eligible bidders
		7. Conduct pre-tender meeting	Pre-tender meeting is not conducted
		8. Submit/receive bids	Bids are not accompanied with all the relevant documents
		9. Manage bidding and tendering quality	Bidding quality is not established
		10. Manage risks	1. Risks are not identified, and response method is evolved 2. Errors in tender documents 3. BOQ not matching with contract documents and drawings
		11. Review bid documents	Bid documents are not properly reviewed
		12. Evaluate bid documents	Checklist is not followed for evaluation of bid documents
		13. Select contractor	Contractor is not selected on procurement strategy
		14. Award contract	Contract award process is not followed

TABLE 5.34

Potential Risks, Probable Effect on Project, and Risk Treatment in Activities Related to Bidding and Tendering Phase

Serial Number	Phase	Major Elements/ Activities	Potential Risk	Probable Effect on Project	Ownership of Risk	Risk Treatment/ Control Measures
1	1.1 Bidding and Tendering Phase					
		1. Organize tender documents	Bid documents are not prepared as per procurement method and contract strategy adopted at the early stage of project	1. Delay to follow tendering process 2. Bid price different than estimated definitive price	Project Manager (Design)	Prepare bid documents as per the procurement method and contract strategy adopted by the owner
		2. Identify project team	Identified project team members may not have direct involvement in the bidding and tendering phase activities	Team members will not be able to follow and perform bidding and tendering process	Project Owner (PMC)	Select team members having knowledge, skills, and experience in bidding and tendering process
		3. Identify tendering procedure	Tendering procedure is not as per procurement strategy	1. Amendment to tendering documents 2. Delay in submission of bid	Project Owner (PMC)	Develop tendering procedure as per procurement strategy selected by the owner
		4. Identify bidders	Short listing to identify bidders is not done with proper system	1. Selected bidders are incompetent	Project Owner (PMC)	Select the bidders based on short listing by pre-qualification questionnaires
		5. Manage tendering process	Tendering process is not followed considering the following activities: • Bid notification • Distribution of tender documents • Pre-tender meeting • Issuing addendum if any • Bid submission	Delay in submission of bids than the notified one	Project Owner (PMC)/ Tendering Committee	Manage the tendering process properly (refer Figure 5.8)
		6. Distribute tendering documents	Tender documents are not distributed to eligible bidders	Selection of incompetent/ unqualified contractor	Project Owner (PMC)/ Tendering Committee	Distribute the bid documents to qualified and short-listed bidders

#	Task	Failure	Consequence	Responsible	Quality action
7.	Conduct pre-tender meeting	Pre-tender meeting is not conducted	Contractor will not get bid/tender clarifications	Project Owner (PMC)	Conduct pre-tender meeting to provide opportunity for the bidders to review and discuss the documents and get clarifications to their queries
8.	Submit/receive bids	Bids are not accompanied with all the relevant documents	Disqualification of bidder(s)	Project Owner (PMC)/ Tendering Committee	Receive the bid documents in accordance with the instructions to the bidders
9.	Manage bidding and tendering quality	Bidding quality is not established	Delay in review of tender documents	Project Owner (PMC)	Establish quality management plan to minimize omissions and mistakes during bidding and tendering process
10.	Manage risks	Risks are not identified, and response method is evolved	Bid value	Project Owner (PMC)	Identify risks and evolve response to ensure that bid value meet the owner's budget
		Errors in tender documents	Delay in tendering submission	Project Owner (PMC)	Review and amend the documents
		BOQ not matching with contract documents and drawings	Correctness of bid value	Bidder	Verify BOQ for correctness with respect to contract documents and drawings
11.	Review bid documents	Bid documents are not properly reviewed	Errors in the selection of qualified contractor	Project Owner (PMC)	Review the bids to ensure compliance to tender documents
12.	Evaluate bid documents	Checklist is not followed for evaluation of bid documents	Verification of bids for the selection of competent bidder	Project Owner (PMC) Project Manager (Design)	Evaluate bids with proper checklist Review bid documents to meet tendering process and requirements
13.	Select contractor	Contractor is not selected on procurement strategy	1. Incompetent contractors 2. Delay in execution of project	Project Owner (PMC)	Select the contractor based on the procurement strategy and competent for successful completion of the project
14.	Award contract	Contract award process is not followed	Selection of unqualified contractor	Project Owner (PMC)	Award the contract to qualified and competent contractor

5.14.3 Risk Monitoring in Activities Related to Quality of Bidding and Tendering Phase

Table 5.35 illustrates an example process of risk monitoring of bidding and tendering phase activity.

5.15 RISK MANAGEMENT IN QUALITY OF EPC PHASE (GENERAL ACTIVITIES)

Figure 5.19 illustrates a typical flowchart for Risk Management in Quality of EPC Phase (general activities).

TABLE 5.35
Risk Monitoring of Bidding and Tendering Phase Activity

Serial Number	Description	Action			
1	Risk ID	Bidding and Tendering—1.1 (5)			
2	Description of Risk	1. Identify project team 1.1. Identified project team members may not have direct involvement in the bidding and tendering phase activities			
3	Response	1.1. Select team members having knowledge, skills, and experience in bidding and tendering process			
4	Strategy of response	Avoidance	Transfer	Mitigation X	Acceptance
5	Monitoring and control	Risk owner	Review date	Critical issue	
		Project Owner (PMC)	D/M/Y	Team members will not be able to follow and perform bidding and tendering process	
6	Estimated impact on the project	Bidding tendering process			
7	Actual impact on the project	Delay in performing bidding and tendering activities			
8	Revised response	Develop FEED taking into consideration TOR requirements and related data			
9	Record update date	D/M/Y			
10	Communication to stakeholders	Information sent to all stakeholders on D/M/Y			

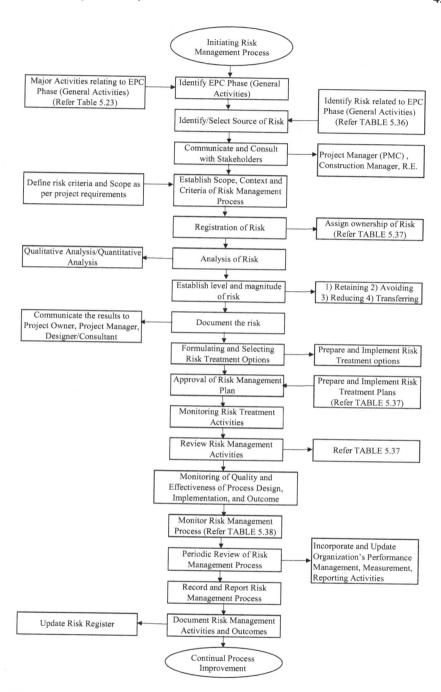

FIGURE 5.19 A typical flowchart for risk management in quality of engineering, procurement, and construction phase (general activities).

5.15.1 Risk in Activities Related to Quality of EPC Phase (General Activities)

Table 5.36 lists potential risks in the activities related to EPC phase (general activities).

5.15.2 Risk Effect and Risk Treatment in Activities Related to Quality of EPC Phase (General Activities)

Table 5.37 lists potential risks, probable effect on project, and risk treatment in activities related to EPC phase (general activities).

5.15.3 Risk Monitoring in Activities Related to Quality of EPC Phase (General Activities)

Table 5.38 is an example process of risk monitoring of EPC phase (general activities) activity.

TABLE 5.36
Potential Risks in Activities Related to EPC Phase (General Activities)

Serial Number	Phase	Major Elements/ Activities	Potential Risk
1	1.1	EPC Phase (General Activities)	
		1. Develop site facilities	Site facilities for all contractors, subcontractors are not developed
		2. Notice to proceed	Delay in issuing notice to proceed
		3. Kick-off meeting	Delay in conducting kick-off meeting
		4. Mobilization	1. Delay in mobilization
			2. All the required activities related to mobilization are not completed as specified
		5. Identify stakeholders	Stakeholders having involvement in the project are not identified
		6. Develop responsibilities matrix	1. Responsibilities matrix is not developed
			2. All the responsibilities are not included in the matrix
		7. Establish construction phase requirements	Construction phase requirements are not developed as specified in the contract documents

Serial Number	Phase	Major Elements/ Activities	Potential Risk
		8. Develop construction phase scope	Scope is not developed taking into consideration contract requirements
		9. Develop contractor's construction schedule	Construction schedule does not match contracted time schedule of the project
		10. Develop project S-Curve	S-Curve is not developed taking into consideration forecasted cash flow based on work activities to be completed as per construction schedule
		11. Develop contractor's quality control plan	Contractor's quality control plan does not contain adequacy and performance for delivering the level of construction quality required by the contract
		12. Develop stakeholder management plan	The needs of relevant stakeholders are not addressed in the contents of stakeholder's plan
		13. Develop resource management plan	Resource management is developed not considering the management of all the project resources (human resources, construction resources (contractor staff), manpower for construction works, material for project)
		14. Develop communication management plan	Communication management plan does include effective communication activities related to all the involved stakeholders
		15. Develop risk management plan	The contents of risk management plan do not include how the risk activities will be performed, recorded, and monitored throughout the project
		16. Develop contract management plan	The contents of contract management plan do not include all the activities for implementation and successful completion of contract
		17. Develop HSE management plan	The guidelines normally specified in the contract documents are not followed for developing HSE management plan

TABLE 5.37
Potential Risks, Probable Effect on Project, and Risk Treatment in Activities Related to EPC Phase (General Activities)

Serial Number	Phase	Major Elements/ Activities	Potential Risk	Probable Effect on Project	Ownership of Risk	Risk Treatment/Control Measures
1	1.1 EPC Phase (General Activities)					
		1. Develop site facilities	Site facilities for all contractors, subcontractors are not developed	Delay in start of activities at construction site	Construction Manager (Contractor)	Develop site facilities as per mobilization plan and contract documents
		2. Notice to proceed	Delay in issuing notice to proceed	Delay in authorizing contractor to proceed with project work	Project Owner (PMC)	Complete all the required documents, permits not to delay the issuance of notice to proceed
		3. Kick-off meeting	Delay in conducting kick-off meeting	Delay in providing an opportunity to concerned project team members to interact and knowing each other	Project Owner (PMC)	Conduct kick-off meeting without any delay to start mobilization
		4. Mobilization	1. Delay in mobilization 2. All the required activities related to mobilization are not completed as specified	1. Delay in the beginning of actual construction works	Construction Manager (Contractor)	1. Start mobilization work as per construction schedule 2. Obtain necessary permits and permission to start mobilization without any delay
		5. Identify stakeholders	Stakeholders having involvement in the project are not identified	Relevant stakeholders not aware of their involvements, roles, and responsibilities and have any influence on the project to address their needs for successful project	Project Owner (PMC)	Identify the relevant stakeholders having involvement, interest, or impact in the project processes
		6. Develop responsibilities matrix	1. Responsibilities matrix is not developed 2. All the responsibilities are not included in the matrix	1. Stakeholders are not aware of roles and responsibilities 2. Related issues/project status will not be addressed and resolved	Project Owner (PMC)	1. Develop responsibilities matrix properly assigning the roles and responsibilities of relevant stakeholders 2. Prepare communication matrix taking into consideration responsibilities of all the related stakeholders

Activity	Potential failure	Consequence	Responsibility	Requirement / Action
7. Establish construction phase requirements	Construction phase requirements are not developed as specified in the contract documents	Delay in the execution of project as specified in the contract documents, drawings, and specifications	Construction Manager (Contractor)	Establish construction phase requirements taking into consideration the following: 1. FEED drawings 2. Contract conditions 3. Technical specifications 4. Other related data
8. Develop construction phase scope	Scope is not developed taking into consideration contract requirements	Contractor will not be fully aware of the scope of works of the project as per contract requirements	Construction Manager (Contractor)	Develop construction phase scope taking into consideration the following major requirements: 1. Development of detailed engineering 2. Procurement of material 3. Construction of the project
9. Develop contractor's construction schedule	Construction schedule does not match contracted time schedule of the project	Project activities will not be executed in an organized and structured manner and the resources will be used to achieve specified objectives	Construction Manager (Contractor)	Develop schedule in an organized and structure manner based on the contracted time schedule considering logical process
10. Develop project S-Curve	S-Curve is not developed taking into consideration forecasted cash flow based on work activities to be completed as per construction schedule	Contractor will not be able to forecast exact cash flow and spending over the established project schedule (time)	Construction Manager (Contractor)	Develop S-Curve for forecasting cash flow taking into considerations the work (activities) which are expected to be completed and how much payments will be received and expenditure over the established project schedule
11. Develop contractor's quality control plan (CQCP)	Contractor's quality control plan does not contain adequacy and performance for delivering the level of construction quality required by the contract	Contractor will not be able to ensure that the day-to-day activities are meeting the standards specified in the contract documents	Construction Manager (Contractor)	Develop CQCP for day-to-day working to ensure meeting the performance standards specified in the contract documents (refer Figure 3.43 in Chapter 3)

(Continued)

TABLE 5.37
(Continued)

Serial Phase Number	Major Elements/ Activities	Potential Risk	Probable Effect on Project	Ownership of Risk	Risk Treatment/Control Measures
	12. Develop stakeholder management plan	The needs of relevant stakeholders are not addressed in the contents of stakeholder's plan	Stakeholders will not be aware as to how their interest will be affected	PMC	Develop stakeholder management plan to address the needs of project stakeholders and their roles and responsibilities
	13. Develop resource management plan	Resource management is developed not considering the management of all the project resources (human resources, construction resources (contractor staff), manpower for construction works, material for project)	Delay in completion of the project as per contract documents and to the satisfaction of owner's requirements	Construction Manager (Contractor)	Develop resource management plan considering the following requirements: 1. Human resources (project teams) 2. Construction resources
	14. Develop communication management plan	Communication management plan does not include effective communication activities related to all the involved stakeholders	Stakeholders will not be provided with timely and accurate information for their contribution to the success of project	Construction Manager (Contractor)	Develop communication management plan for smooth, timely flow of information to the identified stakeholders
	15. Develop risk management plan	The contents of risk management plan do not include how the risk activities will be performed, recorded, and monitored throughout the project	Risk will not be identified, recorded, monitored, and timely response action is taken for successful project	Construction Manager (Contractor)	Identify and register the risks on a regular basis and response action/treated immediately

16. Develop contract management plan	The contents of contract management plan do not include all the activities for implementation and successful completion of contract	Project will not be managed in an organized method, process, and procedure to obtain the required project successfully	Construction Manager (Contractor)	Develop contract management plan in an organizational method, process, and procedure to execute the project for successful completion of project
17. Develop HSE management plan	The guidelines normally specified in the contract documents are not followed for developing HSE management plan	1. Worker's health and safety and safety of worksite/workplace will be in danger 2. More number of occurrences of accidents at site	Construction Manager (Contractor)	Develop HSE management plan considering following the major points: 1. Project scope detailing safety requirements 2. Safety policy 3. Regulatory requirements 4. Roles and responsibilities 5. Hazard identification 6. Emergency execution plan 7. Emergency evacuation plan

TABLE 5.38

Risk Monitoring of Engineering, Procurement, and Construction (EPC) Phase (General Activity)

Serial Number	Description	Action			
1	Risk ID	General Activities—1.1 (9)			
2	Description of risk	1. Develop contractor's construction schedule 1.1. Construction schedule does not match contracted time schedule of the project			
3	Response	1.1. Develop schedule in an organized and structure manner based on the contracted time schedule considering logical process			
4	Strategy of response	Avoidance	Transfer	Mitigation	Acceptance X
5	Monitoring and control	Risk owner	Review date	Critical issue	
		Construction Manager (Contractor)	D/M/Y	Project activities will not be executed in an organized and structured manner and the resources will be used to achieve specified objectives	
6	Estimated impact on the project	Development of management plans			
7	Actual impact on the project	Delay in development and approval of management plans			
8	Revised response	Develop FEED taking into consideration TOR requirements and related data			
9	Record update date	D/M/Y			
10	Communication to stakeholders	Information sent to all stakeholders on D/M/Y			

5.16   RISK MANAGEMENT IN QUALITY OF EPC PHASE (DETAILED ENGINEERING)

Figure 5.20 illustrates a typical flowchart for risk management in quality of EPC phase (detailed engineering).

5.16.1   RISK IN ACTIVITIES RELATED TO QUALITY OF EPC PHASE (DETAILED ENGINEERING)

Table 5.39 lists potential risks in the activities related to EPC phase (detailed engineering).

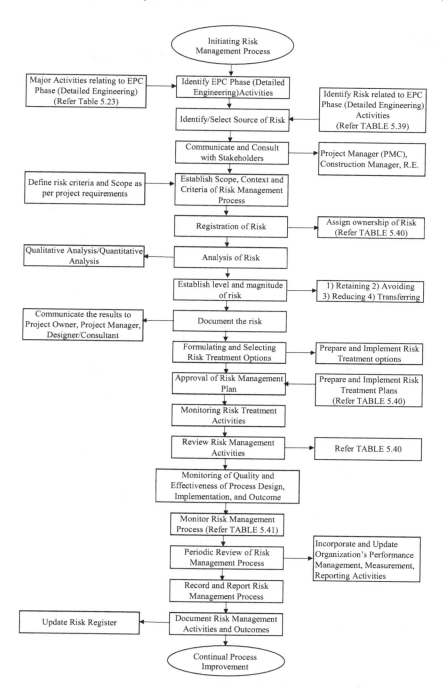

FIGURE 5.20 A typical flowchart for risk management in quality of EPC phase (detailed engineering).

TABLE 5.39
Potential Risks in Activities Related to EPC Phase (Detailed Engineering)

Serial Number	Phase	Major Elements/Activities	Potential Risk
1		1.1 EPC (Detailed Engineering)	
		1. Identify detailed engineering deliverables	1. Engineering requirements of individual components such as size, shape, levels, performance characteristics, and technical details for development of detailed engineering are not identified
			2. Detailed engineering deliverables are identified based on engineering as well contractual requirements
		2. Develop detailed engineering	Detailed engineering is not developed based on the criteria to meet owner's requirements
		3. Perform interdisciplinary coordination	Interdisciplinary coordination is not performed
		4. Develop project schedule	Schedule is not developed as per contracted specification and revised taking into consideration detailed engineering
		5. Estimate project cost	Project cost is not based on detailed engineering requirements
		6. Manage detailed engineering design quality	Quality plan is not developed based on detailed engineering requirements
		7. Estimate project resources	Errors in estimating project resources
		8. Manage project risks	Potential risks are not identified, and response method is evolved
		9. Manage HSE plan	Environmental impact and HAZID report are not considered while developing HSE plan
		10. Review detailed engineering	Detailed engineering drawings are not analyzed and reviewed
		11. Finalize detailed engineering	Review comments are not incorporated prior to finalizing detailed engineering

5.16.2 RISK EFFECT AND RISK TREATMENT IN ACTIVITIES RELATED TO QUALITY OF EPC PHASE (DETAILED ENGINEERING)

Table 5.40 lists potential risks, probable effect on project, and risk treatment in activities related to EPC Phase (detailed engineering).

5.16.3 RISK MONITORING IN ACTIVITIES RELATED TO QUALITY OF EPC PHASE (DETAILED ENGINEERING)

Table 5.41 is an example process of risk monitoring of EPC phase (detailed engineering) activity.

Quality of Project Life Cycle Phases **435**

TABLE 5.40
Potential Risks, Probable Effect on Project, and Risk Treatment in Activities Related to EPC Phase (Detailed Engineering)

Serial Number	Phase	Major Elements/Activities	Potential Risk	Probable Effect on Project	Ownership of Risk	Risk Treatment/Control Measures
1	1.1 EPC (Detailed Engineering)	1. Identify detailed engineering deliverables	1. Engineering requirements of individual components such as size, shape, levels, performance characteristics, technical details for development of detailed engineering are not identified 2. Detailed engineering deliverables are identified based on engineering as well contractual requirements	1. Detailed engineering not in compliance with specification requirements 2. Noncompliance of project quality through various activities 3. Developing a realistic schedule with appropriate milestone to confirm the progress	Construction Manager (Contractor)	1. Develop detailed engineering deliverables as per TOR requirements for a successful completion of the project 2. Identify all the engineering requirements to be included in the detailed engineering
		2. Develop detailed engineering	Detailed engineering is not developed based on the criteria to meet owner's requirements	Project quality not meeting the owner's requirements and the project goals and objectives	Construction Manager (Contractor)	Develop detailed engineering documents taking into considerations following items: 1. Owner's requirements, project goals, and objectives 2. TOR requirements 3. FEED 4. The size, shape, levels, performance characteristics, technical details, and requirements of all the individual components

(Continued)

TABLE 5.40
(Continued)

Serial Number	Phase	Major Elements/Activities	Potential Risk	Probable Effect on Project	Ownership of Risk	Risk Treatment/Control Measures
		3. Perform interdisciplinary coordination	Interdisciplinary coordination is not performed	1. Design errors 2. Conflict in interrelationship of among different trades	Construction Manager (Contractor)	Perform interdisciplinary coordination to incorporate the requirements of all the trades
		4. Develop project schedule	Schedule is not developed as per contracted specification and revised taking into consideration detailed engineering	Schedule will not meet contract requirements accurately to achieve the specified project quality	Construction Manager (Contractor)	Develop the revised schedule taking into considerations approved detailed engineering requirements and contracted schedule
		5. Estimate project cost	Project cost is not based on detailed engineering requirements	1. Errors in exact project financing/funding amount requirements 2. Estimate cost not as per budget	Construction Manager (Contractor)	Estimate the cost taking into consideration approved detailed engineering requirements that do not exceed that contracted project amount
		6. Manage detailed engineering design quality	Quality plan is not developed based on detailed engineering requirements	Errors and omissions in the detailed engineering	Construction Manager (Contractor)	Develop detailed engineering as per specification requirements and review to reduce omissions, errors, and mistakes
		7. Estimate project resources	Errors in estimating project resources	Execution of project within schedule	Construction Manager (Contractor)	Estimate the resources based on approved detailed engineering required to complete execution of the project
		8. Manage project risks	Potential risks are not identified, and response method is evolved	Detailed engineering not meeting contracted specification requirements	Construction Manager (Contractor)	Identify potential risks on a regular basis and develop response for the same

9. Manage HSE plan	Environmental impact and HAZID report are not considered while developing HSE plan	Noncompliance with regulatory requirements and environmental considerations	Construction Manager (Contractor)	Develop detailed engineering taking into account environmental impact analysis HAZOP and HAZID report
10. Review detailed engineering	Detailed engineering drawings are not analyzed and reviewed	1. Errors and omission 2. Non-coordination among various trades	Construction Manager (Contractor)	Review detailed engineering for conflict and coordination with other trades to avoid omissions and errors.
11. Finalize detailed engineering	Review comments are not incorporated prior to finalizing detailed engineering	Errors in the detailed engineering	Construction Manager (Contractor) PMC	Check the detailed engineering prior to submission for approval Review submitted drawings for conformance with owner's requirements

TABLE 5.41

Risk Monitoring of (EPC) Phase (Detailed Engineering) Activity

Serial Number	Description	Action			
1	Risk ID	Detailed Engineering—1.1 (1)			
2	Description of risk	1. Identification of detailed engineering requirements 1.1. Engineering requirements of individual components such as size, shape, levels, performance characteristics, technical details for development of detailed engineering are not identified 1.2. Detailed engineering deliverables are identified based on engineering as well as contractual requirements			
3	Response	1.1. Develop detailed engineering deliverables as per TOR requirements for successful completion of the project 1.2. Identify all the engineering requirements to be included in the detailed engineering			
4	Strategy of response	Avoidance	Transfer	Mitigation X	Acceptance
5	Monitoring and control	Risk owner	Review date	Critical issue	
		Construction Manager (Contractor)	D/M/Y	Detailed engineering not in compliance with specification requirements 2. Noncompliance of project quality through various activities 3. Developing a realistic schedule with appropriate milestone to confirm the progress	
6	Estimated impact on the project	Detailed engineering drawings			
7	Actual impact on the project	Execution of project activities			
8	Revised response	Develop detailed engineering drawings as per TOR requirements for proper functioning of process equipment			
9	Record update date	D/M/Y			
10	Communication to stakeholders	Information sent to all stakeholders on D/M/Y			

5.17 RISK MANAGEMENT IN QUALITY OF EPC PHASE (PROCUREMENT)

Figure 5.21 illustrates a typical flowchart for risk management in quality of EPC Phase (procurement).

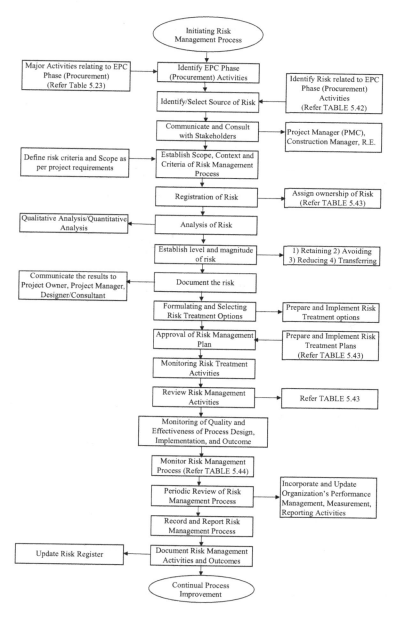

FIGURE 5.21 A typical flowchart for risk management in quality of EPC phase (procurement).

Figure 5.22 illustrates material approval and procurement procedure.

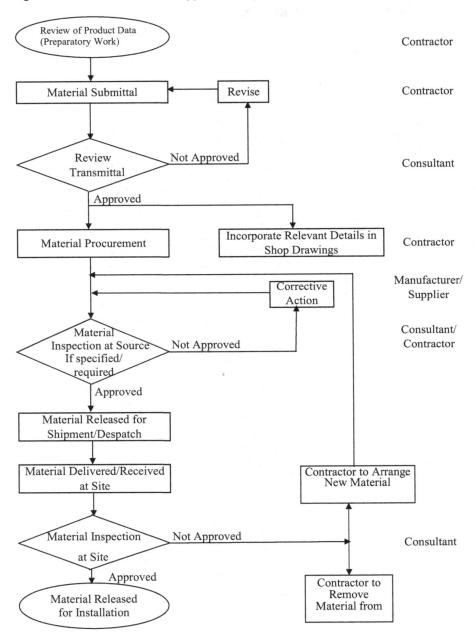

FIGURE 5.22   Material approval and procurement procedure.

5.17.1   RISK IN ACTIVITIES RELATED TO QUALITY OF EPC PHASE (PROCUREMENT)

Table 5.42 lists potential risks in the activities related to EPC Phase (procurement).

5.17.2   RISK EFFECT AND RISK TREATMENT IN ACTIVITIES RELATED TO QUALITY OF EPC PHASE (PROCUREMENT)

Table 5.43 lists potential risks, probable effect on project, and risk treatment in activities related to EPC phase (Procurement).

TABLE 5.42
Potential Risks in Activities related to EPC Phase (Procurement)

Serial Number	Phase	Major Elements/ Activities	Potential Risk
1	1.1 EPC (Procurement)		
		1. Identify procurement deliverable	Procurement deliverables are not identified as per contract documents and approved detailed engineering
		2. Develop procurement documents	Procurements documents do not contain all the related documents as required per procurement strategy
		3. Prepare material procurement list	Material list has major items missing
		4. Identify vendors	Vendors are not identified as per procurement strategies
		5. Manage procurement quality	Specific inspection and tests are not carried out on the materials, equipment, and systems as per contract documents
		6. Finalize procurement	Detailed procedure for submitting materials, samples as specified in the contract is not followed

TABLE 5.43

Potential Risks, Probable Effect on Project, and Risk Treatment in Activities Related to EPC Phase (Procurement)

Serial Number	Phase	Major Elements/Activities	Potential Risk	Probable Effect on Project	Ownership of Risk	Risk Treatment/Control Measures
1	1.1 EPC (Procurement)					
		1. Identify procurement deliverable	Procurement deliverables are not identified as per contract documents and approved detailed engineering	Delay in procurement of required items	Construction Manager (Contractor)	Identify procurement deliverables based on approved detailed engineering, BOQ, and specifications
		2. Develop procurement documents	Procurements documents do not contain all the related documents as required per procurement strategy	Delay in procurement of specified items	Construction Manager (Contractor)	Develop procurement documents on corporate policy taking into consideration detailed engineering and project-specific requirements
		3. Prepare material procurement list	Material list has missing major items	Delay in compiling the list and procurement process	Construction Manager (Contractor)	Prepare procurement list mentioning major items based on the specifications and corporate policy to execute the project successfully as per schedule without any delay for the sake of availability of material
		4. Identify vendors	Vendors are not identified as per procurement strategies	Delay in sourcing vendors able to meet the product specifications	Construction Manager (Contractor)	Identify the vendors taking into consideration procurement strategy and approved list of manufacturers/vendors listed in the contract specifications
		5. Manage procurement quality	Specific inspection and tests are not carried out on the materials, equipment, and systems as per contract documents	1. Rejection/non approval of material 2. Delay in execution of project	Construction Manager (Contractor)	Establish quality management plan for procurement of material clearly defining the inspection stages (refer Figure 5.22)

| 6. Finalize procurement | Detailed procedure for submitting materials, samples as specified in the contract is not followed | 1. Delay in approval of material
2. Delay in execution of works | Construction Manager (Contractor) | 1. Follow procurement procedure to ensure material and sample approval is as per contract requirements
2. Follow procurement schedule and make required material available for execution of site works without any interruption |

5.17.3   RISK MONITORING IN ACTIVITIES RELATED TO QUALITY OF EPC PHASE (PROCUREMENT)

Table 5.44 is an example process of risk monitoring of EPC phase (procurement) activity.

5.18   RISK MANAGEMENT IN QUALITY OF EPC PHASE (CONSTRUCTION)

Figure 5.23 illustrates a typical flowchart for risk management in quality of EPC phase (construction).

TABLE 5.44
Risk Monitoring of (EPC) Phase (Procurement) Activity

Serial Number	Description	Action		
1	Risk ID	Procurement—1.1 (1)		
2	Description of Risk	1. Manage procurement quality 1.1. Specific inspection and tests are not carried out on the materials, equipment, and systems as per contract documents		
3	Response	1.1. Establish quality management plan for procurement of material clearly defining the inspection stages		
4	Strategy of response	Avoidance	Transfer	Mitigation Acceptance X
5	Monitoring and control	Risk owner	Review date	Critical issue
		Construction Manager (Contractor)	D/M/Y	Rejection/non approval of material
6	Estimated impact on the project	Quality of material, equipment and systems		
7	Actual impact on the project	Project quality		
8	Revised response	Develop quality management plan, material management plan		
9	Record update date	D/M/Y		
10	Communication to stakeholders	Information sent to all stakeholders on D/M/Y		

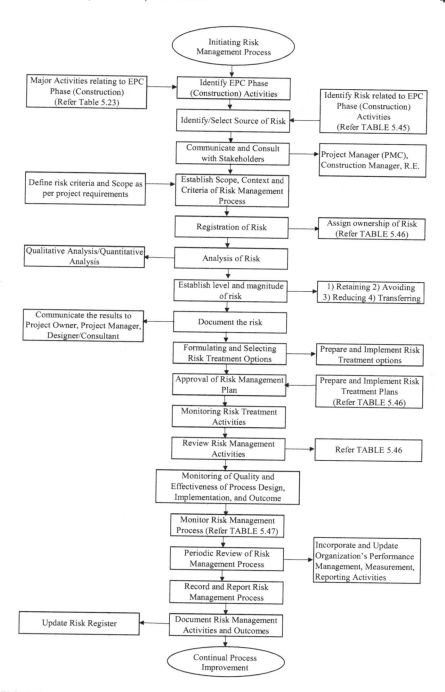

FIGURE 5.23 A typical flow chart for risk management in quality of EPC phase (construction).

5.18.1 Risk in Activities Related to Quality of EPC Phase (Construction)

Table 5.45 lists potential risks in the activities related to EPC phase (construction).

5.18.2 Risk Effect and Risk Treatment in Activities Related to Quality of EPC Phase (Construction)

Table 5.46 lists potential risks, probable effect on project, and risk treatment in activities related to EPC phase (construction)

5.18.3 Risk Monitoring in Activities Related to Quality of EPC Phase (Construction)

Table 5.47 illustrates an example process of risk monitoring of EPC phase (construction) activity.

TABLE 5.45
Potential Risks in Activities Related to EPC Phase (Construction)

Serial Number	Phase	Major Elements/ Activities	Potential Risk
	1.4 EPC (Construction)		
		1. Identify construction works deliverables	1. Construction works deliverables are not as specified in contract documents, owner's requirements 2. Deliverables are not in accordance with approved detailed engineering drawings and specifications
		2. Manage execution/ installation of works	The contractor is not able to manage the works as per contract drawings (approved detailed engineering) and specifications
		3. Manage scope change	1. The contractor is able to manage the scope changes as per contract documents 2. Change requirement are not identified in time to avoid unnecessary disruption of works 3. Delay in approval of scope changes

Serial Number	Phase	Major Elements/ Activities	Potential Risk
		4. Manage construction quality	Construction quality is not managed by regular inspection and testing carried out at following stages: 1. During construction process 2. Receipt of material, equipment, and systems 3. Before final delivery or commissioning and handover
		5. Manage construction resources	Contractor is not able to manage resources mainly consisting of the following: 1. Construction workforce 2. Construction equipment and machinery 3. Construction material, equipment, and systems to be installed 4. Delay in getting the material
		6. Manage communication	Proper communication system is not established clearly identifying the submission process for correspondence, submittals, minutes of meetings and reports
		7. Manage construction risks	Potential risks are not identified and response method is evolved
		8. Manage contract	Contract is not managed in an organized method, process, and procedure
		9. Manage HSE requirements	Contractor not following all the requirements of HSE management plan
		10. Manage project finances	Payments due to subcontractors, vendors are not settled
		11. Manage claims	Claims are not resolved and settled
		12. Monitoring and control of project works	Monitoring and controlling of project is not followed up on a regular basis, and progress of work is not updated
		13. Validate executed works	1. Inspection of works is not performed throughout the execution of project 2. Check lists are not used to check the installed/executed works

TABLE 5.46

Potential Risks, Probable Effect on Project, and Risk Treatment in Activities Related to EPC Phase (Construction)

Serial Number	Phase	Major Elements/ Activities	Potential Risk	Probable Effect on Project	Ownership of Risk	Risk Treatment/Control Measures
		1.4 EPC (Construction)				
		1. Identify construction works deliverables	1. Construction works deliverables are not as specified in contract documents, owner's requirements	1. Contractor not executing the works in accordance with approved detailed engineering and specifications as specified in the contract documents	Construction Manager (Contractor)	1. Identify and execute construction works as per approved detailed engineering documents and specifications
			2. Deliverables are not in accordance with approved detailed engineering drawings and specifications	2. Executed works not meeting the contracted specification		2. Develop project deliverables as per contract documents and approved detailed engineering
		2. Manage execution/ installation of works	The contractor is not able to manage the works as per contract drawings (approved detailed engineering) and specifications	Project quality not meeting owner's requirements, needs	Construction Manager (Contractor)	Execute the works as per approved shop drawings, installing the approved material, equipment, systems
		3. Manage scope change	1. The contractor is not able to manage the scope changes as per contract documents	1. Project objectives may not be achieved	Construction Manager (Contractor)	1. Submit scope change proposal for change in the project scope
			2. Change requirement are not identified in time to avoid unnecessary disruption of works	2. Delay in execution of project		2. Submit RFI immediately after noticing the discrepancies in the contract specification, documents
			3. Delay in approval of scope changes	3. Impact on schedule and cost		

Activity	Risk	Responsible	Action
4. Manage construction quality	Construction quality is not managed by regular inspection and testing carried out at the following stages: 4. During construction process 5. Receipt of material, equipment, and systems 6. Before final delivery or commissioning and handover	PMC	1. Review RFI/change order proposal considering the overall impact on the project performance (refer Figure 3.22 in Chapter 3 and Figure 4.6 as applicable) 2. Take action within specified limit in the contract documents
	1. Works are not accomplished in accordance with the requirements specified in the contract 2. Inspection of construction works is not carried out on regular basis to ensure project quality	Construction Manager (Contractor)	Develop contractor's quality control plan and get approval to manage day-to-day quality of the project (refer Figure 3.43 in Chapter 3)
5. Manage construction resources	Contractor is not able to manage resources mainly consisting of the following: 1. Construction workforce 2. Construction equipment and machinery 3. Construction material, equipment, and systems to be installed 4. Delay in getting the material	Construction Manager (Contractor)	1. Workmanship 2. Delay in completion of project
			1. Manage the construction resources and their availability as per requirements to execute the works to be completed as per schedule 2. Follow resource calendar and procurement log for the availability of material at site

(Continued)

TABLE 5.46 (Continued)

Serial Number	Phase	Major Elements/Activities	Potential Risk	Probable Effect on Project	Ownership of Risk	Risk Treatment/Control Measures
6.		Manage communication	Proper communication system is not established clearly identifying the submission process for correspondence, submittals, minutes of meetings, and reports	1. Implementation of project activities will not be smooth 2. Delay in distribution and submission of reports, control documents, and project-related information	Construction Manager (Contractor) PMC	Prepare communication plan for smooth implementation of communication system clearly identifying the submission and communication process for correspondence, submittals, minutes of meetings, and reports Prepare responsibilities matrix and site administration matrix for smooth communication
7.		Manage construction risks	Potential risks are not identified, and response method is evolved	1. Prevention of unwanted consequences 2. Effect on project quality	Construction Manager (Contractor)	Identify construction risks on regular basis and develop response/risk treatment plan
8.		Manage contract	Contract is not managed in an organized method, process, and procedure	Delay in execution of project works within agreed-upon time and cost in accordance with the contract conditions and specifications	Construction Manager (Contractor) PMC	Manage contract in an organized method, process, and procedures taking into consideration involvement of all stakeholders, owner, contractor, subcontractor, manufacturer, vendors, designer, and PMC Monitor and control the execution works in order to complete the project successfully

Activity	Cause	Effect	Responsible	Action
9. Manage HSE requirements	Contractor not following all the requirements of HSE management plan	1. Occupational health 2. Accidents and injuries at site 3. Project not complying with regulatory requirements and environmental issues	Construction Manager (Contractor)	Follow HSE management plan to avoid accidents and implementation of safety measures
10. Manage project finances	Payments due to subcontractors, vendors and not settled	Effect on project execution schedule due to delay in due payments	Project Manager (PMC)	Manage the due payments and pay on time as per contract documents
11. Manage claims	Claims are not resolved and settled	Delay in project closeout	Project Manager (PMC)	Manage the claims amicably to resolve the issue
12. Monitoring and control of project works	Monitoring and controlling of project is not followed up on regular basis, and progress of work is not updated	1. Effect on project progress and performance 2. Delay in tracking and recognizing any obstacles encountered during execution of works	Construction Manager (Contractor)	Monitor and control the project on a regular basis during execution of project by recognizing any obstacle for smooth progress of works as per schedule
			PMC Project Manager (Supervision)	Monitor the project progress properly Monitor the execution of works properly to complete the project as per schedule and specification requirements
13. Validate executed works	1. Inspection of works is not performed throughout the execution of project 2. Check lists are not used to check the installed/executed works	1. Quality of project 2. Workmanship of executed/installed works and project quality 3. Works are not in full compliance with the contract documents	Project Manager (Supervision) Project Manager (Supervision)	1. Make inspection and monitoring of the progress of execution works regularly 2. Submit check list for all the executed works once completed Monitor the project progress and project status as per approved schedule

TABLE 5.47

Risk Monitoring of (EPC) Phase (Construction) Activity

Serial Number	Description	Action		
1	Risk ID	Construction—1.1 (4)		
2	Description of risk	1. Manage construction quality 1.1. Construction quality is not managed by regular inspection and testing carried out at the following stages: 1. During construction process 2. Receipt of material, equipment, and systems 3. Before final delivery or commissioning and handover		
3	Response	1.1. Develop contractor's quality control plan and get approval to manage day-to-day quality of the project		
4	Strategy of response	Avoidance	Transfer	Mitigation Acceptance X
5	Monitoring and control	Risk owner	Review date	Critical issue
		Construction manager (Contractor)	D/M/Y	1. Works are not accomplished in accordance with the requirements specified in the contract 2. Inspection of construction works is not carried out on regular basis to ensure project quality
6	Estimated impact on the project	Quality of project works		
7	Actual impact on the project	Project quality		
8	Revised response	Develop contractor's quality control plan		
9	Record update date	D/M/Y		
10	Communication to stakeholders	Information sent to all stakeholders on D/M/Y		

5.19   RISK MANAGEMENT IN QUALITY OF TESTING, COMMISSIONING, AND HANDOVER PHASE

Figure 5.24 illustrates a typical flowchart for risk management in quality of testing, commissioning, and handover phase.

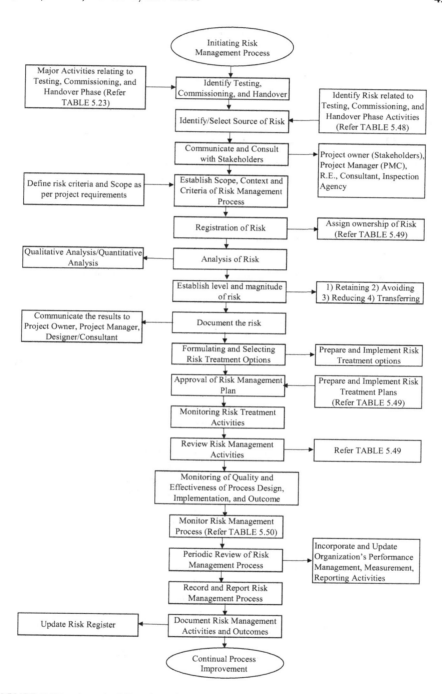

FIGURE 5.24 A typical flowchart for risk management in quality of testing, commissioning, and handover phase.

5.19.1   Risk in Activities Related to Quality of Testing, Commissioning, and Handover Phase

Table 5.48 lists potential risks in the activities related to Testing, Commissioning, and Handover Phase.

5.19.2   Risk Effect and Risk Treatment in Activities Related to Quality of Testing, Commissioning, and Handover Phase

Table 5.49 lists potential risks, probable effect on project, and risk treatment in activities related to testing, commissioning, and handover phase.

TABLE 5.48
Potential Risks in Activities Related to Testing, Commissioning, and Handover Phase

Serial Number	Phase	Major Elements/Activities	Potential Risk
1	1.1	Testing, Commissioning, and Handover Phase	
		1. Identify testing, commissioning, and start-up requirements	All the requirements as specified in the contract documents are not identified and listed
		2. Identify stakeholders	All the relevant stakeholders are not identified
		3. Develop scope of works	1. Scope of works is not properly developed as per contract documents
			2. Errors in scope of works
		4. Develop testing, commissioning, and handover plan	Testing, commissioning, and handover plan is not properly developed with proper sequencing of the activities
		5. Execute testing and commissioning works	The tests are not coordinated with and scheduled with specialist contractors
		6. Manage testing and commissioning quality	The quality plan is not established to follow the specified procedure
		7. As-built drawings	As-built drawings are not developed incorporating all the information from record drawings
		8. Technical manuals and documents	Technical manuals don't consist of all the documents specified in contract documents
		9. Record books (project, construction, manufacturing)	Contract specified record books are not submitted
		10. Train owner's/end user's personnel	Training program as per specifications is not established and approved
		11. Regulatory/authority approval	Regulatory approvals from concerned authorities are not obtained or delayed
		12. Handover the project to the owner/end user	Delay in handover/acceptance of the project
		13. Substantial completion	Delay in issuance of substantial certificate
		14. Settle payments	All the dues/payments are not settled
		15. Settle claims	All the claims are not amicably settled

TABLE 5.49

Potential Risks, Probable Effect on Project, and Risk Treatment in Activities Related to Testing, Commissioning, and Handover Phase

Serial Number	Phase	Major Elements/ Activities	Potential Risk	Probable Effect on Project	Ownership of Risk	Risk Treatment/Control Measures
1		1.1 Testing, Commissioning, and Handover Phase				
		1. Identify testing, commissioning, and start-up requirements	All the requirements as specified in the contract documents are not identified and listed	Missing installed/ executed works may not be functional and suitable for usage to meet owner's requirements, needs	Construction Manager (Contractor)	Identify testing, commissioning, and start-up requirements as per TOR requirements and contract documents
		2. Identify stakeholders	All the relevant stakeholders are not identified	Testing and commissioning not fully satisfying the requirements needs of missing stakeholders	PMC Project Manager (Supervision)	Identify relevant stakeholders having interest in the testing and commissioning Select team members having experience in testing and commissioning of major projects
					Construction Manager (Contractor)	1. Select testing and commissioning team members as per contract documents 2. Engage specialist agency for testing and commissioning purpose if required as per contract documents
		3. Develop scope of works	1. Scope of works is not properly developed as per contract documents 2. Errors in scope of works	Testing and commissioning not meeting the requirements as specified in contract documents	Construction Manager (Contractor)	1. Develop scope of work as per requirements of contract documents and specifications 2. Review scope of work to match contract requirements and to minimize errors

(Continued)

TABLE 5.49
(Continued)

Serial Number	Phase	Major Elements/ Activities	Potential Risk	Probable Effect on Project	Ownership of Risk	Risk Treatment/Control Measures
		4. Develop testing, commissioning, and handover plan	Testing, commissioning, and handover plan is not properly developed with proper sequencing of the activities	Testing and commissioning will not be systematic approach to ensure proper function of the system, equipment	Construction Manager (Contractor)	Develop testing, commissioning, and handover plan as per schedule and contract documents (refer Figure 5.25)
		5. Execute testing, and commissioning works	The tests are not carried out and coordinated as per approved schedule with specialist contractors	Test results not accurate to meet owner's requirements, usage	Construction Manager (Contractor)	Engage specialist contractor to perform testing, commissioning of systems/ equipment as per contract requirements as per approved schedule
		6. Manage testing and commissioning quality	The quality plan is not established to follow the specified procedure	Specified procedures are not followed	Construction Manager (Contractor)	Perform testing and commissioning as per the procedure in contract documents
		7. As-built drawings	As-built drawings are not developed incorporating all the information from record drawings	As-built drawings not indicating the actual installation as installed/ executed at site	Construction Manager (Contractor)	Prepare as-built drawings based on record drawings and actual installation/executed works
		8. Technical manuals and documents	Technical manuals don't consist of all the documents specified in contract documents	Incomplete information about installed equipment/ systems	Construction Manager (Contractor)	Develop technical manuals consisting of all the information about installed equipment/systems as per contract documents
		9. Record books (project, construction, manufacturing)	Contract-specified record books are not submitted	Record not available during maintenance period	Construction Manager (Contractor)	Submit the record books as per contract requirements

10. Train owner's/end user's personnel	Training program as per specifications is not established and approved	Owner's personnel will not be fully aware of maintenance and operation procedures	Construction Manager (Contractor)	Conduct training for owner's/user's personnel to get familiarized with the installed works and their operations
11. Regulatory/authority approval	Regulatory approvals from concerned authorities are not obtained or delayed	Delay in handover/takeover/start-up of project	Construction Manager (Contractor)	Obtain necessary regulatory approvals from the respective concerned authorities
12. Handover the project to the owner/end user	Delay in handover/acceptance of the project	Delay in start-up of the project	Construction Manager (Contractor)	Handover the project/facility as per approved schedule
13. Substantial completion	Delay in the issuance of substantial certificate	Delay in project closeout	Project Owner (PMC)	Issue substantial completion certificate once it is established that contractor has completed the works as per contract documents and to the satisfaction of owner
14. Settle payments	All the dues/payments are not settled	Delay in project closeout	Project Owner (PMC)	Settle the approved and due payments to close the project
15. Settle claims	All the claims are not amicably settled	Delay in project close out	Project Owner (PMC)	Settle the claims amicably

Figure 5.25 illustrates a flowchart for the development of inspection and test plan.

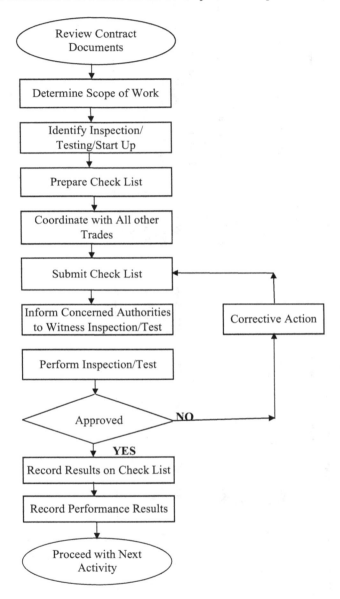

FIGURE 5.25 A logic flowchart for the development of inspection and test plan.

Source: Abdul Razzak Rumane. (2016). *Handbook of Construction Management*. Reprinted with permission of Taylor & Francis Group.

5.19.3 Risk Monitoring in Activities Related to Quality of Testing, Commissioning, and Handover Phase

Table 5.50 illustrates an example process of risk monitoring of testing, commissioning, and handover phase activity.

TABLE 5.50
Risk Monitoring of Testing, Commissioning, and Handover Phase Activity

Serial Number	Description	Action			
1	Risk ID	Testing, Commissioning—1.1 (4)			
2	Description of risk	1. Execute testing and commissioning works 1.1. The tests are not carried out and coordinated as per approved scheduled with specialist contractors			
3	Response	1.1. Engage specialist contractor to perform testing, commissioning of systems/equipment as per contract requirements as per approved schedule			
4	Strategy of response	Avoidance	Transfer	Mitigation	Acceptance X
5	Monitoring and control	Risk owner	Review date	Critical issue	
		Construction Manager (Contractor)	D/M/Y	Test results not accurate to meet owner's requirements, usage	
6	Estimated impact on the project	Testing and commissioning			
7	Actual impact on the project	Accuracy of test results to meet project specification requirements			
8	Revised response	Perform tests as per contract requirements			
9	Record update date	D/M/Y			
10	Communication to stakeholders	Information sent to all stakeholders on D/M/Y			

Bibliography

AACE International Recommended Practice No. 18R-97 (2010)

AACE International Recommended Practice No. 27R-03 (2010)

AACE International Recommended Practice No. 37R-06 (2010)

Rumane, Abdul Razzak (2013), *Quality Tools for Managing Construction Projects*, CRC Press, Taylor & Francis Group, Boca Raton, FL.

Rumane, Abdul Razzak (2016), *Handbook of Construction Management: Scope, Schedule, and Cost Control*, CRC Press, Boca Raton, FL.

Rumane, Abdul Razzak (2017), *Quality Management in Construction Projects*, Second Edition, CRC Press, Taylor & Francis Group, Boca Raton, FL.

Rumane, Abdul Razzak (2021), *Quality Management in Oil and Gas Projects*, CRC Press, Taylor & Francis Group, Boca Raton, FL.

www.iso.org (ISO 31000 Risk Management)

Index

Printed in the United States
by Baker & Taylor Publisher Services